Buried viable seed banks are a fundamental aspect of seed plant biology. They play a key role in the conservation and restoration of plant communities and the response of plants to changing land use and climate. There is almost no area of plant ecology in which seed banks are not implicated.

Despite several recent reviews of the ecology of seed banks, there has previously been no single source of data on seed persistence in individual species. This volume, which compiles the available data from the nineteenth century up to the end of 1993, provides this source for 1189 members of the North West European flora. The text describes the criteria for inclusion of data and discusses seed bank classification systems, the relative representation of different habitats, methods and taxa, and challenges for future research.

The soil seed banks of North West Europe:
methodology, density and longevity

The soil seed banks of North West Europe:
methodology, density and longevity

Ken Thompson
NERC Unit of Comparative Plant Ecology
Department of Animal and Plant Sciences
The University
Sheffield S10 2TN
UK

Jan P. Bakker and Renée M. Bekker
Laboratory of Plant Ecology
University of Groningen
PO Box 14
9750 AA Haren
The Netherlands

CAMBRIDGE
UNIVERSITY PRESS

PUBLISHED BY THE PRESS SYNDICATE OF THE UNIVERSITY OF CAMBRIDGE
The Pitt Building, Trumpington Street, Cambridge CB2 1RP, United Kingdom

CAMBRIDGE UNIVERSITY PRESS
The Edinburgh Building, Cambridge CB2 2RU, United Kingdom
40 West 20th Street, New York, NY 10011-4211, USA
10 Stamford Road, Oakleigh, Melbourne 3166, Australia

First published 1997

Printed in the United Kingdom at the University Press, Cambridge

Typeset in Minion 9/13pt and Univers 6/9½pt

A catalogue record for this book is available from the British Library

Library of Congress Cataloguing in Publication data

Thompson, Ken, 1954–
      The soil seed banks of north west Europe : methodology, density
and longevity / Ken Thompson, Jan P. Bakker, and Renée M. Bekker.
         p.   cm.
      Includes bibliographical references.
      ISBN 0 521 49519 9 (hardcover)
      1. Soil seed banks–Europe.   I. Bakker, Jan Pouwel.   II. Bekker,
Renée M.   III. Title.
   QK281.T48 1997
   582'.05–DC20   96-21788 CIP

ISBN 0 521 49519 9

Contents

Preface

There is a rapidly increasing demand for standardised information on the ecological attributes of a wide range of plant species, both as a tool of vegetation management for conservation and restoration, and as a data source for academic work in plant biology and ecology. For the British flora (or at least large parts of it) standardised accounts already exist on geographical distribution (Perring and Walters 1962; Stewart, Pearman and Preston 1994), seed germination requirements (Grime *et al.* 1981) and mycorrhizas (Harley and Harley 1986). The *Ecological Flora Database* (Fitter and Peat 1994) has assembled much of the available ecological information on British vascular plants, while the *Dutch Botanical Database* (Netherlands Bureau of Statistics 1994) does the same for the Netherlands. Hodgson *et al.* (1995) provide autecological accounts for a large number of common British species. As more data accumulate, and as computerised methods of data storage and retrieval become more widely available, we expect to see an increasing proliferation of specialised databases.

One aspect of plant ecology which has seen a recent explosion of new data is buried viable seed banks (Vyvey 1989a, b; Bernhardt and Poschlod 1993), and the time is ripe to attempt a synthesis of what is now available. Such syntheses are valuable not only as a source of information, but also as a guide to where important gaps in our knowledge remain to be filled. A number of generalisations have emerged from the hitherto very scattered literature on seed longevity in the soil; for example, Harper (1977) suggested that (i) long-lived seeds are characteristic of disturbed habitats, (ii) most long-lived seeds are annuals or biennials, (iii) small seeds tend to have much greater longevity than large ones, (iv) aquatic plants may have great

seed longevity, and (v) seeds of mature tropical forest trees have very short lives. We hope to provide the means of testing many of these hypotheses.

Buried viable seed banks are a fundamental aspect of seed plant biology, play an important role in the conservation and restoration of plant communities (Bakker 1989), and are important predictors of plant response to changing land use and climate (Hodgson and Grime 1990). There is hardly a single area of modern plant ecology in which seed banks are not suspected of involvement. A random trawl through the recent literature reveals studies of the role of seed banks in vegetation dynamics (Milberg and Hansson 1994, Milberg and Persson 1994), prediction of marsh vegetation after a drawdown (Ter Heerdt and Drost 1994, Haukos and Smith 1993), recolonisation after forest fire (Clark and Wilson 1994) and volcanic eruption (Tsuyuzaki 1994), sensitivity of wetlands to pollution (Leck and Simpson 1992), rangeland improvement (Bertiller and Coronato 1994), weed biology (Granados and Torres 1993; Marshall and Arnold 1994), persistence of transgenes (Linder and Schmitt 1994), succession (Kiirikki 1993), endangered species conservation (Milberg 1994), plant community restoration (Bakker *et al.* 1996) and spread of invasive aliens (Lonsdale 1993).

There are several recent reviews of the ecology of seed banks (Roberts 1981; Leck, Parker and Simpson 1989; Thompson 1992), but until now there has been no single source of data on seed persistence in individual species. In this volume we hope to provide this source, at least for North West Europe. We also draw attention to some of the more significant features of the data, including habitats and taxa where we know very little, and provide guidelines for future work on seed banks. Lack of standardisation of methods remains a barrier to progress (Leck, Parker and Simpson 1989). This book is essentially a compilation of available knowledge. We would be pleased if readers would feel stimulated to check whether our interpretation is justified, rather than take our conclusions for granted.

The collaboration which led to this database began with KT requesting unpublished data from various colleagues. When one of these requests reached JPB early in 1989, he suggested working together on the project. Very soon agreement was reached that KT would collate data in English, while JPB would deal with sources in Dutch, German and French. Later, RMB became involved as part of her PhD study of seed bank dynamics.

We would like now to thank all those who have provided access to unpublished data, including Harald Albrecht, Gilbert Barralis, Karl-Georg Bernhardt, Peter Csontos, John Hodgson, T. Jayasingam, Tim King, Per Milberg, Begoña Peco, Peter Poschlod, Gert Rosenthal, Henk Schat, Herman Stieperaere, Kees Vegelin and Arnout Zwaenepoel. Richard Pankhurst kindly provided access to the *Flora Europaea Database*. We are grateful to the British Council, the Royal Society, the UK Natural Environment Research Council and the Directorate of Science and Technics from the Dutch Ministry of Agriculture, Nature Conservation and Fisheries for financial support. We are particularly indebted to Han Olff for discussions about the structure of the database and for introducing two of us (JPB and KT) to computers, and to Ric Colasanti for continuing the tuition. David Corker also helped with computing and figure drawing. Phil Grime and Jelte Van Andel made valuable comments on the text. Thanks also to David Lewis, for suggesting the whole crazy enterprise, and to the manufacturers of hagelslag, the cement in the Anglo–Dutch relationship. Finally we thank Pat Thompson, Suus Bakker and Marius Schwartz for putting up with us during the long gestation of this project.

This book is accompanied by an electronic version of the database on a standard 3.5 inch disk. The electronic version is identical to the printed version, with the single exception of an extra 'Family' field, which was omitted from the printed version in order to save space.

The disk contains a single file: seedbank.csv. This comma-delimited format should be accessible by all modern spreadsheet packages, including Microsoft Excel, Borland Quattro Pro, Lotus 1-2-3 and dBase. Users of modern Macintosh computers should also experience no difficulty. However, the publishers can make no guarantees about the functioning of the file. In the event of a problem please contact the authors.

Seed banks: coverage, criteria and classification

1

In this chapter we briefly describe how the database was constructed, which countries, taxa and data sources are included, and the criteria which we have employed to classify seed banks.

Geographical and taxonomic coverage

Only in North West Europe has the union of a relatively impoverished flora and an abundance of botanical research over many years resulted in a reasonably complete coverage of seed persistence for at least the commoner elements in the flora. The boundaries of the area to be included were governed by the availability of the two pocket floras by Richard and Alastair Fitter (Fitter, Fitter and Blamey 1985; Fitter, Fitter and Farrer 1984), which together provide a reasonably complete list of most native and commonly naturalised vascular plants in North West Europe. The area included in this account is consequently that covered by these two volumes: the whole of North West Europe extending roughly from the Loire in the south to the tip of arctic Norway in the north (Figure 1.1).

With few exceptions, we have stuck closely to Fitters' species list; the only species omitted are a few of the less thoroughly naturalised aliens. Inevitably, however, much of Fitters' taxonomy is out of date. In our account, nomenclature of British natives and naturalised aliens, about two-thirds of the total, is taken from Kent (1992). Nomenclature of the remaining species is derived from the *Flora Europaea Database*. Our final list contains 2568 taxa, but seed bank data are available for only 1189. The quantity of data available for individual species also varies enormously; 250 species are represented by a single record each. We have taken the simplest possible taxonomic view; aggregates or groups of closely related species have usually been collapsed into a single entity. Thus, our account contains data for several species, such as *Taraxacum officinale, Rubus fruticosus* and *Euphrasia officinalis*, which have no real taxonomic validity. This may offend some purists but has been done for entirely pragmatic reasons. Only very rarely do sources of seed bank data make fine taxonomic distinctions, and these and other 'species' continue to have real meaning for ecologists, even if not for professional taxonomists.

Many seed bank data are available for European species from outside the area defined above, for instance elsewhere in Europe, North America, Australia and Japan. Also, for obvious reasons much of the best data for naturalised species are to be found in the literature of their countries of origin. We have therefore included all sources of data, regardless of origin. For nearly all species, the great majority of the data originate from within the area defined in Figure 1.1, so we have not felt it worthwhile to make the distinction in the database. For a few cosmopolitan species (e.g. *Chenopodium album, Trifolium repens*), however, a substantial fraction of the data arises from outside North West Europe, and one cannot assume that these 'alien' data correspond in all respects to those from Europe.

Despite a determination to use data from all available sources, we have not been able to abstract data from all Eastern European, Russian and Japanese sources. Data contained in papers entirely in Russian or Japanese, without English summaries and sometimes even without scientific

Figure 1.1 The area covered by the database is Europe to the north and west of the bold line.

names, have eluded us. Our attempts to obtain funding to translate some of the more important Russian papers were unsuccessful. Otherwise, however, we feel confident that our coverage of the available literature is reasonably complete up to the end of 1993. We owe a particular debt to the publications of Vyvey (1989a, b), Milberg (1990) and Bernhardt and Poschlod (1993), and to the assistance of H. A. Roberts, in locating sources of seed bank data.

Seed bank classification

It has become customary for seed banks, at least in temperate regions, to be classified with reference to the scheme proposed by Thompson and Grime (1979), and it might therefore seem natural to employ that scheme in this account. There are good reasons for not doing so, however, some of which have been given by Bakker (1989), and it seems worthwhile briefly to discuss the subject here. The Thompson and Grime scheme was proposed on the basis of the observed behaviour of seeds in the soil as revealed by a programme of seasonal sampling over a period of one year (Figure 1.2). Species were classified first as persistent or transient, according to whether their germinable seeds were detected throughout the year or not, respectively. Transient seed banks were then further divided into those with seeds present only in the summer (Type I) or only in the winter (Type II). Persistent seed banks were also further divided into those with (Type III) and those without (Type IV) a pronounced seasonal peak. The ecological usefulness of these distinctions is confirmed by the enduring use of the system by many ecologists; furthermore, the different types of seed bank are correlated with significant differences in

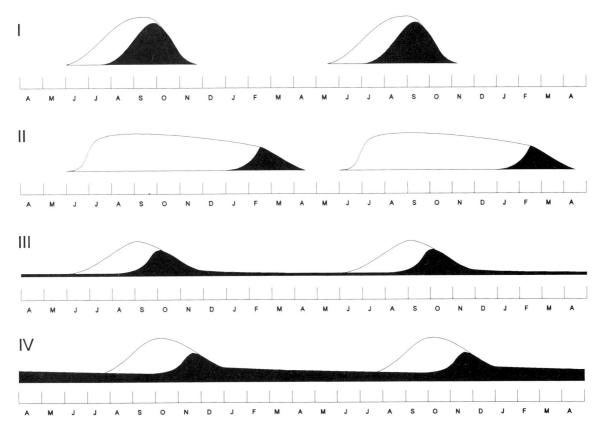

Figure 1.2 Four types of seed bank of common occurrence in temperate regions (after Thompson and Grime 1979). The curves illustrate the seasonal abundance of immediately germinable (shaded area) and viable but dormant (unshaded area) seeds both in the soil and on the soil surface.

geographical distribution and germination physiology. For the purposes of this account, however, this approach suffers from three disadvantages:

1. In the overwhelming majority of seed bank studies the detailed knowledge of seasonal dynamics necessary to separate the four types is not available. Admittedly species can often be classified partly on the basis of seed morphology and germination requirements (e.g. Thompson, Band and Hodgson 1993), but these data too are unavailable for most species.

2. It was always recognised that Types III and IV are ends of a continuum, but there is accumulating evidence that the same species may behave as Type III or IV at different times and in different places (Cresswell 1982; Miller and Cummins 1987).

3. Perhaps most seriously, the four types tell us too little about longevity. For practical purposes, as Vyvey (1986), Bakker (1989) and Pfadenhauer and Maas (1987) have pointed out, we would like to know whether seeds are likely to persist beneath plant communities which have been destroyed or degraded at some time in the past.

We have therefore adopted a modified version of the seed bank classification proposed by Bakker (1989; Bakker *et al.* 1991) and described in Thompson (1992, 1993), which defines three types:

Transient   Species with seeds which persist in the soil for less than one year, often much less. This corresponds directly to Thompson and Grime's Types I and II, and acknowledges that for many species the two types are inseparable.

Short-term persistent   Species with seeds which persist in the soil for at least one year, but less than five years. This type, originally described by Bakker (1989) as 'persistent', may play a role in the maintenance of plant populations after poor seed setting in a dry year or after cutting too early.

Long-term persistent   Species with seeds which persist in the soil for at least five years. This type, originally termed 'permanent' by Bakker (1989), is the only one likely to contribute to the regeneration of destroyed or degraded plant communities.

The cut-off point of five years between the latter two types is

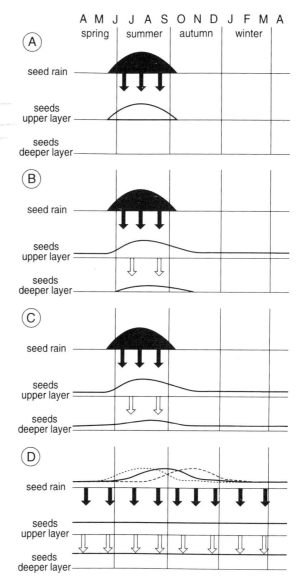

Figure 1.3 Four types of seed bank (after Poschlod and Jackel 1993), defined on the basis of seasonal dynamics of the seed rain and seeds in the upper and deeper soil layers. For more detailed description see text.

admittedly arbitrary, and was chosen largely because it is the end point of a significant proportion of burial experiments.

While we were working on the database, Poschlod and Jackel (1993) published an elaboration of the classification into transient, short-term persistent and long-term persistent types, which relies on the dynamics of the seed bank and seed rain (Figure 1.3). They recognise four types:

A   Transient   Seeds confined to the upper soil layer, and only for a short period after seed rain (persistent for < 1 year).
B   Transient   Seeds in the surface soil all year, with a distinct peak following seed rain, and some seeds in the lower soil layer (persistent for one or two years).
C   Persistent   Many seeds in the surface soil and some seeds in the lower layer all year, with a distinct peak following seed rain in the upper layer and a much smaller peak in the lower layer (persistent for some years to some decades).
D   Persistent   At least as many seeds in the lower soil layer as in the upper layer all year, and no distinct peak after seed rain (persistent for several decades).

Persistence was estimated from the date of afforestation of former chalk grassland, assuming that species from the latter habitat rapidly disappear after canopy closure. This scheme, by combining seasonal behaviour with depth distribution, is certainly a refinement of previous seed bank classification systems. Unfortunately, like all such systems, it suffers from the drawback that the data needed to apply it to most species are simply not available.

Classification criteria

Seed bank studies are exceptionally variable, and there is therefore a wide range of possible criteria which could be employed in allocating species to seed bank types. Without doubt the simplest type of data (though not necessarily the least ambiguous) derives from experiments in which seeds are artificially buried. The results of such studies are theoretically capable of providing an exact upper limit to the longevity of seeds under the particular set of conditions employed. The chief sources of variability between different studies are soil type, depth of burial and degree of

disturbance. Of these the first is relatively unimportant, but the other two can both significantly affect the outcome. Deeply buried seeds generally persist longer than shallow ones, and seeds persist longer in undisturbed soil. The database identifies data from artificially buried seeds, the depth of burial and whether the soil was disturbed after burial, but for more information on the methods used in particular studies the reader must consult the original reference.

The results of artificial burial experiments have a beguiling air of authority and precision, but must be interpreted with caution. Artificial burial bypasses the crucial role of natural burial mechanisms, and therefore is prone to serious exaggeration of seed longevity. It is probably safe to say that if a species proves to be short-lived when artificially buried, then it will also be short-lived under more natural conditions. The converse, however, cannot be relied upon. Pons (1989) found that artificially buried seeds of *Molinia caerulea* survived for at least three years, yet he remarks that he knows 'of no records of persistent seed banks of this species'. The database confirms that most of the data from naturally buried seed suggest that *Molinia* is short-lived, implying that under normal conditions *Molinia* seeds, for whatever reason, rarely become buried. Numerous similar examples will be noted in the database, and it is generally true that the results of long-term burial experiments are close to the upper limit of longevity recorded for many species.

A second difficulty with artificial burial experiments is that, with a very few notable exceptions, the period of burial is relatively short. This is inevitable; the funding problems encountered by seed burial experiments are no different from those which beset all long-term ecological research. For many species, meaningful burial experiments need to continue beyond the working life of a single experimenter. For these reasons, burial experiments are much better at separating short-lived from long-lived seeds than they are at determining the potential longevity of the latter.

Artificially buried seeds are also, to varying degrees, protected from the attentions of potential predators. It is uncertain how far seed banks are depleted by predation, but buried seeds are ingested in large numbers by earthworms, and many are either killed by this treatment, or are exhumed and stimulated to germinate (Van Tooren and During 1988; Thompson, Green and Jewels 1994).

Studies of naturally buried seeds also frequently provide direct evidence of seed longevity. This evidence usually takes the form of species which are no longer present in the community but are still present as seeds in the soil. Provided the last time the species grew at the site can be ascertained with reasonable certainty, buried seeds can often be accurately dated. Two common examples of communities in which this is normally possible are weed seeds beneath formerly arable grasslands, and seeds of light-demanding species beneath woodlands and plantations of known age. Other direct sources of evidence of seed longevity include seeds buried beneath volcanic ash or buildings of known age, seeds in stored topsoil and prevention of fresh seed input by close cutting or the application of herbicides. All these sources, of course, normally provide evidence only of *minimum* potential longevity.

Some of these kinds of sources occasionally provide evidence of apparently very great longevity in buried seeds, and great care must be taken to guarantee that such reports are not the result of contamination by seeds of more recent origin. Contamination of soil samples by wind-borne or surface seed is always a potential problem, of course, but it becomes much more serious when, as with very old seeds, the likely density of genuine seed is low. A good discussion of records of great longevity in buried seeds can be found in Priestley (1986). We return to this point later when we discuss admissibility of data.

Another valuable but less direct source of evidence is the vertical distribution of seeds in the soil. There is abundant evidence that deeply buried seeds are older than shallow ones, allowing the ratio of deeply buried to shallow seeds to be used as an index of seed longevity. This approach, used with care, has been shown to produce results which are broadly comparable with those of more direct measures of longevity (Bakker 1989). Poschlod (1993), however, feels that too many discrepancies occur, perhaps partly due to the vertical transport of seeds by earthworms (Willems and Huijsmans 1994). A potential obstacle to the general application of the method is the wide variety of soil depths employed by different investigators. Wherever possible we have taken 'surface soil' to mean the top 5 cm, but other depths have been used where this could not be avoided.

Direct evidence of longevity is reported in the database, and is expressed simply as the maximum number of years

persistence in the soil recorded in a particular study. Where this number of years is less than the maximum which could have been recorded (e.g. a species survives for four years in a burial experiment lasting five years), the number of years for which the species survived is given. Where the species survives to the end of an experiment, the number is preceded by >, in the above example >5. Note that although these examples refer to artificial burial experiments, studies of naturally buried seed can generate longevities in exactly the same way. Indirect measurements of longevity, derived from vertical distribution in the soil, can of course only allocate species to a persistence class and cannot be more precise.

A depressingly large quantity of seed bank data is not amenable to any of the above methods of measuring longevity. If the soil is sampled on a single occasion and is not vertically subdivided, then seeds of species which are present in the vegetation clearly cannot be dated. In order to accommodate these rather low-grade data we have defined an extra seed bank category: 'present', defined as seeds in the soil but of unknown longevity. There are huge numbers of these not very useful data, but for many species they are the only data available; we have included them to allow interested readers to consult the original reference and draw their own conclusions.

In an attempt to formalise the above criteria into a more usable form, we have devised a key to seed bank types (Figure 1.4). The key applies only to naturally buried seeds and to data of the most common type, that is, an enumeration of seeds in soil sampled on a single occasion. The key uses both direct and indirect evidence of longevity, but gives priority to direct evidence. The key deals with the small quantity of incompletely or inadequately described vegetation by assuming that everything in the seed bank is also in the vegetation. While not really satisfactory, there seems little alternative.

A few points of explanation are necessary. Any species in the vegetation but not detected in the seed bank is considered to be transient. Occasionally a species does not produce seed at a particular site and will then appear as transient even if its seeds are in fact persistent. In reality, of course, there is little practical distinction between a transient seed bank and a failure to produce seed at all.

A more serious potential problem is the failure to detect seeds present in the soil. Germination methods employed in

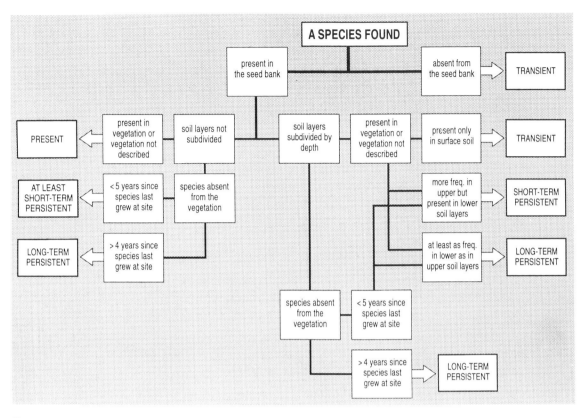

Figure 1.4 A dichotomous key to the three seed bank types employed in the database: **transient**, **short-term persistent** and **long-term persistent**. For definitions of seed bank types see text.

the majority of seed bank studies may fail to detect seeds with unusual or 'difficult' germination requirements. Nakagoshi (1985), using extraction methods, reported persistent seed banks in a range of genera which are normally regarded as having transient seeds, e.g. *Viburnum, Ligustrum, Ilex* and *Sorbus*. Whether the European and North American members of these genera have persistent seed banks which have so far escaped attention remains to be determined. Only artificial burial experiments or extraction of surviving seeds from the soil are likely to resolve this problem. A related problem concerns the Orchidaceae, a family not recorded in the seed bank by any source. Whether this reflects the genuine absence of the family from the seed bank, or a failure to provide the right germination conditions, is unknown.

Seed bank data of other types are dealt with according to criteria consistent with those above, as far as possible. For instance, seasonal sampling without subdivision by depth

(e.g. Thompson and Grime 1979) is incapable of distinguishing short-term from long-term persistence. The solution adopted in this case has been to allocate all persistent species to the short-term persistent category. Of course many of them are in fact long-term persistent, but the only alternatives would have been to ignore the data altogether or use the almost valueless 'present' category. The guiding principle in dealing with all data sources has been to use the data if at all possible, while making the fewest assumptions and using the 'present' category only as a last resort.

Data collection and admissibility

The data presented in this account are a condensed version of a larger database, which records all relevant information about each data source. Data were allocated to one of a pair of linked databases, which contain data about each *source* and each *species* respectively. Much of this information is not directly reported here and was collected primarily in order to determine whether the data were admissible.

We have had to exercise our judgement on the subject of how few naturally buried seeds we were prepared to accept as evidence of persistence in the soil. Ideally, one would arrive at a separate decision for each data source, based on an assessment of the thoroughness of the sampling, but in practice we have attributed **one** or **two** buried seeds to the possible consequences of contamination or recent dispersal. In order to apply this criterion we had to know exactly how many seeds of each species were actually recovered. The most frequent single reason for rejecting data was an inability to discover this information, usually because it was impossible to work out the actual area sampled. Some classic papers had to be omitted on account of this problem (e.g. Milton 1943, 1948).

Other frequent causes of rejection were: burial experiments conducted for too short a period (often measured only in months); data for two or more sites, treatments or taxa pooled; data presented only as frequencies or graphs; species identification poor, e.g. only to genus. For example, the classic paper of Van Altena and Minderhoud (1972), describing the established vegetation and seed bank of over 70 meadows, could not be used since all data were condensed to frequencies. The second most frequent cause of rejection was that no attempt was made to determine the viability of seeds extracted from soil. Much has been written on the relative merits of extraction or germination of seeds from soil samples (e. g. Gross 1990; see also Chapter 3, this volume), but germination has the undeniable advantage of guaranteeing that the seeds recovered are alive. Symonides (1978) found that fewer than 10% of seeds of some species extracted from the seed bank were capable of germination. It therefore seemed prudent to reject data where no effort was made to determine if seeds extracted from the soil were viable. Germination, staining with tetrazolium and a firm or white embryo were all accepted as evidence of viability.

Some of the problems we encountered are a consequence of the inevitable condensation of large amounts of data necessary to meet the demands of journal editors. A further difficulty we encountered was inconsistency between methods and results. For every source we attempted to work out if the methods as described could have produced the stated results. To give a simplified and hypothetical example, if a total area of 0.1 m² of soil was sampled, and the data expressed on a m⁻² basis, then (a) the minimum possible density was 10, and (b) all densities should be multiples of 10. Surprisingly frequently, calculations of this sort revealed data which could not have been obtained from the methods as described. Wherever possible, we tried to correct methodological problems or abbreviated data by contacting the authors. Following up publications in this way sometimes led us to useful unpublished data. Inevitably, however, some problems remained unresolved and the sources had to be rejected.

Data collection in the future

We summarise the criteria to be satisfied before a paper can be incorporated into the database:

1. Full and correct identification of species.
2. Seeds germinated or tested for viability.
3. Calculation of seed density per square metre possible.
4. Sampling depth reported.
5. Data reported for individual treatments, without pooling of sites or taxa.
6. Burial experiments exceed one year.

We are aware of large numbers of new papers published since the beginning of 1994, and of course we have overlooked many, old and new. We have hardly exploited the huge Russian literature at all. We would be pleased to hear from authors of new papers or overlooked old ones, from those with unpublished seed bank data, and especially from Russian-speaking ecologists with time on their hands.

Principal features of the data 2

Previous reviews

Before discussing our dataset, we refer briefly to three earlier published compilations of seed bank literature.

The first is the bibliographical review by the Belgian Quirin Vyvey published in *Excerpta Botanica* section B (1989a, b). This review covers 845 publications on natural buried viable seed banks as well as studies of artificially buried seeds, and provides a good summary of the general seed bank literature up to 1989. We quote the introduction by Vyvey: 'Publications of interest for nature conservation (e.g. natural revegetation after disturbance, survival of endangered plant populations) and publications with agricultural interest are included. More than one third (35%) of the publications are dealing with weed seeds in the soil, mainly in Europe, North America and Australia. Seed bank studies of natural or semi-natural grasslands (18%) and forests (16%) are less numerous and are mostly restricted to Europe, North America and Japan. The oldest mentioned study is Darwin's (1859). In the 19th century 9 more studies have been published and from 1900 until 1959 I found 144 titles. In the 1960's 99 studies and in the 1970's already 176 papers have been published. From 1980 until about half 1988 416 papers came out, of which about 58 in 1988. This illustrates the increasing interest in study of the soil seed bank.'

The second is the review 'Diasporenbank im Boden als Vegetationsbestandteil' (Soil seed bank as part of the vegetation) by the Germans Karl-Georg Bernhardt and Peter Poschlod, also published in *Excerpta Botanica* section B (1993). The review includes 193 publications. The far lower number of publications than in Vyvey's review is due to the selection made by Bernhardt and Poschlod. The latter authors only mentioned publications dealing with the dynamics and interaction of the established vegetation and the soil seed bank, and omitted papers on methodological problems. A considerable part of their publications are from Russia. Unfortunately, the great majority of Russian papers, mostly published in the 1950s and 1960s, are not accessible as they are only in Russian. The majority deal with grassland and its restoration, and our knowledge of grassland seed banks would be much improved if translations of these papers were available. Like Vyvey, Bernhardt and Poschlod found an increase in the number of publications during recent decades, particularly in the 1980s (Figure 2.1), and that the majority involve arable land (Figure 2.2). The discrepancy in sample number between Figures 2.1 and 2.2 arises from population dynamic studies not attributed to a specific habitat.

The third bibliography is a review by the Swede Per Milberg on the longevity of seeds as indicated from old seed collections and from burial trials including the percentage of viable seeds (Milberg 1990). It summarises the results of over 200 publications.

Number of publications, source records and species records

The present North West European seed bank database is derived from 275 publications up to the beginning of 1994. This is a far lower number than included by Vyvey (1989a, b). The latter, if continued, would have reached beyond

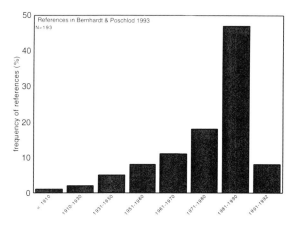

Figure 2.1 Dates of origin of the references reviewed by Bernhardt and Poschlod (1993).

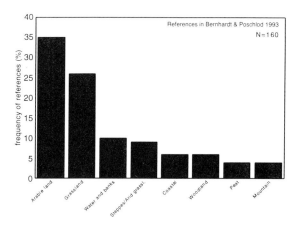

Figure 2.2 Subdivision of the references reviewed by Bernardt and Poschlod (1993) by plant communities.

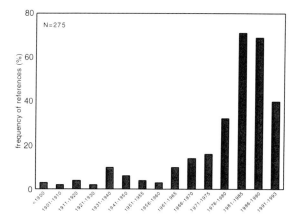

Figure 2.3 Dates of origin of publications included in the database.

1000 publications by 1994. There are two reasons for the relatively small number of publications taken into account in our database. Firstly, Vyvey cites publications from all over the world including those with species which do not occur in North West Europe. Secondly, we used only those publications in which the methods are clear and the seed density can be calculated as described in Chapter 1. This latter reason also accounts for the absence of some publications cited by Bernhardt and Poschlod (1993) from our database, although we cite many publications lacking from their review. About 90% of their publications are from continental Europe including Russia.

Many of the publications in our database include data from more than one site, the same site sampled in different years, or burial experiments at various depths or different soils. We treat each such record separately, and thus our total of 275 publications contains 1936 *source records*. Since each source record normally contains data on more than one species, the database contains a total of 21 071 s*pecies records*.

Period and origin of publications

Our database covers the period 1882 until the beginning of 1994. The number of publications remained low until the 1970s and increased rapidly in the 1980s (Figure 2.3).

The true numbers are almost certainly higher; we must have missed publications (+ unpublished data) matching our criteria. We would be grateful if readers would alert us to studies lacking from the database, and of course new papers. Our data suggest that seed bank studies concerning the North West European flora derive particularly from Great Britain, USA, Germany, The Netherlands, Canada and Sweden (Figure 2.4). It is possible that this pattern conceals some (perhaps linguistic) bias; certainly we did not expect to find quite so few records from France and Belgium. The majority of the studies included in the database (68%) were carried out in North West Europe (as defined in Chapter 1).

Methods used in publications

The method adopted in most source records (70%) is sampling soil from natural vegetation without extraction

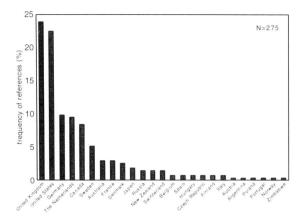

Figure 2.4 Countries of origin of publications included in the database.

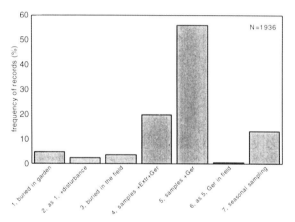

Figure 2.5 Methods adopted in the seed bank studies included in the database. Extr: seeds are extracted from the soil sample. Ger: germination or viability of the seeds is tested. See Chapters 1 and 3 for further explanation.

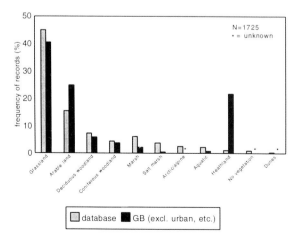

Figure 2.6 Relative representation of habitats in the database compared with the abundance of habitats in Great Britain, according to the 1990 satellite land cover map.

of seeds or sample reduction, followed by germination in the glasshouse or outside (Figure 2.5). Methods involving reduction of sample volume or seed extraction prior to germination or viability testing account for 20% of source records. About 10% of the source records involve deliberate burial of seeds in an experimental garden or in the field. As both soil seed bank analysis (comprising methods 4, 5, 6, and 7) and seed burial experiments (comprising methods 1, 2 and 3) are rather distinct methods, we discuss each separately below. See the description of the database (Chapter 4) for an explanation of the different methods.

Analysis of natural soil seed banks

The total number of source records involving the enumeration of natural soil seed banks is 1725. The majority of the studies were carried out in grassland (Figure 2.6), of which 75% received no fertiliser application. The majority of these records also concern managed grasslands, most often by grazing and less often by hay-making (Figure 2.7).

Arable fields comprise the only other large habitat category, reflecting the continuing interest in the behaviour of buried weed seeds, although studies of arable fields are not quite as common as might be expected from the abundance of the habitat in Britain. There are relatively few data available from woodland, heathland, dunes, (salt) marshes, arctic/alpine and aquatic communities. However, when we compare these data with the relative abundance of these habitats in Great Britain, all these communities turn out to be relatively well represented. On a European scale, arctic/alpine habitats may turn out to be underrepresented, but we do not have data for the distribution of habitats in North West Europe generally. Heathlands (including upland dwarf shrub communities) seem to be underrepresented, probably reflecting their low species richness and economic value.

The total depth of sampling of natural seed banks shows that very few records were taken from the uppermost 2 cm. The majority of the samples were taken from the top 5 or 10 cm (Figure 2.8). The great majority of source records (80%) derive from studies in which a single layer was tested for the presence of viable seeds, while two or three layers were sampled in only 7% or 5% of cases (Figure 2.9). In more than half the cases in which more than one layer was

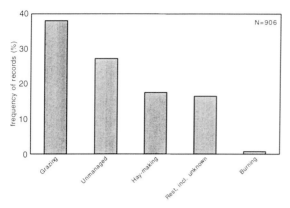

Figure 2.7 Management of grassland habitats included in the database.

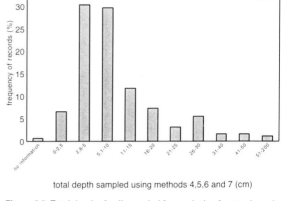

Figure 2.8 Total depth of soil sampled for analysis of natural seed banks (i.e. methods 4, 5, 6 and 7).

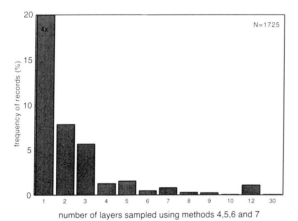

Figure 2.9 Number of layers of soil sampled for analysis of natural seed banks (i.e. methods 4, 5, 6 and 7). 80% of records sampled only one layer, but for clarity the scale has been condensed.

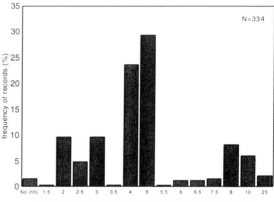

Figure 2.10 The thickness of the top layer in cases where more than one layer is sampled for analysis of natural seed banks (i.e. methods 4, 5, 6 and 7).

sampled, the thickness of the uppermost layer was 4 or 5 cm (Figure 2.10). The total volume of soil sampled covers a large range; more than half the source records relate to studies in which less than 5 litres was collected (Figure 2.11). We discuss soil volume in more detail in Chapter 3.

The sampling period clearly shows two peaks (Figure 2.12), firstly in spring (March – May) and secondly in autumn (October). Apparently, many authors are interested in the persistent viable seed bank after stratification of dormant seeds during the winter period and before the input of fresh seeds. The peak in autumn reflects the interest of many authors in the viable seed bank shortly

after fresh seed rain. Very few samples are collected in January and February, a period when the ground is frozen or waterlogged in many north temperate countries. Seed bank investigators probably also respond poorly to being frozen or waterlogged.

The period judged to be adequate for germination of seeds in soil samples seems to depend on the method adopted. Studies in which seeds were extracted generally employed a shorter period than those in which there was no extraction (Figure 2.13). This difference is discussed in Chapter 3.

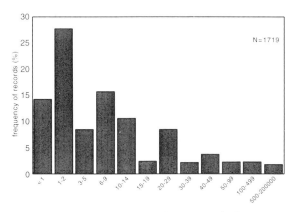

Figure 2.11 Number of litres of soil sampled for analysis of natural seed banks (i.e. methods 4, 5, 6 and 7).

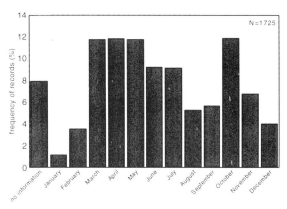

Figure 2.12 Timing of soil sampling for analysis of natural seed banks (i.e. method 4, 5, 6 and 7).

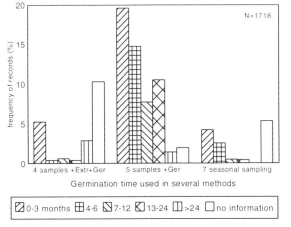

Figure 2.13 Duration of germination period used in several methods of natural seed bank analysis. Method 6 is excluded owing to low sample size (see also Figure 2.5).

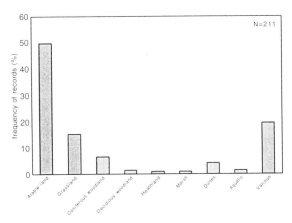

Figure 2.14 Distribution of burial experiments among different habitats.

Artifical seed burial experiments

The database contains 211 source records on seed burial experiments. Half were carried out on arable fields, but 40% of the experiments were carried out in natural or semi-natural landscapes. In a total of 20% of the records seeds were buried in grassland habitats (Figure 2.14). Seeds were buried within 10 cm of the soil surface in nearly 70 % of burial experiments (Figure 2.15), although in exceptional cases seeds were buried up to 2 metres below the soil surface.

The duration of burial experiments ranges from 1–100 years, but 75% of the records relate to a burial period of less than 6 years (Figure 2.16). For more than half of the records the period of burial is unknown, which explains the low number of records in Figure 2.16. More than 95% of the publications on seed burial experiments provided no information about the time of year when seeds were exhumed or the period during which germination of exhumed seeds was tested.

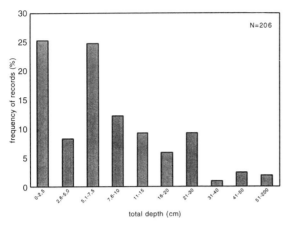

Figure 2.15 Distribution of different depths applied in burial experiments.

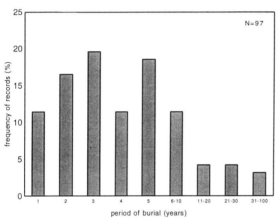

Figure 2.16 Burial experiments subdivided by duration.

Taxonomic relationships of the database

The North West European region harbours species from 120 higher plant families, of which 98 are represented in our database (Table 2.1). About half of the 22 families for which we lack any information are primarily woody. For reasons discussed in Chapter 1, the Orchidaceae were omitted from the database at the outset. Otherwise, no large family is absent from the database. Other than those families which are wholly absent, the main lacunae in the database can be derived from Table 2.1 by comparing the number of species occurring in North West Europe with the number of species and records present in the database. For a typical large family, we have some data for about half of the species present in North West Europe. This is true, for instance, of the Caryophyllaceae, Compositae, Gramineae, Leguminosae, Ranunculaceae, Scrophulariaceae and Umbelliferae. Representation of smaller families depends to a large extent on their abundance in the agricultural habitats which have been the main focus of seed bank investigation. Not surprisingly therefore, the Urticaceae, Plantaginaceae and Polygonaceae are the top three families in terms of records per species. In contrast, some medium-sized families which are mostly confined to semi-natural, unproductive habitats, such as the Liliaceae, Gentianaceae and Saxifragaceae, are underrepresented in the database. Aquatic families, such as the Alismataceae and Potamogetonaceae, are generally poorly represented. There is a clear tendency for small-seeded families to have the highest proportion of long-term persistent records.

The top-100 species ranked by their number of records in the database amplifies the pattern shown in Figure 2.6; they are almost entirely species from productive grassland and arable habitats (Table 2.2). Extremely few species from heathland and woodland are represented in the top-100, and one has to travel to number 45 before encountering the first of these (*Calluna vulgaris*). The great majority of species in the top-100 tend to have a long-term persistent seed bank.

Table 2.1 Families occurring in North West Europe and their representation in the database by species numbers and the number of records per seed bank type. The database contains information only on the families in bold

Family	Total number of species in NW Europe	Number of species represented in the database	Number of *transient* records	Number of *short-term persistent* records	Number of *long-term persistent* records	Total number of records in the database
Aceraceae	5	3	23	0	0	24
Adoxaceae	1	1	4	0	0	4
Aizoaceae	1	0	0	0	0	0
Alismataceae	10	1	0	6	6	18
Amaranthaceae	4	3	3	7	16	54
Amaryllidaceae	7	0	0	0	0	0
Apocynaceae	2	0	0	0	0	0
Aquifoliaceae	1	1	13	0	0	13
Araceae	4	2	13	0	0	13
Araliaceae	1	1	31	0	0	31
Aristolochiaceae	2	1	5	0	0	5
Asclepiadaceae	1	1	2	0	0	2
Balsaminaceae	4	4	14	0	0	14
Berberidaceae	1	0	0	0	0	0
Betulaceae	7	5	48	43	20	131
Boraginaceae	46	17	79	29	16	195
Buddlejaceae	1	0	0	0	0	0
Butomaceae	1	0	0	0	0	0
Buxaceae	1	0	0	0	0	0
Callitrichaceae	4	1	1	2	0	5
Campanulaceae	27	17	55	23	7	138
Cannabaceae	2	2	4	0	0	4
Caprifoliaceae	12	7	38	8	3	49
Caryophyllaceae	118	55	283	382	166	1240
Celastraceae	2	1	2	0	0	2
Ceratophyllaceae	2	1	9	0	0	9
Chenopodiaceae	33	14	30	139	78	373
Cistaceae	6	3	8	3	0	16
Compositae	244	134	1029	517	218	2249
Convolvulaceae	8	4	39	2	10	52
Cornaceae	5	1	3	0	0	3
Corylaceae	2	2	12	0	0	14
Crassulaceae	23	7	7	2	0	17
Cruciferae	145	54	140	288	168	861
Cucurbitaceae	3	0	0	0	0	0
Cupressaceae	3	1	12	0	0	12
Cyperaceae	174	78	290	132	62	653
Diapensiaceae	1	1	1	0	0	1
Dioscoraceae	1	1	3	0	0	3
Dipsacaceae	9	6	61	10	4	84
Droseraceae	3	2	0	2	0	3
Elaeagnaceae	1	0	0	0	0	0
Elatinaceae	3	1	0	0	1	1
Empetraceae	2	2	9	0	1	17
Ephedraceae	1	0	0	0	0	0
Ericaceae	26	16	123	38	34	279
Eriocaulaceae	1	1	0	1	0	2
Euphorbiaceae	22	12	34	23	33	166
Fagaceae	8	4	29	0	0	29
Frankeniaceae	1	0	0	0	0	0
Fumariaceae	19	6	6	3	17	31
Gentianaceae	29	12	19	8	3	44
Geraniaceae	22	13	67	15	12	105
Globulariaceae	1	0	0	0	0	0
Gramineae	241	130	2033	855	270	4237

Table 2.1 Continued

Family	Total number of species in NW Europe	Number of species represented in the database	Number of *transient* records	Number of *short-term persistent* records	Number of *long-term persistent* records	Total number of records in the database
Grossulariaceae	5	1	3	0	0	3
Guttiferae	16	9	34	66	48	212
Haloragaceae	3	1	3	0	0	3
Hippocastanaceae	1	0	0	0	0	0
Hippuridaceae	1	0	0	0	0	0
Hydrocharitaceae	8	1	1	0	0	1
Iridaceae	14	3	10	0	1	12
Juglandaceae	1	0	0	0	0	0
Juncaceae	53	29	151	344	225	968
Juncaginaceae	3	1	5	10	0	15
Labiatae	72	39	243	104	67	581
Leguminosae	138	60	654	139	153	1384
Lemnaceae	6	2	13	0	0	13
Lentibulariaceae	8	4	4	3	0	13
Liliaceae	60	24	104	5	0	116
Linaceae	9	2	19	16	6	66
Loranthaceae	1	0	0	0	0	0
Lythraceae	4	2	19	12	8	51
Malvaceae	13	5	3	10	10	23
Menyanthaceae	2	2	3	2	0	7
Monotropaceae	1	1	1	0	0	1
Myricaceae	2	1	1	0	0	2
Najadaceae	3	2	3	0	2	6
Nymphaeaceae	3	2	10	0	0	10
Oleaceae	4	2	25	1	0	26
Onagraceae	34	16	80	118	26	299
Orobanchaceae	17	0	0	0	0	0
Oxalidaceae	7	3	22	12	2	75
Papaveraceae	11	8	22	22	17	95
Parnassiaceae	1	1	6	1	0	8
Pinaceae	16	9	26	9	0	35
Plantaginaceae	7	6	150	112	44	443
Plumbaginaceae	7	3	17	0	0	22
Polemoniaceae	2	0	0	0	0	0
Polygalaceae	6	5	28	2	0	34
Polygonaceae	42	21	282	196	157	910
Portulaceae	4	3	7	12	17	46
Potamogetonaceae	23	4	24	2	0	38
Primulaceae	30	17	124	55	41	337
Pyrolaceae	8	3	9	0	0	9
Ranunculaceae	81	40	363	116	101	724
Resedaceae	5	3	0	5	1	9
Rhamnaceae	3	1	4	0	0	4
Rosaceae	101	47	497	112	64	853
Rubiaceae	27	17	272	43	8	466
Ruppiaceae	2	2	1	0	1	3
Salicaceae	33	13	47	0	8	56
Santalaceae	7	0	0	0	0	0
Saxifragaceae	29	9	13	11	2	33
Scrophulariaceae	100	52	206	182	108	772
Solanaceae	11	8	15	23	31	89
Sparganiaceae	7	1	6	0	0	6
Tamaricaceae	2	0	0	0	0	0
Taxaceae	1	0	0	0	0	0
Thymelaeaceae	3	0	0	0	0	0
Tiliaceae	3	1	1	0	0	1
Typhaceae	5	2	1	9	4	17

Table 2.1 Continued

Family	Total number of species in NW Europe	Number of species represented in the database	Number of *transient* records	Number of *short-term persistent* records	Number of *long-term persistent* records	Total number of records in the database
Ulmaceae	4	2	0	0	0	0
Umbelliferae	105	46	324	98	34	525
Urticaceae	5	2	28	82	25	182
Valerianaceae	9	6	36	2	1	47
Verbenaceae	1	1	1	1	0	7
Violaceae	24	11	49	37	19	182
Zannichelliaceae	1	1	1	8	0	16
Zosteraceae	3	2	5	0	0	5

Table 2.2 Top-100 species ranked by the number of records in the database

Species	Total number of records	Species	Total number of records	Species	Total number of records
Trifolium repens	326	*Fallopia convolvulus*	104	*Festuca ovina*	70
Juncus effusus	278	*Thlaspi arvense*	103	*Myosotis arvensis*	70
Holcus lanatus	267	*Arrhenatherum elatius*	102	*Veronica arvensis*	70
Cerastium fontanum	265	*Alopecurus myosuroides*	101	*Persicaria maculosa*	69
Poa trivialis	254	*Hypericum perforatum*	101	*Sinapis arvensis*	69
Ranunculus repens	250	*Filipendula ulmaria*	100	*Galium verum*	68
Stellaria media	247	*Galium aparine*	100	*Papaver rhoeas*	68
Poa pratensis	233	*Vicia cracca*	100	*Trisetum flavescens*	67
Festuca rubra	224	*Alopecurus pratensis*	97	*Veronica chamaedrys*	67
Taraxacum officinale	221	*Festuca pratensis*	96	*Alopecurus geniculatus*	66
Chenopodium album	207	*Calluna vulgaris*	95	*Centaurea jacea*	66
Rumex acetosa	204	*Luzula campestris*	95	*Angelica sylvestris*	65
Plantago major	198	*Potentilla erecta*	95	*Chamerion angustifolium*	65
Poa annua	197	*Cirsium palustre*	93	*Glyceria fluitans*	65
Plantago lanceolata	194	*Rumex obtusifolius*	92	*Sonchus asper*	65
Agrostis capillaris	187	*Deschampsia cespitosa*	91	*Agrostis canina*	64
Agrostis stolonifera	180	*Viola arvensis*	89	*Arenaria serpyllifolia*	63
Juncus bufonius	179	*Lychnis flos-cuculi*	87	*Campanula rotundifolia*	63
Anthoxanthum odoratum	172	*Bellis perennis*	85	*Heracleum sphondylium*	63
Dactylis glomerata	166	*Lotus pedunculatus*	84	*Linum catharticum*	63
Urtica dioica	164	*Medicago lupulina*	84	*Veronica hederifolia*	63
Capsella bursa-pastoris	146	*Prunella vulgaris*	83	*Aphanes arvensis*	62
Anagallis arvensis	140	*Lotus corniculatus*	82	*Deschampsia flexuosa*	62
Polygonum aviculare	139	*Rumex acetosella*	82	*Vaccinium myrtillus*	62
Sagina procumbens	135	*Juncus articulatus*	81	*Euphorbia exigua*	61
Achillea millefolium	129	*Veronica serpyllifolia*	79	*Rubus fruticosus*	61
Ranunculus acris	127	*Rubus idaeus*	78	*Galium saxatile*	59
Cirsium arvense	125	*Lathyrus pratensis*	77	*Betula pubescens*	58
Trifolium pratense	120	*Phleum pratense*	77	*Ajuga reptans*	56
Lolium perenne	119	*Gnaphalium uliginosum*	75	*Arabidopsis thaliana*	56
Elytrigia repens	117	*Stellaria graminea*	75	*Avena fatua*	56
Trifolium dubium	117	*Rumex crispus*	74	*Epilobium ciliatum*	56
Leucanthemum vulgare	114	*Veronica persica*	74		
Cardamine pratensis	106	*Cynosurus cristatus*	72		

Table 2.3 Species ranked by highest average density m^{-2}. The top-25 species are listed for each of three depths

Depth	Lowest density (m^{-2})	Highest density (m^{-2})	Average density (m^{-2})	Number of records
0–3 cm				
Stellaria uliginosa	46433	46433	46433	1
Urtica dioica	2264	6509	4924	3
Vaccinium oxycoccos	1667	7400	4534	2
Epilobium hirsutum	3917	3917	3917	1
Juncus effusus	3833	3833	3833	1
Mentha arvensis	3783	3783	3783	1
Poa trivialis	113	7584	3707	6
Juncus inflexus	3350	3350	3350	1
Crassula tillaea	2125	3313	2719	2
Mentha aquatica	2517	2517	2517	1
Galium verum	2350	2350	2350	1
Epilobium obscurum	2167	2167	2167	1
Veronica beccabunga	2067	2067	2067	1
Hypericum perforatum	283	6300	1970	10
Minuartia verna	800	2700	1806	3
Matricaria discoidea	867	2300	1584	2
Hypericum humifusum	1567	1567	1567	1
Scirpus sylvaticus	538	2547	1543	2
Cerastium semidecandrum	250	3867	1522	3
Sagina procumbens	1383	1383	1383	1
Leucanthemum vulgare	1238	1238	1238	1
Carex echinata	1200	1200	1200	1
Arenaria serpyllifolia	1150	1150	1150	1
Cardamine flexuosa	1150	1150	1150	1
Ranunculus repens	396	2038	1146	6
0–5 cm				
Lythrum salicaria	410000	410000	410000	1
Spergularia marina	376234	376234	376234	1
Cytisus scoparius	56834	56834	56834	1
Sagina procumbens	49408	49408	49408	1
Calluna vulgaris	28000	68000	46167	6
Ulex europaeus	29493	29493	29493	1
Medicago arabica	11787	11787	11787	1
Lychnis flos-cuculi	9125	9125	9125	1
Agrostis capillaris	48	16544	6118	5
Holcus lanatus	600	16900	5957	7
Trifolium dubium	161	38427	5689	81
Poa annua	48	18176	5582	10
Galega officinalis	5167	5167	5167	1
Juncus effusus	4700	4700	4700	1
Hypericum humifusum	560	8736	4648	2
Torilis japonica	3556	3986	3771	2
Trifolium subterraneum	269	10333	3703	20
Lotus subbiflorus	161	12702	3601	12
Stellaria media	3504	3504	3504	1
Trifolium pratense	215	14101	3475	14
Trifolium repens	161	54035	3257	69
Atriplex prostrata	2655	2655	2655	1
Juncus bufonius	64	6368	2645	3
Erica tetralix	357	6010	2433	5
Poa trivialis	2400	2400	2400	1
0–10 cm				
Spergularia marina	488708	488708	488708	1
Juncus effusus	2444	97032	41522	8

Table 2.3 Continued

Depth	Lowest density (m⁻²)	Highest density (m⁻²)	Average density (m⁻²)	Number of records
Calluna vulgaris	26702	26702	26702	1
Urtica dioica	526	67979	23819	5
Bromus hordeaceus	18110	18110	18110	1
Glyceria fluitans	1248	38376	17957	6
Cardamine pratensis	468	32448	11249	3
Lolium multiflorum	11200	11200	11200	1
Salicornia europaea	3360	18600	10980	2
Typha latifolia	9510	9510	9510	1
Hypericum maculatum	1052	31944	7806	10
Hypericum perforatum	1404	12090	6747	2
Poa trivialis	2704	11180	6479	5
Alopecurus geniculatus	1508	16328	6023	6
Agrostis capillaris	88	17750	5237	34
Ranunculus repens	260	10920	5170	7
Juncus bufonius	38	89063	5083	23
Sagina procumbens	4576	4576	4576	1
Digitalis purpurea	4256	4256	4256	1
Hypochaeris glabra	4140	4140	4140	1
Stellaria media	4110	4110	4110	1
Lychnis flos-cuculi	3171	4472	3822	2
Galium palustre	572	6344	3458	2
Poa pratensis	3213	3213	3213	1
Juncus conglomeratus	88	16350	3196	6

Seed density

Data on seed density show a large variation. The majority of species have seed densities below 500 seeds m⁻² (Figure 2.17). Average densities up to 5000 seeds m⁻² are, however, not exceptional, and maximal densities of 300 000–500 000 seeds m⁻² of a single species have been reported. A summary of the top-25 species according to their average seed densities is given in Table 2.3. The summary is given for three frequently sampled depths (0–3 cm, 0–5 cm, 0–10 cm) since density obviously depends on the total volume sampled. Not surprisingly therefore, the average densities of the top-25 species tend to increase if a greater depth is taken into account. This also holds for several species sampled at more than one depth, e.g. *Juncus effusus, Poa trivialis, Spergularia marina, Hypericum perforatum*. Some species, however, do not show such a clear relationship, e.g. *Calluna vulgaris, Sagina procumbens*. Also the lowest and highest densities recorded for a particular depth can vary enormously e.g. 88–17 750 m⁻² for *Agrostis capillaris* and 38–89 063 m⁻² for *Juncus bufonius* in the 10 cm layer (Table 2.3). This variation, which has already been noted in Chapter 1 and will be discussed further in Chapter 3, is an

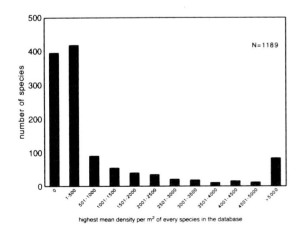

Figure 2.17 Distribution of highest mean density of seeds m⁻² irrespective of volume sampled.

Table 2.4 Top 100 species ranked by their maximum recorded longevity

Species	Method [1]	Maximum longevity (years)	Species	Method [1]	Maximum longevity (years)
Lamium album	5	>660	Viola canina	5	>40
Lamium purpureum	5	>660	Ambrosia artemisiifolia	3	40
Taraxacum officinale	5	>660	Lepidium virginicum	3	40
Glechoma hederacea	5	>460	Portulaca oleracea	3	40
Trifolium repens	5	>460	Apium graveolens	1	>39
Carex bigelowii	5	>200	Arctium lappa	1	>39
Luzula parviflora	5	200	Calystegia sepium	1	>39
Sambucus nigra	5	>160	Chenopodium hybridum	1	>39
Galium saxatile	5	>120	Leucanthemum vulgare	1	>39
Juncus conglomeratus	5	>100	Nicotiana tabacum	1	>39
Malva pusilla	3	>100	Onopordum acanthium	1	>39
Verbascum blattaria	3	>100	Poa pratensis	1	>39
Verbascum thapsus	3	>100	Robinia pseudoacacia	1	>39
Veronica officinalis	5	>100	Setaria verticillata	1	>39
Hyoscyamus niger	5	>90	Solanum nigrum	1	>39
Malva sylvestris	5	>90	Ballota nigra	5	>35
Rubus idaeus	5	>87	Chenopodium polyspermum	5	>35
Medicago lupulina	5	>80	Daucus carota	5	>35
Sinapis arvensis	5	>80	Juncus articulatus	5	>35
Oenothera biennis	3	80	Juncus bufonius	5	>35
Rumex crispus	3	80	Juncus squarrosus	5	>35
Juncus effusus	5	>73	Knautia arvensis	5	>35
Calluna vulgaris	5	>68	Luzula campestris	5	>35
Carex pilulifera	5	>68	Potentilla reptans	5	>35
Atriplex patula	5	>58	Vicia sativa	5	>35
Chenopodium rubrum	5	>54	Potentilla erecta	5	>32
Elatine triandra	3	>50	Buphthalmum salicifolium	5	>30
Gnaphalium uliginosum	5	>50	Chaenorhinum minus	5	>30
Ranunculus sceleratus	3	>50	Erica tetralix	5	>30
Rorippa islandica	5	>50	Euphorbia cyparissias	5	>30
Brassica nigra	3	50	Euphorbia exigua	5	>30
Persicaria hydropiper	3	50	Geranium pusillum	5	>30
Hypericum perforatum	5	>45	Linum catharticum	5	>30
Prunella vulgaris	5	>45	Mentha arvensis	5	>30
Ranunculus repens	5	>45	Mercurialis annua	5	>30
Sinapis alba	5	>41	Potentilla neumanniana	5	>30
Agrostis capillaris	5	>40	Ranunculus bulbosus	5	>30
Campanula patula	5	>40	Sanguisorba minor	5	>30
Carduus crispus	5	>40	Setaria pumila	1	>30
Cerastium fontanum	5	>40	Valerianella dentata	5	>30
Clinopodium acinos	5	>40	Phalaris arundinacea	1	30
Epilobium montanum	5	>40	Stellaria media	3	30
Filago minima	5	>40	Anagallis arvensis	4	>26
Juncus inflexus	5	>40	Rumex acetosella	1	>26
Plantago major	5	>40	Lamium amplexicaule	5	>25
Potentilla argentea	5	>40	Convolvulus arvensis	6	>22
Potentilla norvegica	5	>40	Achillea millefolium	5	>20
Sagina procumbens	5	>40	Aethusa cynapium	5	>20
Veronica polita	5	>40	Arenaria serpyllifolia	5	>20
Veronica serpyllifolia	5	>40	Capsella bursa-pastoris	5	>20

[1]For details of method codes see text p.31.

inevitable consequence of the compilation of records with varying seed rain, soils and different storage conditions for the survival of seeds. Most of the species in Table 2.3 have small or very small seeds.

Seed longevity

The maximum longevity of seeds is known to exceed 100 years for only a few species (Table 2.4). It must be emphasised, however, that data on longevities measured in centuries or many decades are available only exceptionally, and seem not always to be reliable (see Chapter 1). The top-100 species ranked by their maximum longevity in the database indicates that only 27 data are derived from burial experiments. The majority of data have been derived from historical records, frequently from soil underneath buildings, requiring an assumption that no fresh seed input has taken place. If we accept this assumption, the maximal longevity of many species is higher than presented in the review by Milberg (1990).

A point not addressed by the present database is interspecific variability in viability of seeds after exhumation, e.g. *Verbascum blattaria* had 42% viable seeds after 100 years of burial, whereas for *Verbascum thapsus* this was only 2% (Kivilaan & Bandurski 1981); *Datura stramonium* had 91% viable seeds after 39 years of burial, *Apium graveolens* only 1% (Toole & Brown 1946). Different authors often report wide variability in the longevity of a single species, e.g. for *Solanum nigrum*, 4% after 11 years (Salzmann 1954) and 83% after 39 years (Toole & Brown 1946).

Challenges for further research

Apart from an overview of principal features of the data, this chapter draws attention to the more obvious lacunae in the data and suggests challenges for further research. The gaps in our knowledge, even for a flora as small and well-studied as that of North West Europe, are remarkable. For many species we have no information at all, and many of the data we possess are of poor quality. The data suggest that attention has been concentrated on the productive agricultural habitats which make up most of the landscape, while semi-natural habitats have received much less attention. This undoubtedly reflects the relative abundance of the habitats, but has led to a serious imbalance in our knowledge. We know a great deal about a handful of important species of arable fields and fertile grassland, and almost nothing about most species (which are often declining or endangered) of less productive, semi-natural habitats. Given the importance of the latter for nature conservation and habitat restoration, it would be helpful if future seed bank studies could begin to redress the balance.

In addition to the obvious taxonomic holes in our knowledge, we suggest the following as topics worthy of future study:

1. The relationship between seed rain and seed bank.
2. Mechanisms of incorporation of seeds into the different layers of the seed bank.
3. Effects on seed viability and longevity of storage in different types of soil, e.g. sand versus peat, and under different types of management, e.g. drained versus waterlogged, fertilised versus unfertilised.
4. The physiological basis of seed longevity in soil.
5. The role of seed banks in ecosystem dynamics.

Methods of seed bank analysis

3

As indicated in the description of the database in Chapter 4, various methods of separation, extraction or emergence exist for the analysis of the soil seed bank:

(i) soil samples can be taken from the natural vegetation and the sample volume reduced, followed by extraction of seeds by flotation or washing.

(ii) soil samples can be taken from the natural vegetation and the seeds germinated inside a glasshouse or outside, without extraction or sample reduction.

(iii) the soil can be thoroughly dug to a known depth in the natural vegetation and the seeds germinated in the field.

The third method is used only rarely (e.g. Chancellor 1986; Graham and Hutchings 1988b) and we do not consider it further. We do, however, consider a similar method, in which seedling emergence is monitored *in situ* in the field after removal of the established vegetation. Sources employing this method were not included in the database since no depth and density of seeds can be determined. Nevertheless we discuss it in this chapter as it raises interesting methodological questions in comparison with other seedling emergence methods.

We comment on the (dis)advantages of these methods with respect to the number of species and individuals found, and technical points such as consumption of time and space. Finally, we conclude with some recommendations for standardising methods of seed bank analysis.

Seed separation methods

Extraction of seeds by flotation

In this method soil samples are washed with a variety of different salt solutions of a specific density. Seeds can than be separated from the other organic and mineral parts of the sample through their own species-specific density.

The flotation method was successfully applied by Malone (1967) for arable weeds and seems adequate if one is interested in the size of the seed bank of a single species. If the aim is the study of the seed bank as part of a plant community the method is not adequate. The main difficulty is that the density (= salt concentration) at which viable seeds can be separated varies among species (Gross 1990). To separate the seeds of all the species present in soil samples requires repeated washing and rinsing, resulting in considerable loss of the sample (Gross 1990).

Extraction of seeds by washing and sieving

Seeds are concentrated by washing soil samples through sieves of various mesh sizes. The finest mesh size, intended to catch the smallest seeds, measured 0.2 mm (Barralis *et al.* 1988), 0.1 mm (Bernhardt and Hurka 1989), 0.243 mm (Gross 1990), or 0.212 mm (Ter Heerdt *et al.* 1996). The 0.212 mm sieve has proved to be fine enough to catch seeds of *Juncus* species. The advantage of washing and sieving is reduction of the volume of the soil sample, which makes it easier to find seeds under the microscope. Since the concentrated sample still contains soil material, examination under a dissecting microscope is necessary. Separation on different sieve (mesh) sizes facilitates the counting and sorting of seeds by their size (Gross 1990).

Both seed separation methods, by flotation and by washing, seem very effective in finding large-seeded species (Malone 1967; Fay and Olsen 1978; Benz *et al.* 1984). These methods are, however, ineffective and time consuming for small-seeded species and for soil including much organic matter (Ter Heerdt *et al.* 1996). Finlayson, Cowie and Bailey (1990) were not able to sort the many small seeds from the organic matter that remained after sieving.

In order to identify the seeds of different species a reliable key at the species level is needed and/or a reference collection of seeds collected from the study sites.

Identification and testing of the viability of seeds after extraction

Even successful separation of seeds provides no information about viability. Since only living seeds are normally considered to contribute to the seed bank, their viability has to be tested. Olmsted and Curtis (1947) found a five- to hundred-fold higher number of seeds by microscopic examination than by germination of the extracted seeds. This large discrepancy suggested either that a vast number of seeds remain dead in the soil but readily recognisable, or that the germination test failed to break the dormancy of the supposedly non-viable seeds. Seeds showing external evidence of damage or disease can be discarded (Gross 1990), and viability can also be tested by pressing the seeds with a needle to find if the embryo is firm (Bakker *et al.* 1996). A tetrazolium test (Moore 1972) is often considered to provide more objective evidence of viability, but not all species from a mixed community can be stained reliably with tetrazolium. In that case, individual species can be tested for germinability in an environmental growth chamber (Gross 1990).

In general, we conclude that methods based on separation and identification of seeds are costly, time consuming and rather inaccurate. In our opinion the disadvantages mentioned outweigh the advantage that flotation and washing are not influenced by differences in germination requirements.

Seedling emergence methods

Treatment of samples prior to germination under controlled conditions

In seedling emergence methods, the samples are spread in trays and kept under conditions known (or suspected) to promote the germination of as many species and individuals as possible. The samples are usually spread on top of sterile subsoil or sand. The subsoil is needed to provide the nutrients necessary for the emerged seedlings to grow, although this is not necessary if seedlings are removed soon after germination. Sometimes the plants have to reach maturity or flower before they can be identified. Often unidentified seedlings are removed from the experimental trays, planted into separate pots and grown to flowering so that they can be identified. Plants are counted, identified and removed as soon as possible to prevent competition for light with new seedlings (Thompson and Grime 1979; Roberts 1981; Bigwood and Inouye 1988; Gross 1990).

Many authors quote Thompson and Grime's (1979) assertion that the seedling emergence method is 'not designed to provide a complete assessment of the seed flora present'. In reality, such methods can come close to this ideal if properly designed and executed. According to Major and Pyott (1966) and Galinato and Van der Valk (1986), species differ greatly in germination requirements, and therefore glasshouse conditions are not always suitable for the germination of all species. In order to break dormancy some authors stratify the seeds by putting the samples into a refrigerator at 5 °C for three weeks (Gross 1990) or by just placing them outside the glasshouse during the winter period (Hoogveld and Muller 1985). These authors found more species and individuals after cold-stratification than with non-stratified treatments. This may also be the reason why some authors put the trays in open cages which only prevent animals from entering (Pfadenhauer and Maas 1987; Poschlod *et al.* 1991). Bakker (1989) put soil samples in the glasshouse but collected them only in early spring when cold-stratification had taken place naturally during the winter period.

Period of germination

The period of seedling emergence in the trays largely depends on the thickness of the layer spread on the sterilised subsoil. Often the samples are crumbled and spread in a layer of one centimetre or more after the removal of vegetative parts of plants. Pfadenhauer and Maas (1987) report about 5 cm, Poschlod and Jackel (1993) 2–3 cm. Several authors have shown that only the seeds at the surface of the sample will germinate (Brenchley and Warington 1930; Bakker 1960; Williams 1969; Galinato and Van der Valk 1986). Seeds deeper in the soil may not germinate because the amount of light reaching the seeds is too low (Fenner 1985). To overcome this problem, after a certain period when no new seedlings emerge the sample is turned upside down or stirred to promote new emergence. Therefore, the period of seedling emergence is inevitably prolonged by thick sample layers. Roberts (1981) suggested that a period of two years would be reasonable. According to our experience, the longer the period of germination, the higher the chance that mosses (which impede seedling emergence and survival) cover the tray. Such long seedling emergence periods also imply large glasshouses or open cages, which are not always available. Poschlod and Jackel (1993) ended the seedling emergence period after about one year because of lack of space.

The space problem in the glasshouse or open cage can be countered by a reduction of the bulk of the sample by sieving (Brenchley and Warington 1930; Kropác 1966; Barralis and Chadoeuf 1980). The records in the database (see Figure 2.13) reveal that up to three months was judged adequate for most viable seeds to germinate from reduced samples, whereas up to one year was necessary when the bulk of the samples was not reduced. Concentrating the samples and using thin layers in the germination trays ensures that all seeds are exposed to light and suitable temperatures.

Bulk reduction

A comparison of concentrated and unconcentrated soil samples was carried out by Ter Heerdt et al. (1996). Sieving with a 0.2 mm mesh width reduced the bulk of the soil by 85% for clay, 70% for peaty soil, and 55% for sandy soil, indicating the potential area saving in the glasshouse. After concentration no new seedlings emerged after 6 weeks, whereas 4 to 6 months were necessary for the unconcentrated samples. Numbers of both species and individuals were higher in the concentrated than in the unconcentrated samples. Handsorting of the remainder of the sample when no new seedlings emerged revealed that no viable seeds remained; in other words all viable seeds had germinated. Rubbing the seeds over the sieve apparently scarified the seed coat to such an extent that all viable seeds germinated. The number of seeds found after seedling emergence plus handsorting in unconcentrated samples was lower than the total number found after concentration. Apparently, up to 20% of the viable seeds did not germinate, and up to 30% may be missed by handsorting, depending on the individual

species. Poiani and Johnson (1988) and Brown (1992) handsorted much smaller volumes of soil than they used in their emergence method, which makes a comparison of methods impossible.

Seedling emergence under field conditions as compared with controlled conditions

Here we discuss three papers in which the authors compare the results of seedling emergence from soil samples in the glasshouse or in an open cage with the results from seedling emergence in the field. A variety of practices has been applied to remove the established vegetation and stimulate seedling emergence in the field: Pfadenhauer and Maas (1987) removed the established vegetation in fen meadows by using herbicides which did not affect seeds; Graham and Hutchings (1988b) thoroughly dug the topsoil of improved grassland on a former arable field; Bakker et al. (1996) removed the established vegetation by cutting scrub grown over calcareous grassland and raking the layer of mosses and litter.

Pfadenhauer and Maas (1987) found that more seedlings of most species emerged from soil samples than in the field. For some species, however, the opposite was true. Similar results applied to the number of species emerged. Species that emerged in larger numbers in the field might represent the transient seed bank found in the top soil, in which no light could penetrate, after fresh seed rain, e.g. *Anthoxanthum odoratum, Plantago lanceolata, Parnassia palustris, Primula farinosa*, which have many records in the transient category. Species which emerged in higher numbers from soil samples seem to represent the persistent seed bank in deeper soil layers, which are suddenly exposed to light in the open cage, e.g. *Poa pratensis, Cerastium fontanum, Juncus articulatus, Carex* spp., which have many records in the long-term persistent category according to the criteria mentioned in Chapter 1.

Graham and Hutchings (1988b) found more species and individuals germinating from soil samples in the glasshouse than emerged under field conditions. These results seem in agreement with those of Pfadenhauer and Maas (1987), since most of the arable species found by Graham and Hutchings (1988b) have a persistent seed bank. The low levels of germination in the field as compared with the glasshouse are similar to results obtained by Kropác (1966), Roberts and Ricketts (1979) and Froud-Williams et al. (1984). All these authors worked with arable weeds.

Bakker *et al.* (1996) specifically addressed several methodological points. Several species were not found as viable seeds in samples from sites where the mature plant was present, yet did germinate in soil samples from other sites. These findings imply that the conditions in the glasshouse were suitable, and their absence may be attributable to the small sampling core area in combination with the patchy distribution of seeds. Some other species only found as emerged seedlings in the field were classified as having a transient seed bank, according to the criteria mentioned in Chapter 1. It is, therefore, unlikely that they have germinated from the seed bank underneath the moss and litter layer. It is possible that they were recently spread onto the bare soil or dropped during the process of raking litter and mosses. This suggests that the method of assessing the viable seed bank by exposing the soil is less reliable than sampling soil cores and growing viable seeds in the glasshouse. The number of seedlings emerging in the field also depends strongly on weather conditions, as in this study most seedlings appeared after a rain period. Unfavourable conditions in the field might explain why seeds of some species were found only in the glasshouse.

We conclude that collecting soil samples and germinating the seeds in the glasshouse under controlled conditions is capable of revealing all (or nearly all) species and individuals in a relatively small soil volume. However, rare species or those with a patchy distribution can easily be missed. Germination in the field following removal of the established vegetation reveals only those species and individuals which emerge under the prevailing weather conditions from a thin layer of soil in which light can penetrate, but can cover a relatively large area and include both rare species and those with a patchy distribution. The heterogeneity of the soil seed bank is further discussed below.

Sample size, volume and statistics

Many different opinions have been expressed concerning the number of samples that one should take and the total area that these samples should cover (see also Chapter 2). Without trying to solve this problem definitively, some remarks seem to be appropriate. It is an axiom of sampling populations that the lower limit of sample size depends on the size of the organisms being studied and on their sensitivity to damage. Since buried seeds can be considered to be almost infinitely tiny, the sample size for seed bank studies could theoretically be very small. However, if individual samples are to be processed separately, which is time-consuming, it is not practical to use very small sample sizes. On the other hand, relatively few large samples often produce unacceptably large sample variances. The two possible ways out of this dilemma are to make the samples as small as practical, or take very small samples and pool them. As Bigwood and Inouye (1988) have shown, either strategy is effective in reducing variance, and the choice between them is often best decided on practical grounds. Thompson and Grime (1979) pooled all 100 individual cores collected from a single site, while five (Gross 1990) or ten (Ter Heerdt *et al.* 1996) cores are often pooled to form a single sample. Bigwood and Inouye (1988) combined 37 samples, but subsequent analysis showed that precision would have been maximised by reducing substantially the number of subsamples and increasing the number of whole units accordingly.

The number of soil samples is particularly important if the aim is a statistical comparison of different sites. A higher number of replicate samples enables statistical analysis of the samples in order to compare the seed banks of, for instance, fertilised versus unfertilised sites or stages of a successional series (e.g. Bakker *et al.* 1996). As sample number is increased, sampling variance declines; Benoit *et al.* (1989) found that doubling the sample number reduced variance by half. However, the relationship was not linear; beyond about 75 samples, the gain in precision did not justify the increase in sampling effort. Owing to the time-consuming method of handsorting, Bernhardt and Hurka (1989) were forced to sample their sites with only two replicates, losing the possibility of statistical analysis. Bernhardt (1992) had to pool his 60 cores per sample and only examined two small subsamples by handsorting.

The distribution of seed numbers in soil samples is rarely normal, and normality can often not be achieved even by a log ($x + 1$) transformation. The application of parametric tests is limited by the requirement for normal distribution of the data, and therefore non-parametric tests are frequently to be preferred. However, even non-parametric statistics can only be applied to those species that occur relatively frequently in the seed bank, and some authors have preferred a descriptive multivariate presentation of the data (Warr *et al.* 1994). Bakker *et al.* (1996) were able to analyse statistically only 18 out of 65 species represented in the top soil using 10 replicates. The deeper soil layers still cause problems, having too many zeros, which causes the distribution of the data to be skewed.

Our general conclusion about the size and number of samples appropriate for seed bank studies does not differ from that expressed by several previous authors, namely that we can only recommend many small samples above a few bigger ones, depending on the aim of the study. We can, however, attempt to make this recommendation more specific. The critical question is: just how representative do we want a sample to be? Looking towards the French–Swiss tradition in vegetation science (Westhoff and Van der Maarel 1978), a homogeneous plant community can representatively be described by taking the minimum area, in which 80% of the species occurring in the established plant community can be found. We are in favour of adopting a similar approach for the representativeness of the seed bank of a particular plant community, and not worrying too much about the few species which are too rare to be detected. Of course, this approach does require that the complete seed bank of a plant community is known, or suspected, in advance.

Gross (1990) calculated the cumulative number of species in 25 unconcentrated soil samples (each consisting of five cores of 2.5 cm diameter and 15 cm depth). From the saturation curve, she concluded that 20 samples were sufficient to include 90% of all species present in the seed bank of the study site. Ter Heerdt *et al.* (1996) found that analysing 10 concentrated samples (each pooled from 10 cores of 4 cm diameter and 10 cm depth) guaranteed (with 99% probability) that every species would occur at least once in the samples from 53 to 96 % of the time, according to the soil type. In contrast, only 18–70% of the species–sample combinations assured finding a species at least once in 10 unconcentrated samples. This means that in the concentrated samples the chance of finding a species was 10–50 % higher than the unconcentrated samples.

A more theoretical approach is to calculate the minimal density of seeds that can be detected with a 95% confidence level. Considering a Poisson distribution (Zar 1984) of seeds at a site we can calculate the minimal detectable seed density per volume (or area) as follows:

N = the number of cores
A = the volume (area) of each core in litres (m^2)
$N \times A$ = total volume (area) of sampling
Q = the density of seeds (unknown)
$m = N \times A \times Q$ = the expected number of seeds in a sample
The chance of finding x seeds in a sample is:
$$p(x;m) = e^{-m} \times (m^x/x!)$$
The chance of finding 0 seeds is then:
$$p(0;m) = e^{-m}$$
With a confidence level of 0.95:
$$e^{-m} < 0.05$$
So: $-m < \ln(0.05) \qquad \rightarrow \qquad -m < -\ln(20)$

Then the value for Q can be calculated as follows:

$Q > \ln(20)/(N \times A)$
where Q is the minimal detectable density of seeds per litre (m^2).

For the method used by Bakker *et al.* (1996) we calculated a minimal seed density of 0.24 seeds l^{-1} or 24 seeds m^{-2} in order to have a 95 % chance of finding one seed of a species by the method adopted. Clearly, therefore, the volume of soil needed to reveal the majority of seeds present will increase as the seed density declines. Several authors in Hutchings (1986) recommended that, if the objective is to determine the species composition of the seed bank, 0.8 litre was necessary for early successional vegetation, 1–1.2 litre for grasslands, and 8–12 litres for climax woodland, reflecting the generally much lower seed density beneath the latter community. These volumes, which can be compared with those collected in published studies (Figure 2.11), are of course quite inadequate for density estimation, because of the high spatial heterogeneity of seed distribution (Thompson 1986). Here, different criteria apply. Schenkeveld and Verkaar (1984) demonstrated a more or less clumped distribution of viable seeds of some (but not all) species within a 0.1 m^2 patch of limestone grassland. Rusch (1992) demonstrated a patchy distribution of limestone grassland species by sampling at 1 metre intervals along a transect. In fact, depending on abundance, densities of populations of individual species in the soil can be described by any of three types of spatial distribution: Poisson distribution, negative binomial (aggregated) distribution, and normal distribution (Goyeau and Fablet

1982). Elliott (1977) describes how to estimate the sample number required for a given precision. In the most general case, where the distribution is known (or suspected) to be normal:

$$\text{Precision (D)} = \text{standard error / mean}$$
$$= \sqrt{(s^2/n)} / x$$

Therefore, the number of sampling units in a random sample is given by:

$$n = \frac{s^2}{D^2 x^2}$$

$$= \frac{s^2}{0.2^2 x^2}$$

(for a standard error of 20 % of the mean)

$$= \frac{25s^2}{x^2}$$

Usually, of course, the distribution will not be normal. If a Poisson series is known to apply, then of course n depends only on the mean:

$$n = \frac{25}{x}$$ (for a standard error of 20 % of the mean)

If a negative binomial distribution applies, then:

$$n = 25(1/x + 1/k)$$
(for a standard error of 20 % of the mean)

where:

n = number of samples required,
s^2 = sample variance,
x = mean density,
k is the exponent in the binomial expansion $(q-p)^{-k}$, and can be estimated by:
$$k = \frac{x^2}{s^2 - x}$$

See Elliott (1977) for a fuller description of methods of estimating k.

Mean and variance (for a normal distribution), mean (for Poisson) or mean and k (for negative binomial) must first be guessed from previous experience or established from a pilot sample. In every case the number of samples required is strongly influenced by the mean density; sparse species require more samples for any given level of precision. This is consistent with the advice from other authors. Zanin et al. (1989) suggested taking 200 cores (each 3.4 cm diameter, 15 cm depth) in order to reach a 95% confidence level for the majority of the species present in the soil seed bank of arable fields, which should contain 10 000–15 000 seeds m^{-2}. Goyeau and Fablet (1982) found that the number of cores (each 5 cm diameter, 30 cm depth) must be > 100 for a majority of species to achieve a 95% confidence level. For the same precision the number of cores should be 50–100 if the distribution is normal and the species is abundant, i.e. > 2500 seeds m^{-2}. The more the distribution of seeds deviates from a Poisson distribution, the more cores are necessary to estimate the mean number of seeds with acceptable accuracy (Röttele and Koch 1981). Lopez et al. (1988) summarised the data of arable field studies from several authors and suggested in order to reach a 20% precision at 95% confidence level: (i) > 200 or 300 cores if < 500 seeds m^{-2} are present, (ii) 100–300 cores if 500–2500 seeds m^{-2} are present, (iii) 50–100 cores if 2500–5000 seeds m^{-2} are present, (iv) about 50 cores if > 5000 seeds m^{-2} of an individual species are present.

Precision can be maximised by an optimal sampling strategy, but it is worth remembering that *accuracy* is still largely determined by the volume of soil sampled. The aggregation of seed banks makes them inherently difficult to sample, and the only way to overcome this problem involves a lot of sampling effort. For rare species, probably no realistic sampling programme can be expected to produce an acceptably small standard error.

Standardisation of seed bank sampling

In summarising recommended methods of seed bank sampling, we largely follow the conclusions of Ter Heerdt et al. (1996), who used a combined method of concentrating soil samples and germination in the glasshouse. The advantage of standardised sampling of the seed bank is the possibility of comparing the results from different studies. Depending on the aim of the study one should take the following points into account:

1. Use a preliminary study of the vegetation and soil seed bank to get an impression of the composition of the

seed bank and to learn to identify the seedlings as soon as possible.

This study should also provide insights into the abundance, distribution and patchiness of the species present. The space needed in the glasshouse can be estimated at this stage.

Favourable germination conditions of many species can be derived from the literature.

2. Deciding whether the species found are persistent or transient is much simpler if at least two layers of soil are sampled separately.

3. To avoid stratification problems, collect soil samples in early spring. Natural stratification has already taken place in the field.

4. Wash the soil samples with water on a coarse sieve to remove roots, pebbles, etc., and on a fine sieve to remove all clay and silt. A mesh size of 0.2 mm will retain seeds of most species.

5. Spread the concentrated sample on a sterilised medium in a layer as thin as possible, and certainly not thicker than 5 mm.

6. If germination is carried out in a glasshouse or open cage, prepare control trays to record contamination by wind-borne seeds.

7. Remove emerging seedlings as soon as possible. When germination has stopped, we recommend further disturbance of the sample to enable seeds deeper in the sample to germinate. Keep a careful watch for signs of herbivore activity and take appropriate action if any are seen.

8. Presence of remaining seeds should be checked with a seed separation method followed by handsorting.

9. Try to give a complete description of the vegetation where the soil samples were taken.

10. Adequate replication is essential in order to be able to perform any statistical analysis of the data. The degree of replication will depend on the density and patchiness of the seed bank.

Pooling of small individual cores into larger samples is advocated both for statistical reasons and for ease of handling. If the soil is stony it is often easier to take small cores.

Description and guide to interpretation of the database

4

Each row of the database contains some or all of the attributes listed below, and represents any *unique* combination of *species, seed bank type, longevity (if known), depth of sampling (or burial), method* and *source.* Thus, a single source and species may contribute several rows to the database, as long as each row records different seed behaviour or methods. The electronic version of the database, which is supplied as a .csv file, is identical to the printed version, except that it contains an extra **Family** field.

Species name.

Authority   Names and authorities of British natives and naturalised aliens are derived from the BSBI database (essentially Kent 1992), the remainder from the Flora Europaea database.

Seed bank type:
1   **Transient**   Seeds which persist in the soil for less than one year.
2   **Short-term persistent** Seeds which persist in the soil for at least one year, but less than five years.
3   **Long-term persistent**   Seeds which persist in the soil for at least five years.
4   **Present**   Seeds present but cannot be assigned to one of the three seed bank types.

Number of records: the number of records (separate sampling sites or occasions) from which data are summarised in that row of the database.

Longevity (if known)   The maximum length of time (in years) that the species has survived in the soil (< 1 year is 0). Where seeds are known for certain to have survived for a specific number of years but may have survived for longer, the number is preceded by $>$.

Minimum seed density m^{-2}   The minimum seed density among the records summarised by that row of the database.

Maximum seed density m^{-2}   The maximum seed density among the records summarised by that row of the database.

Mean seed density m^{-2}   The mean seed density of all the records summarised by that row of the database. (For artificial burial experiments all seed density fields are zero)

Depth   Total depth of sampling, in cm. For artificial burial experiments, the depth at which seeds were buried.

Method code   Seven method categories are recognised:
1   Seeds deliberately buried in a garden plot without subsequent disturbance.
2   Seeds deliberately buried in a garden plot with subsequent disturbance.
3   Seeds deliberately buried in the field.
4   Soil samples taken from natural vegetation, seeds extracted and germination or viability tested (includes methods involving any reduction of sample volume other than just discarding part of sample).

5   Soil samples taken from natural vegetation, seeds germinated in a glasshouse or outdoors, without extraction or sample reduction.

6   As 5, but germination in the field, i.e. soil was excavated, mixed and replaced.

7   Sequential sampling of natural seed banks on at least SIX occasions per year.

Source code   Corresponds to one of the source references listed at the end of the database. Nearly always a single code refers to a single source. In a few cases (for example, where two or more phases of a burial experiment are reported separately, or data are reported briefly in an accessible location and more extensively in a less accessible location) one code refers to more than one source. Source codes to which this applies are 35, 64, 81, 162, 177, 188, 257 (each 2 sources per code) and 125 (3 sources). Where a single source combines published and unpublished data, this is noted in the reference list.

Use of the database

The data in two extensive databases (concerning *sources* and *species* respectively) were reorganised and abstracted to produce the species accounts which form the bulk of this book. Some of the more significant features of the data have been described in Chapter 2. The rest of the book consists of the data for individual species. The data are arranged alphabetically by family, and by genus and species within families.

Even a brief glance at the database reveals that for nearly every species, the data are extremely variable. This is inevitable; the data have been collected by hundreds of separate investigators in dozens of countries, employing a huge variety of methods in a wide range of environments. This variability raises the question of how far it is possible to allocate a species to a single persistence category. To a large extent our ability to do this depends on the number of records available for a species. Species with many records can often be allocated unambiguously to a persistence category. *Juncus effusus*, with a total of 219 records (excluding type 4), is clearly persistent, and the small minority of records which suggest otherwise can be discounted.

Confidence in describing a species as long-term persistent, however, declines rapidly as the number of records decreases. Only 12 out of 102 records for *Agrostis capillaris* are type 3, yet most would accept this species as long-term persistent. It is not so evident, however, that *Trientalis europaea* is persistent, with a similar proportion but a much smaller absolute number (2 out of 18) of type 3 records. In contrast, the data for genuinely transient species are frequently less ambiguous; all 41 records confirm that the seeds of *Anemone nemorosa* are short-lived, while only one of 104 records for the genus *Bromus* (including *Bromopsis* and *Anisantha*) is long-term persistent. In part, the relative reliability of data for persistent and transient species is an inevitable consequence of the methodological and statistical problems discussed in Chapter 3. Persistent species may fail to be recorded in the soil for a variety of reasons, especially if patchily distributed at low density. Spurious presence of genuinely transient species is much less likely to occur.

Two further complications arise in attempts to interpret the data. The first is that some variation in the data probably reflects real variability, both environmental and genetic, in the behaviour of seeds in the soil. Thus, while *Anthoxanthum odoratum* seems to be mostly short-lived, there is enough evidence to suggest that it is occasionally persistent. This type of variability appears to be particularly prevalent among moderate-sized seeds from families with a wide range of seed sizes, e.g. Compositae and Gramineae.

The second is wind-dispersal of seeds. One category of evidence for persistence is the presence of seeds at a site where the vegetative plant is absent. Clearly dispersal can also be responsible for this. We could have eliminated this problem by removing all examples of this type of evidence for wind-dispersed species during the creation of the database. However, we preferred not to make such 'value judgements', and in any case the definition of 'wind-dispersed' is not without ambiguity. Evidence that wind-dispersal is a real problem is not hard to find. Perhaps the most obvious example is *Chamerion angustifolium*, known certainly to be transient (Thompson and Grime 1979), yet with a majority of type 2 records. Almost without exception, these record the presence of seeds at sites where the plant is absent, and are most plausibly accounted for by effective wind-dispersal. Sometimes the solution is less obvious. Most *Epilobium* species have a few type 3 records, but their true status is not clear. Unfortunately there is no burial experiment for any member of the genus (but see Myerscough and Whitehead 1966).

When interpreting the data, adhere to the following guidelines:

1. Species with many records, especially if derived from several sources, are more reliably classified than those with few records. Single records should be treated with extreme caution.

2. Erroneous persistent records for transient species are much rarer than erroneous transient records for persistent species. Thus any species with several type 3 records probably is long-term persistent, even if type 1 and 2 records are more numerous.

3. Remember that for a number of environmental or genetic reasons, seeds of different populations of the same species may behave differently in the soil.

4. Remember that some sources (method 7) can allocate persistent species (however long-lived) only to type 2. Also remember that in the case of ambiguous or unknown data, the database makes the most conservative assumption about seed persistence. For both these reasons, type 3 records are the best, but not the *only* evidence of long-term persistence.

5. Do not accept a single persistent record from an artificial burial experiment at face value if it is inconsistent with numerous field data.

6. Treat type 3 records for wind-dispersed species with circumspection.

7. If in doubt, consult the original source.

The database   5

Species	Authority	Seed bank type	Number of records	Longevity (y)	Minimum density (seeds m⁻²)	Maximum density (seeds m⁻²)	Mean density (seeds m⁻²)	Depth (cm)	Method	Source code
Aceraceae										
Acer campestre	L.	1	2	—	0	0	0	6.5	7	182
Acer campestre	L.	1	1	—	0	0	0	10	5	255
Acer campestre	L.	1	1	—	0	0	0	15	5	257
Acer platanoides	L.	1	1	—	0	0	0	4	5	176
Acer platanoides	L.	4	1	—	1	1	1	4	5	176
Acer pseudoplatanus	L.	1	2	—	0	0	0	20	5	67
Acer pseudoplatanus	L.	1	1	—	0	0	0	1	5	96
Acer pseudoplatanus	L.	1	4	—	0	0	0	3	5	96
Acer pseudoplatanus	L.	1	6	—	0	0	0	3	5	208
Acer pseudoplatanus	L.	1	1	—	0	0	0	3	7	235
Acer pseudoplatanus	L.	1	3	—	0	0	0	10	5	255
Acer pseudoplatanus	L.	1	1	—	0	0	0	15	5	257
Adoxaceae	—		—		—	—	—			
Adoxa moschatellina	L.	1	1	—	0	0	0	20	5	67
Adoxa moschatellina	L.	1	2	—	0	0	0	4	5	176
Adoxa moschatellina	L.	1	1	—	0	0	0	15	5	257
Alismataceae	—		—		—		—	—	—	
Alisma plantago-aquatica	L.	2	1	—	36	36	36	16	5	216
Alisma plantago-aquatica	L.	2	1	—	84	84	84	3	7	235
Alisma plantago-aquatica	L.	2	3	—	80	220	173	6.5	5	128
Alisma plantago-aquatica	L.	2	1	—	42021	42021	42021	50	5	216
Alisma plantago-aquatica	L.	3	6	—	185	4893	1586	35	5	247
Alisma plantago-aquatica	L.	4	1	—	67	67	67	3	5	96
Alisma plantago-aquatica	L.	4	2	—	180	1380	780	6.5	5	128
Alisma plantago-aquatica	L.	4	3	—	780	6757	3379	50	5	216
Amaranthaceae	—		—		—		—	—	—	
Amaranthus albus	L.	4	2	—	104	378	241	7.5	5	10
Amaranthus hybridus	L.	1	1	—	0	0	0	10	4	22
Amaranthus hybridus	L.	1	1	—	0	0	0	5	5	189
Amaranthus hybridus	L.	2	2	—	50	90	70	10	4	22
Amaranthus hybridus	L.	4	1	—	6	6	6	5	5	189
Amaranthus retroflexus	L.	1	1	—	43	43	43	15	5	175
Amaranthus retroflexus	L.	2	3	>3	0	0	0	7	2	44
Amaranthus retroflexus	L.	2	1	—	37	37	37	7.5	5	10
Amaranthus retroflexus	L.	2	1	—	172	172	172	15	5	175
Amaranthus retroflexus	L.	3	1	>5	0	0	0	30	3	14
Amaranthus retroflexus	L.	3	1	>10	0	0	0	23	1	38
Amaranthus retroflexus	L.	3	3	>5	0	0	0	7	2	44

Species	Authority									
Amaranthus retroflexus	L.	3	3	>6	0	0	0	7	2	44
Amaranthus retroflexus	L.	3	1	>5	0	0	0	23	1	64
Amaranthus retroflexus	L.	3	1	40	0	0	0	—	3	124
Amaranthus retroflexus	L.	3	1	>5	0	0	0	7.5	2	195
Amaranthus retroflexus	L.	3	1	10	0	0	0	15	1	240
Amaranthus retroflexus	L.	3	1	10	0	0	0	50	1	240
Amaranthus retroflexus	L.	3	1	10	0	0	0	100	1	240
Amaranthus retroflexus	L.	3	1	—	301	301	301	15	5	175
Amaranthus retroflexus	L.	3	1	—	1075	1075	1075	30	4	12
Amaranthus retroflexus	L.	4	1	—	8	8	8	10	5	8
Amaranthus retroflexus	L.	4	1	—	420	420	420	10	5	15
Amaranthus retroflexus	L.	4	5	—	30	3435	863	7.5	5	10
Amaranthus retroflexus	L.	4	18	—	20	82470	5541	30	4	13
Aquifoliaceae										
Ilex aquifolium	L.	1	13	—	0	0	0	15	5	257
Araceae										
Acorus calamus	L.	1	3	—	0	0	0	10	5	131
Acorus calamus	L.	1	3	—	0	0	0	32	5	131
Arum maculatum	L.	1	2	—	0	0	0	6	5	49
Arum maculatum	L.	1	2	—	0	0	0	20	5	67
Arum maculatum	L.	1	3	—	0	0	0	15	5	257
Araliaceae										
Hedera helix	L.	1	6	—	0	0	0	3	5	96
Hedera helix	L.	1	1	—	0	0	0	10	5	126
Hedera helix	L.	1	1	—	0	0	0	20	4	225
Hedera helix	L.	1	3	—	0	0	0	3	7	235
Hedera helix	L.	1	20	—	0	0	0	15	5	257
Aristolochisceae										
Asarum europaeum	L.	1	1	—	0	0	0	20	5	67
Asarum europaeum	L.	1	2	—	0	0	0	5	6	170
Asarum europaeum	L.	1	2	—	0	0	0	12	5	171
Asclepiadaceae										
Vincetoxicum hirundinaria	Medicus	1	1	—	0	0	0	6	5	49
Vincetoxicum hirundinaria	Medicus	1	1	—	0	0	0	6	5	181
Balsaminaceae										
Impatiens capensis	Meerb.	1	3	—	0	0	0	5	5	19
Impatiens capensis	Meerb.	1	1	—	0	0	0	10	5	131
Impatiens capensis	Meerb.	1	3	—	0	0	0	5	4	215
Impatiens capensis	Meerb.	1	1	—	830	830	830	32	5	131
Impatiens capensis	Meerb.	2	1	—	530	530	530	32	5	131
Impatiens capensis	Meerb.	4	1	—	100	100	100	5	4	215

Species	Authority	Seed bank type	Number of records	Longevity (y)	Minimum density (seeds m^{-2})	Maximum density (seeds m^{-2})	Mean density (seeds m^{-2})	Depth (cm)	Method	Source code
Impatiens glandulifera	Royle	1	1	—	0	0	0	3	5	96
Impatiens noli-tangere	L.	1	1	—	0	0	0	5	5	213
Impatiens noli-tangere	L.	1	1	—	0	0	0	20	4	225
Impatiens noli-tangere	L.	4	2	—	1	12	7	4	5	176
Impatiens parviflora	DC.	1	2	—	0	0	0	20	5	67
Impatiens parviflora	DC.	1	1	—	0	0	0	20	4	225
Betulaceae										
Alnus glutinosa	—	—	—	—	—	—	—	—	—	
Alnus glutinosa	(L.) Gaertner	1	1	—	0	0	0	1	5	96
Alnus glutinosa	(L.) Gaertner	1	3	—	0	0	0	3	5	208
Alnus glutinosa	(L.) Gaertner	1	1	—	0	0	0	20	4	225
Alnus glutinosa	(L.) Gaertner	1	1	—	0	0	0	5	5	248
Alnus glutinosa	(L.) Gaertner	1	3	—	0	0	0	10	5	255
Alnus glutinosa	(L.) Gaertner	1	2	—	26	39	33	2	7	99
Alnus glutinosa	(L.) Gaertner	1	1	—	53	53	53	8	7	275
Alnus glutinosa	(L.) Gaertner	1	1	—	171	171	171	50	5	107
Alnus glutinosa	(L.) Gaertner	1	1	—	228	228	228	25	5	107
Alnus glutinosa	(L.) Gaertner	2	1	—	6	6	6	3	7	235
Alnus glutinosa	(L.) Gaertner	2	1	—	67	67	67	3	5	96
Alnus glutinosa	(L.) Gaertner	2	1	—	85	85	85	3	5	208
Alnus glutinosa	(L.) Gaertner	2	2	—	1200	2000	1600	20	4	225
Alnus glutinosa	(L.) Gaertner	3	1	—	198	198	198	3	5	208
Alnus glutinosa	(L.) Gaertner	4	3	—	50	183	111	3	5	96
Alnus glutinosa	(L.) Gaertner	4	1	—	509	509	509	3	5	208
Alnus glutinosa	(L.) Gaertner	4	1	—	1200	1200	1200	20	4	225
Alnus incana	(L.) Moench	3	1	>5	0	0	0	2	3	83
Betula nana	L.	1	2	—	0	0	0	10	5	63
Betula pendula	Roth	1	2	—	0	0	0	10	5	255
Betula pendula	Roth	1	1	—	64	64	64	9.5	5	84
Betula pendula	Roth	1	1	—	298	298	298	10	5	84
Betula pendula	Roth	2	2	>3	0	0	0	0	3	85
Betula pendula	Roth	2	1	2	0	0	0	0	3	85
Betula pendula	Roth	2	1	—	5	5	5	3	7	235
Betula pendula	Roth	2	3	—	38	75	54	10	5	255
Betula pendula	Roth	2	5	—	50	183	93	3	5	96
Betula pendula	Roth	2	6	—	300	3780	1707	15	5	257
Betula pendula	Roth	2	2	—	6000	12800	9400	20	4	225
Betula pendula	Roth	3	1	>5	0	0	0	2	3	83
Betula pendula	Roth	3	10	—	88	789	213	10	5	150
Betula pendula	Roth	4	2	—	320	500	410	15	5	257
Betula pendula	Roth	4	1	—	35	35	35	20	4	111
Betula pendula	Roth	4	4	—	38	63	54	10	5	255

Species	Authority									
Betula pendula	Roth	4	3	—	224	412	297	5	5	92
Betula pubescens	Ehrh.	1	4	—	0	0	0	6	5	99
Betula pubescens	Ehrh.	1	1	0	0	0	0	10	3	216
Betula pubescens	Ehrh.	1	3	—	0	0	0	5	5	222
Betula pubescens	Ehrh.	1	2	—	4	12	8	3	7	235
Betula pubescens	Ehrh.	1	1	—	33	33	33	24	5	169
Betula pubescens	Ehrh.	1	2	—	33	44	39	32	5	169
Betula pubescens	Ehrh.	1	1	—	53	53	53	7.5	5	84
Betula pubescens	Ehrh.	1	1	—	56	56	56	16	4	51
Betula pubescens	Ehrh.	1	5	—	26	117	69	2	7	99
Betula pubescens	Ehrh.	1	3	—	53	210	105	8	7	275
Betula pubescens	Ehrh.	1	1	—	138	138	138	9.5	5	84
Betula pubescens	Ehrh.	1	2	—	156	156	156	6	5	11
Betula pubescens	Ehrh.	1	1	—	424	424	424	10	4	51
Betula pubescens	Ehrh.	1	1	>3	617	617	617	10	5	84
Betula pubescens	Ehrh.	2	2	2	0	0	0	0	3	85
Betula pubescens	Ehrh.	2	1	—	0	0	0	0	3	85
Betula pubescens	Ehrh.	2	1	—	5	5	5	3	7	235
Betula pubescens	Ehrh.	2	1	—	26	26	26	5	5	82
Betula pubescens	Ehrh.	2	1	—	51	51	51	3	5	82
Betula pubescens	Ehrh.	2	1	—	67	67	67	24	5	168
Betula pubescens	Ehrh.	2	3	—	83	133	100	3	5	96
Betula pubescens	Ehrh.	2	1	—	156	156	156	16	5	168
Betula pubescens	Ehrh.	2	2	—	208	208	208	6	5	11
Betula pubescens	Ehrh.	2	1	—	208	208	208	6	5	99
Betula pubescens	Ehrh.	2	2	—	230	230	230	30	5	180
Betula pubescens	Ehrh.	2	1	—	423	423	423	10	5	93
Betula pubescens	Ehrh.	2	1	—	2166	2166	2166	25	5	107
Betula pubescens	Ehrh.	3	2	>4	0	0	0	10	3	216
Betula pubescens	Ehrh.	3	1	—	228	228	228	30	5	107
Betula pubescens	Ehrh.	3	2	—	110	430	270	30	5	180
Betula pubescens	Ehrh.	4	1	—	50	50	50	3	5	96
Betula pubescens	Ehrh.	4	1	—	551	551	551	5	5	222
Betula pubescens	Ehrh.	4	1	—	780	780	780	6	5	99
Betula pubescens	Ehrh.	4	4	—	341	1810	1177	5	5	92
Boraginaceae				—						
Anchusa arvensis	(L.) M. Bieb.	1	1	—	0	0	0	25	5	2
Anchusa arvensis	(L.) M. Bieb.	1	1	—	0	0	0	3	5	96
Anchusa arvensis	(L.) M. Bieb.	1	1	—	0	0	0	10	5	255
Anchusa arvensis	(L.) M. Bieb.	2	1	>2	0	0	0	2	3	245
Anchusa arvensis	(L.) M. Bieb.	2	1	>2	0	0	0	15	3	245
Anchusa arvensis	(L.) M. Bieb.	2	1	—	112	112	112	12	5	162
Anchusa officinalis	L.	2	1	—	188	188	188	17	5	162
Anchusa officinalis	L.	2	1	—	3070	3070	3070	22	5	162
Borago officinalis	L.	1	1	—	0	0	0	10	4	22
Borago officinalis	L.	1	2	—	0	0	0	10	5	184

Species	Authority	Seed bank type	Number of records	Longevity (y)	Minimum density (seeds m⁻²)	Maximum density (seeds m⁻²)	Mean density (seeds m⁻²)	Depth (cm)	Method	Source code
Cynoglossum officinale	L.	1	2	—	0	0	0	10	5	184
Cynoglossum officinale	L.	1	1	0	0	0	0	2	3	245
Cynoglossum officinale	L.	1	1	0	0	0	0	15	3	245
Cynoglossum officinale	L.	3	1	>5	0	0	0	7.5	2	198
Echium vulgare	L.	1	3	—	0	0	0	3	5	132
Echium vulgare	L.	2	1	>2	0	0	0	2	3	245
Echium vulgare	L.	2	1	>2	0	0	0	15	3	245
Echium vulgare	L.	2	1	—	6	6	6	6	5	211
Heliotropium europaeum	L.	1	2	—	0	0	0	3	5	132
Heliotropium europaeum	L.	2	1	—	538	538	538	3	5	132
Heliotropium europaeum	L.	4	1	—	113	113	113	30	4	13
Lithospermum arvense	L.	1	1	—	0	0	0	25	4	36
Lithospermum arvense	L.	2	1	3	0	0	0	25	3	209
Myosotis alpestris	F.W. Schmidt	1	1	—	0	0	0	6	5	39
Myosotis alpestris	F.W. Schmidt	4	2	—	13	30	22	6	5	39
Myosotis arvensis	(L.) Hill	1	3	—	0	0	0	25	4	36
Myosotis arvensis	(L.) Hill	1	1	—	0	0	0	20	4	225
Myosotis arvensis	(L.) Hill	1	1	—	368	368	368	8	7	275
Myosotis arvensis	(L.) Hill	2	1	>3	0	0	0	20	3	268
Myosotis arvensis	(L.) Hill	2	1	—	80	80	80	25	4	36
Myosotis arvensis	(L.) Hill	2	5	—	67	150	103	3	5	96
Myosotis arvensis	(L.) Hill	2	5	—	63	200	150	10	5	255
Myosotis arvensis	(L.) Hill	2	1	—	608	608	608	38	5	186
Myosotis arvensis	(L.) Hill	3	1	>11	0	0	0	25	3	209
Myosotis arvensis	(L.) Hill	3	1	—	194	194	194	30	5	29
Myosotis arvensis	(L.) Hill	4	1	—	9	9	9	10	5	110
Myosotis arvensis	(L.) Hill	4	6	—	8	76	25	25	4	36
Myosotis arvensis	(L.) Hill	4	1	—	108	108	108	15	4	31
Myosotis arvensis	(L.) Hill	4	3	—	100	267	183	3	5	96
Myosotis arvensis	(L.) Hill	4	6	—	9	850	325	15	4	30
Myosotis arvensis	(L.) Hill	4	1	—	339	339	339	20	4	111
Myosotis arvensis	(L.) Hill	4	12	—	20	2138	565	30	4	13
Myosotis arvensis	(L.) Hill	4	16	—	63	1313	649	10	5	255
Myosotis arvensis	(L.) Hill	4	1	—	975	975	975	20	5	118
Myosotis arvensis	(L.) Hill	4	2	—	580	2350	1465	25	4	3
Myosotis arvensis	(L.) Hill	4	1	—	1852	1852	1852	25	5	2
Myosotis laxa	Lehm.	1	2	—	0	0	0	10	5	27
Myosotis laxa	Lehm.	1	1	—	0	0	0	20	5	67
Myosotis laxa	Lehm.	1	1	—	0	0	0	3	5	96
Myosotis laxa	Lehm.	2	1	—	156	156	156	10	5	27
Myosotis laxa	Lehm.	3	2	—	156	260	208	10	5	27
Myosotis ramosissima	Rochel	1	1	—	0	0	0	6	5	67

Taxon	Authority									
Myosotis ramosissima	Rochel	1	1	—	0	0	0	3	5	96
Myosotis ramosissima	Rochel	1	1	—	0	0	0	3	5	132
Myosotis ramosissima	Rochel	1	2	—	0	0	0	10	5	152
Myosotis ramosissima	Rochel	2	1	—	38	38	38	3	5	132
Myosotis ramosissima	Rochel	2	1	—	183	183	183	16	5	96
Myosotis ramosissima	Rochel	3	1	>20	56	56	56	24	5	168
Myosotis ramosissima	Rochel	3	1	>20	56	56	56	24	5	168
Myosotis ramosissima	Rochel	3	1	>15	89	89	89	6	5	168
Myosotis ramosissima	Rochel	4	1	—	76	76	76	3	5	67
Myosotis ramosissima	Rochel	4	1	—	250	250	250	3	5	96
Myosotis ramosissima	Rochel	4	3	—	400	1100	733	15	5	249
Myosotis scorpioides	L.	1	3	—	0	0	0	6	5	96
Myosotis scorpioides	L.	1	1	—	0	0	0	6.5	5	99
Myosotis scorpioides	L.	1	3	—	0	0	0	12	5	128
Myosotis scorpioides	L.	1	6	—	0	0	0	3	5	174
Myosotis scorpioides	L.	1	7	—	0	0	0	20	5	208
Myosotis scorpioides	L.	1	1	—	0	0	0	3	4	225
Myosotis scorpioides	L.	1	3	—	0	0	0	10	7	235
Myosotis scorpioides	L.	1	5	—	0	0	0	2	5	255
Myosotis scorpioides	L.	1	1	—	52	52	21	6	7	99
Myosotis scorpioides	L.	1	1	—	85	85	85	10	5	208
Myosotis scorpioides	L.	1	3	—	156	156	156	6	5	27
Myosotis scorpioides	L.	3	1	—	156	156	156	6	5	99
Myosotis scorpioides	L.	3	5	—	85	1104	583	5	5	208
Myosotis scorpioides	L.	4	1	—	10	10	10	5	5	213
Myosotis scorpioides	L.	4	2	—	25	67	46	12	5	98
Myosotis scorpioides	L.	4	4	—	12	559	246	3	5	174
Myosotis scorpioides	L.	4	3	—	85	509	255	3	5	208
Myosotis secunda	Al. Murray	2	5	—	717	717	717	2	7	96
Myosotis stricta	Link ex Roemer & Schultes	1	1	—	0	0	0	3	7	99
Myosotis stricta	Link ex Roemer & Schultes	2	1	—	0	0	0	3	5	132
Myosotis stricta	Link ex Roemer & Schultes	2	2	—	63	63	63	20	5	132
Myosotis stricta	Link ex Roemer & Schultes	2	1	—	121	121	121	32	5	111
Myosotis stricta	Link ex Roemer & Schultes	3	2	>18	44	167	106	3	5	169
Myosotis stricta	Link ex Roemer & Schultes	4	1	—	63	63	63	25	5	132
Myosotis stricta	Link ex Roemer & Schultes	4	2	—	267	467	367	6	5	260
Pulmonaria officinalis	L.	1	1	—	0	0	0	20	5	49
Pulmonaria officinalis	L.	1	1	—	0	0	0	12.5	5	67
Pulmonaria officinalis	L.	1	1	—	0	0	0	5	5	126
Symphytum officinale	L.	1	1	—	0	0	0	5	4	88
Symphytum officinale	L.	1	2	—	53	105	79	8	7	275
Symphytum tuberosum	L.	1	2	—	0	0	0	6	5	49
Callitrichaceae	—			—					—	
Callitriche spp.	L.	1	5	—	0	65	13	2	7	99
Callitriche stagnalis	Scop.	1	1	—	27	27	27	3	7	235
Callitriche stagnalis	Scop.	2	1	—	41	41	41	5	5	222

Species	Authority	Seed bank type	Number of records	Longevity (y)	Minimum density (seeds m^{-2})	Maximum density (seeds m^{-2})	Mean density (seeds m^{-2})	Depth (cm)	Method	Source code
Callitriche stagnalis	Scop.	2	1	—	67	67	67	3	5	96
Callitriche stagnalis	Scop.	4	1	—	92	92	92	5	5	222
Callitriche stagnalis	Scop.	4	1	—	100	100	100	3	5	96
Campanulaceae	—		—	—	—		—	—	—	
Campanula barbata	L.	1	1	—	0	0	0	5	5	55
Campanula barbata	L.	4	2	—	632	1656	1144	5	4	91
Campanula cochleariifolia	Lam.	4	1	—	120	120	120	5	5	55
Campanula patula	L.	1	1	—	0	0	0	10	5	149
Campanula patula	L.	1	1	—	0	0	0	6.5	7	182
Campanula patula	L.	1	1	—	0	0	0	10	5	255
Campanula patula	L.	2	1	—	258	258	258	13	5	68
Campanula patula	L.	3	1	—	75	75	75	6.5	5	68
Campanula patula	L.	3	1	>40	80	80	80	20	5	173
Campanula persicifolia	L.	1	3	—	0	0	0	6	5	49
Campanula persicifolia	L.	1	1	—	0	0	0	10	5	152
Campanula persicifolia	L.	2	3	—	234	312	286	12	5	171
Campanula persicifolia	L.	3	2	—	88	88	88	10	5	152
Campanula persicifolia	L.	4	4	—	88	263	161	10	5	150
Campanula persicifolia	L.	4	2	—	390	546	468	10	5	150
Campanula rapunculoides	L.	1	1	—	0	0	0	5	5	152
Campanula rapunculoides	L.	2	1	—	188	188	188	17	5	69
Campanula rapunculoides	L.	4	2	—	10	21	16	6	5	162
Campanula rapunculoides	L.	4	1	—	250	250	250	10	5	49
Campanula rapunculus	L.	1	4	—	0	0	0	10	5	255
Campanula rapunculus	L.	4	1	—	152	152	152	6	5	255
Campanula rapunculus	L.	4	1	—	2850	2850	2850	10	5	67
Campanula rotundifolia	L.	1	3	—	0	0	0	3	5	96
Campanula rotundifolia	L.	1	3	—	0	0	0	20	5	97
Campanula rotundifolia	L.	1	2	—	0	0	0	10	5	110
Campanula rotundifolia	L.	1	1	—	0	0	0	5	5	122
Campanula rotundifolia	L.	1	2	—	0	0	0	10	5	150
Campanula rotundifolia	L.	1	1	—	0	0	0	10	5	152
Campanula rotundifolia	L.	1	1	—	0	0	0	2	5	153
Campanula rotundifolia	L.	1	2	—	0	0	0	13	5	183
Campanula rotundifolia	L.	1	6	—	0	0	0	10	5	184
Campanula rotundifolia	L.	1	1	—	0	0	0	10	5	255
Campanula rotundifolia	L.	1	1	—	8	8	8	6.5	7	182
Campanula rotundifolia	L.	1	1	—	240	240	240	6	5	48
Campanula rotundifolia	L.	2	5	>1	0	0	0	—	1	275
Campanula rotundifolia	L.	2	3	—	5	30	14	3	7	235

Species	Authority									
Campanula rotundifolia	L.	2	1	—	27	27	27	6.5	7	182
Campanula rotundifolia	L.	2	4	—	50	425	153	10	5	255
Campanula rotundifolia	L.	2	2	—	234	546	390	10	5	152
Campanula rotundifolia	L.	3	1	—	136	136	136	10	5	181
Campanula rotundifolia	L.	4	1	—	23	23	23	7	5	58
Campanula rotundifolia	L.	4	1	—	83	83	83	3	5	96
Campanula rotundifolia	L.	4	3	—	9	204	89	10	5	110
Campanula rotundifolia	L.	4	2	—	151	252	202	17.8	5	155
Campanula rotundifolia	L.	4	2	—	24	432	228	6	5	181
Campanula rotundifolia	L.	4	9	—	38	1188	243	10	5	255
Campanula rotundifolia	L.	4	1	—	253	253	253	17.8	5	156
Campanula rotundifolia	L.	4	3	—	500	1000	800	15	5	249
Campanula rotundifolia	L.	4	1	—	1056	1056	1056	5	5	69
Campanula scheuchzeri	Vill.	1	1	—	0	70	0	5	4	91
Campanula scheuchzeri	Vill.	4	1	—	70	70	70	5	4	91
Campanula scheuchzeri	Vill.	4	2	—	114	155	135	5	5	55
Campanula trachelium	L.	3	1	—	51	51	51	12.5	5	126
Jasione montana	L.	1	1	—	0	0	0	3	5	132
Jasione montana	L.	4	1	—	4	4	4	20	4	111
Jasione montana	L.	4	1	—	151	151	151	17.8	5	155
Legousia hybrida	(L.) Delarbre	1	1	—	0	0	0	10	4	22
Legousia hybrida	(L.) Delarbre	2	1	>3	0	0	0	7.5	2	195
Legousia hybrida	(L.) Delarbre	3	1	>20	82	82	82	20	5	40
Legousia hybrida	(L.) Delarbre	4	2	—	28	74	51	15	4	31
Legousia hybrida	(L.) Delarbre	4	2	—	110	141	126	10	4	22
Legousia hybrida	(L.) Delarbre	4	7	—	26	971	361	15	4	30
Legousia speculum-veneris	(L.) Chaix	1	1	—	0	0	0	25	5	2
Lobelia dortmanna	L.	2	1	—	41	41	41	0.538	5	269
Phyteuma nigrum	F.W. Schmidt	1	1	0	0	0	0	10	5	149
Phyteuma nigrum	F.W. Schmidt	1	3	0	0	0	0	10	5	255
Phyteuma orbiculare	L.	1	1	0	0	0	0	5	5	55
Phyteuma orbiculare	L.	1	1	—	0	0	0	10	5	255
Phyteuma spicatum	L.	1	7	—	0	0	0	10	5	255
Wahlenbergia hederacea	(L.) Reichb.	2	1	—	474	474	474	30	5	45
Cannabaceae										
Cannabis sativa	L.	1	1	—	0	0	0	15	1	240
Cannabis sativa	L.	1	1	—	0	0	0	50	1	240
Cannabis sativa	L.	1	1	—	0	0	0	100	1	240
Humulus lupulus	L.	1	1	—	53	53	53	8	7	275
Caprifoliaceae										
Linnaea borealis	L.	1	1	—	0	0	0	10	5	7
Linnaea borealis	L.	1	1	—	0	0	0	3	5	82
Linnaea borealis	L.	1	3	—	0	0	0	5	5	82
Linnaea borealis	L.	1	1	—	0	0	0	3	4	158
Linnaea borealis	L.	1	1	—	0	0	0	5	5	264

Species	Authority	Seed bank type	Number of records	Longevity (y)	Minimum density (seeds m⁻²)	Maximum density (seeds m⁻²)	Mean density (seeds m⁻²)	Depth (cm)	Method	Source code
Linnaea borealis	L.	2	1	—	4	4	4	0	5	157
Lonicera nigra	L.	1	1	—	0	0	0	10	5	255
Lonicera periclymenum	L.	1	1	—	0	0	0	15	5	32
Lonicera periclymenum	L.	1	1	—	0	0	0	10	5	126
Lonicera periclymenum	L.	1	1	—	0	0	0	12.5	5	126
Lonicera periclymenum	L.	1	1	—	0	0	0	17.5	5	126
Lonicera periclymenum	L.	1	1	—	0	0	0	3	7	235
Lonicera periclymenum	L.	1	13	—	0	0	0	15	5	257
Lonicera xylosteum	L.	1	1	—	0	0	0	10	5	181
Lonicera xylosteum	L.	1	1	—	0	0	0	10	5	255
Sambucus nigra	L.	1	1	—	6	6	6	10	5	93
Sambucus nigra	L.	1	1	—	53	53	53	8	7	275
Sambucus nigra	L.	1	1	—	160	160	160	10	5	181
Sambucus nigra	L.	1	1	—	200	200	200	13	5	183
Sambucus nigra	L.	2	1	—	5	5	5	15	5	32
Sambucus nigra	L.	2	2	—	5	8	7	5	5	122
Sambucus nigra	L.	2	1	—	27	27	27	6	5	49
Sambucus nigra	L.	3	1	>160	94	94	94	57	5	161
Sambucus racemosa	L.	1	1	—	0	0	0	7.5	5	84
Sambucus racemosa	L.	1	1	—	0	0	0	10	5	255
Sambucus racemosa	L.	2	2	—	0	4	2	5	6	170
Sambucus racemosa	L.	2	1	—	90	90	90	10	5	119
Sambucus racemosa	L.	3	1	>10	7	7	7	10	4	241
Sambucus racemosa	L.	3	1	—	44	44	44	32	5	169
Viburnum opulus	L.	1	1	—	0	0	0	10	5	181
Viburnum opulus	L.	1	2	—	0	0	0	16	5	216
Viburnum opulus	L.	1	1	—	0	0	0	5	5	248
Caryophyllaceae	—		—		—		—		—	
Agrostemma githago	L.	1	1	0	0	0	0	25	1	125
Agrostemma githago	L.	1	1	0	0	0	0	25	3	209
Agrostemma githago	L.	1	1	0	0	0	0	20	1	219
Agrostemma githago	L.	1	1	0	0	0	0	15	1	240
Agrostemma githago	L.	1	1	0	0	0	0	50	1	240
Agrostemma githago	L.	2	3	1	0	0	0	100	1	44
Arenaria serpyllifolia	L.	1	2	—	0	0	0	7	2	36
Arenaria serpyllifolia	L.	1	1	—	0	0	0	25	4	69
Arenaria serpyllifolia	L.	1	2	—	0	0	0	5	5	96
Arenaria serpyllifolia	L.	1	2	—	0	0	0	3	5	152
Arenaria serpyllifolia	L.	2	1	>2	0	0	0	10	5	179
Arenaria serpyllifolia	L.	2	2	—	7	79	43	6	3	235

Species	Author									
Arenaria serpyllifolia	L.	2	1	—	75	75	75	10	5	255
Arenaria serpyllifolia	L.	2	9	—	50	3533	1009	3	5	96
Arenaria serpyllifolia	L.	2	3	—	240	2096	1477	18	5	162
Arenaria serpyllifolia	L.	3	1	>5	0	0	0	7.5	2	195
Arenaria serpyllifolia	L.	3	1	>11	0	0	0	25	3	209
Arenaria serpyllifolia	L.	3	1	>20	6	6	6	20	5	40
Arenaria serpyllifolia	L.	3	2	—	1976	1976	1976	6	5	48
Arenaria serpyllifolia	L.	4	1	—	194	4564	2379	30	5	29
Arenaria serpyllifolia	L.	4	1	—	6	6	6	25	4	36
Arenaria serpyllifolia	L.	4	2	—	50	50	50	25	5	2
Arenaria serpyllifolia	L.	4	1	—	7	104	56	15	4	31
Arenaria serpyllifolia	L.	4	1	—	143	143	143	5	5	122
Arenaria serpyllifolia	L.	4	7	—	176	176	176	6	5	67
Arenaria serpyllifolia	L.	4	12	—	32	437	190	15	4	30
Arenaria serpyllifolia	L.	4	1	—	20	1498	217	30	4	13
Arenaria serpyllifolia	L.	4	1	—	233	233	233	20	4	111
Arenaria serpyllifolia	L.	4	1	—	333	333	333	25	5	260
Arenaria serpyllifolia	L.	4	3	—	432	432	432	5	5	69
Arenaria serpyllifolia	L.	4	1	—	53	1290	470	10	5	110
Arenaria serpyllifolia	L.	4	1	—	900	900	900	15	5	249
Arenaria serpyllifolia	L.	4	1	—	1150	1150	1150	3	5	96
Arenaria serpyllifolia	L.	4	1	—	2245	2245	2245	8.5	4	254
Arenaria serpyllifolia subsp. *leptoclados*	(Reichb.) Nyman	3	1	>5	0	0	125	7.5	2	195
Arenaria serpyllifolia subsp. *leptoclados*	(Reichb.) Nyman	4	4	—	63	225	87	3	5	132
Cerastium alpinum	L.	4	2	—	41	132	0	2	5	73
Cerastium arvense	L.	1	1	—	0	0	0	25	4	36
Cerastium arvense	L.	1	1	—	0	0	0	5	5	69
Cerastium arvense	L.	2	1	—	38	38	38	10	5	255
Cerastium arvense	L.	2	3	—	63	183	108	6	5	39
Cerastium arvense	L.	2	1	—	1527	1527	1527	2.5	5	114
Cerastium arvense	L.	4	1	—	120	120	120	30	4	13
Cerastium arvense	L.	4	3	—	143	200	167	6	5	39
Cerastium arvense	L.	4	5	—	300	400	340	15	5	249
Cerastium arvense	L.	4	4	—	573	2083	1368	2.5	5	114
Cerastium brachypetalum	Pers.	1	1	—	0	0	0	3	5	132
Cerastium brachypetalum	Pers.	1	2	—	0	0	0	6.5	7	182
Cerastium brachypetalum	Pers.	4	1	—	63	63	63	3	5	132
Cerastium diffusum	Pers.	1	3	—	0	0	0	30	5	138
Cerastium fontanum	Baumg.	1	7	—	0	0	0	6	5	11
Cerastium fontanum	Baumg.	1	5	—	0	0	0	25	4	36
Cerastium fontanum	Baumg.	1	1	—	0	0	0	1	5	57
Cerastium fontanum	Baumg.	1	1	—	0	0	0	3	5	59
Cerastium fontanum	Baumg.	1	3	—	0	0	0	6	5	67
Cerastium fontanum	Baumg.	1	3	—	0	0	0	5	5	69
Cerastium fontanum	Baumg.	1	2	—	0	0	0	3	5	96
Cerastium fontanum	Baumg.	1	3	—	0	0	0	20	5	97
Cerastium fontanum	Baumg.	1	2	—	0	0	0	6	5	99

Species	Authority	Seed bank type	Number of records	Longevity (y)	Minimum density (seeds m^{-2})	Maximum density (seeds m^{-2})	Mean density (seeds m^{-2})	Depth (cm)	Method	Source code
Cerastium fontanum	Baumg.	1	4	—	0	0	0	6.5	5	128
Cerastium fontanum	Baumg.	1	1	—	0	0	0	12	5	174
Cerastium fontanum	Baumg.	1	1	—	0	0	0	3	5	208
Cerastium fontanum	Baumg.	1	3	—	0	0	0	6	5	208
Cerastium fontanum	Baumg.	1	2	—	0	0	0	5	5	222
Cerastium fontanum	Baumg.	1	4	—	0	0	0	20	4	225
Cerastium fontanum	Baumg.	1	1	—	0	0	0	5	5	250
Cerastium fontanum	Baumg.	1	2	—	0	0	0	5	5	251
Cerastium fontanum	Baumg.	1	5	—	0	0	0	10	5	255
Cerastium fontanum	Baumg.	1	1	—	226	226	226	30	5	45
Cerastium fontanum	Baumg.	1	7	—	53	1788	372	8	7	275
Cerastium fontanum	Baumg.	1	1	—	375	375	375	10	4	88
Cerastium fontanum	Baumg.	1	3	—	468	468	468	10	5	27
Cerastium fontanum	Baumg.	1	1	—	1248	1248	1248	6	5	28
Cerastium fontanum	Baumg.	2	5	>1	0	0	0	—	1	275
Cerastium fontanum	Baumg.	2	2	—	7	15	11	3	7	235
Cerastium fontanum	Baumg.	2	1	—	17	17	17	7	5	58
Cerastium fontanum	Baumg.	2	2	—	25	25	25	12	5	174
Cerastium fontanum	Baumg.	2	2	—	41	56	49	5	5	92
Cerastium fontanum	Baumg.	2	2	—	49	62	56	11.5	5	134
Cerastium fontanum	Baumg.	2	1	—	57	57	57	15	5	62
Cerastium fontanum	Baumg.	2	1	—	64	64	64	9	5	84
Cerastium fontanum	Baumg.	2	6	—	38	138	75	10	5	255
Cerastium fontanum	Baumg.	2	2	—	85	85	85	3	5	208
Cerastium fontanum	Baumg.	2	1	—	92	92	92	5	5	222
Cerastium fontanum	Baumg.	2	1	—	100	100	100	15	5	257
Cerastium fontanum	Baumg.	2	2	—	75	140	108	30	5	29
Cerastium fontanum	Baumg.	2	1	—	128	128	128	13	5	162
Cerastium fontanum	Baumg.	2	2	—	135	137	136	10	5	252
Cerastium fontanum	Baumg.	2	1	—	180	180	180	18	5	162
Cerastium fontanum	Baumg.	2	1	—	190	190	190	6	5	151
Cerastium fontanum	Baumg.	2	6	—	67	383	217	3	5	96
Cerastium fontanum	Baumg.	2	5	—	91	393	283	2	7	99
Cerastium fontanum	Baumg.	2	1	—	417	417	417	20	5	97
Cerastium fontanum	Baumg.	2	2	—	500	600	550	5	5	251
Cerastium fontanum	Baumg.	2	1	—	570	570	570	20	5	107
Cerastium fontanum	Baumg.	2	1	—	875	875	875	15	5	162
Cerastium fontanum	Baumg.	2	7	—	875	1765	976	30	5	45
Cerastium fontanum	Baumg.	2	2	—	312	1768	1040	10	5	27
Cerastium fontanum	Baumg.	2	4	—	156	2288	1092	6	5	11
Cerastium fontanum	Baumg.	2	1	—	1100	1100	1100	20	5	69
Cerastium fontanum	Baumg.	2	1	—	1430	1430	1430	12	5	81

Taxon	Author									
Cerastium fontanum	Baumg.	2	1	—	1500	1500	1500	10	4	88
Cerastium fontanum	Baumg.	2	1	—	1567	1567	1567	10	5	184
Cerastium fontanum	Baumg.	2	17	—	390	3120	1620	10	5	152
Cerastium fontanum	Baumg.	2	7	>5	1404	2470	1872	6	5	28
Cerastium fontanum	Baumg.	3	1	>10	0	0	0	7.5	2	195
Cerastium fontanum	Baumg.	3	1	—	7	7	7	10	4	241
Cerastium fontanum	Baumg.	3	1	>20	20	20	20	6	5	211
Cerastium fontanum	Baumg.	3	1	>40	24	24	24	13	5	183
Cerastium fontanum	Baumg.	3	1	—	33	33	33	24	5	168
Cerastium fontanum	Baumg.	3	1	—	60	60	60	15	5	257
Cerastium fontanum	Baumg.	3	3	—	170	170	170	3	5	208
Cerastium fontanum	Baumg.	3	3	—	156	520	312	10	5	27
Cerastium fontanum	Baumg.	3	4	—	172	484	328	30	5	29
Cerastium fontanum	Baumg.	3	2	—	400	400	400	5	5	251
Cerastium fontanum	Baumg.	3	2	—	75	1238	657	30	5	45
Cerastium fontanum	Baumg.	3	1	—	313	1042	678	20	5	97
Cerastium fontanum	Baumg.	3	2	—	750	750	750	10	4	88
Cerastium fontanum	Baumg.	3	1	—	312	1404	858	6	5	99
Cerastium fontanum	Baumg.	3	2	—	1048	1048	1048	8	5	108
Cerastium fontanum	Baumg.	4	1	—	18	18	18	25	4	36
Cerastium fontanum	Baumg.	4	2	—	2	46	24	5	5	19
Cerastium fontanum	Baumg.	4	1	—	29	29	29	5	5	55
Cerastium fontanum	Baumg.	4	1	—	31	31	31	10	5	33
Cerastium fontanum	Baumg.	4	2	—	27	45	36	13.5	5	163
Cerastium fontanum	Baumg.	4	1	—	38	38	38	10	5	255
Cerastium fontanum	Baumg.	4	1	—	82	82	82	5	5	222
Cerastium fontanum	Baumg.	4	2	—	85	85	85	3	5	208
Cerastium fontanum	Baumg.	4	1	—	110	110	110	10	5	110
Cerastium fontanum	Baumg.	4	1	—	112	112	112	6	5	181
Cerastium fontanum	Baumg.	4	1	—	137	137	137	15	5	267
Cerastium fontanum	Baumg.	4	3	—	35	299	153	15	5	62
Cerastium fontanum	Baumg.	4	1	—	182	182	182	20	5	111
Cerastium fontanum	Baumg.	4	1	—	196	196	196	6	5	67
Cerastium fontanum	Baumg.	4	1	—	200	200	200	6.5	5	12b
Cerastium fontanum	Baumg.	4	5	—	117	380	222	10	5	150
Cerastium fontanum	Baumg.	4	2	—	143	367	255	5	5	92
Cerastium fontanum	Baumg.	4	1	—	337	337	337	5	5	244
Cerastium fontanum	Baumg.	4	3	—	172	549	341	3	5	59
Cerastium fontanum	Baumg.	4	1	—	341	341	341	7	5	58
Cerastium fontanum	Baumg.	4	1	—	375	375	375	5	4	88
Cerastium fontanum	Baumg.	4	2	—	210	600	405	5	5	250
Cerastium fontanum	Baumg.	4	1	—	538	538	538	15	4	31
Cerastium fontanum	Baumg.	4	4	—	200	920	553	15	5	94
Cerastium fontanum	Baumg.	4	10	—	112	1840	608	5	5	69
Cerastium fontanum	Baumg.	4	12	—	67	1600	631	3	5	96
Cerastium fontanum	Baumg.	4	1	—	762	762	762	5	5	234
Cerastium fontanum	Baumg.	4	1	—	1200	1200	1200	1	5	57

Species	Authority	Seed bank type	Number of records	Longevity (y)	Minimum density (seeds m⁻²)	Maximum density (seeds m⁻²)	Mean density (seeds m⁻²)	Depth (cm)	Method	Source code
Cerastium fontanum	Baumg.	4	15	—	260	5304	1453	6	5	99
Cerastium fontanum	Baumg.	4	4	—	505	2377	1641	12	5	174
Cerastium fontanum	Baumg.	4	1	—	1650	1650	1650	10	5	184
Cerastium fontanum	Baumg.	4	3	—	1404	2496	1950	10	5	152
Cerastium fontanum	Baumg.	4	3	—	857	4224	3087	17.8	5	155
Cerastium glomeratum	Thuill.	1	2	—	0	0	0	10	4	22
Cerastium glomeratum	Thuill.	1	1	—	0	0	0	25	4	36
Cerastium glomeratum	Thuill.	1	2	—	0	0	0	3	5	132
Cerastium glomeratum	Thuill.	2	3	—	6	14	9	25	4	36
Cerastium glomeratum	Thuill.	3	1	>5	0	0	0	7.5	2	195
Cerastium glomeratum	Thuill.	4	1	—	173	173	173	15	5	207
Cerastium pumilum	Curtis	1	2	—	0	0	0	6	5	67
Cerastium pumilum	Curtis	4	1	—	48	48	48	5	5	69
Cerastium semidecandrum	L.	1	1	—	0	0	0	3	5	96
Cerastium semidecandrum	L.	1	1	—	0	0	0	3	5	132
Cerastium semidecandrum	L.	2	2	—	117	250	184	3	5	96
Cerastium semidecandrum	L.	4	1	—	500	500	500	15	5	249
Cerastium semidecandrum	L.	4	3	—	250	3867	1522	3	5	96
Dianthus carthusianorum	L.	1	2	—	0	0	0	6	5	67
Dianthus carthusianorum	L.	1	1	—	0	0	0	13	5	183
Dianthus caryophyllus	L.	2	1	—	3278	3278	3278	50	5	260
Dianthus superbus	L.	1	1	—	0	0	0	10	5	255
Gypsophila muralis	L.	1	1	—	0	0	0	6.5	5	128
Gypsophila muralis	L.	2	2	—	12	28	20	6.5	7	182
Herniaria ciliolata	Meld.	4	1	—	156	156	156	6	5	99
Herniaria glabra	L.	4	1	—	400	400	400	15	5	249
Herniaria hirsuta	L.	1	1	—	0	0	0	3	5	132
Herniaria hirsuta	L.	4	3	—	75	175	133	3	5	132
Lychnis flos-cuculi	L.	1	1	—	0	0	0	10	5	27
Lychnis flos-cuculi	L.	1	1	—	0	0	0	20	5	67
Lychnis flos-cuculi	L.	1	2	—	0	0	0	6	5	99
Lychnis flos-cuculi	L.	1	1	—	0	0	0	6.5	5	128
Lychnis flos-cuculi	L.	1	5	—	0	0	0	10	5	149
Lychnis flos-cuculi	L.	1	3	—	0	0	0	6	5	208
Lychnis flos-cuculi	L.	1	4	—	0	0	0	20	4	225
Lychnis flos-cuculi	L.	1	7	—	0	0	0	10	5	255
Lychnis flos-cuculi	L.	1	3	—	0	52	26	2	7	99
Lychnis flos-cuculi	L.	1	1	—	156	156	156	6	5	11
Lychnis flos-cuculi	L.	2	5	>1	0	0	0	—	1	275
Lychnis flos-cuculi	L.	2	1	—	31	31	31	5	5	222
Lychnis flos-cuculi	L.	2	3	—	170	283	236	3	5	208
Lychnis flos-cuculi	L.	2	2	—	85	650	368	2	7	99

Species	Author									
Lychnis flos-cuculi	L.	2	1	—	849	849	849	6	5	208
Lychnis flos-cuculi	L.	2	1	—	1616	1616	1616	13	5	68
Lychnis flos-cuculi	L.	2	6	—	38	4563	1688	10	5	255
Lychnis flos-cuculi	L.	2	1	—	1950	1950	1950	6.5	5	68
Lychnis flos-cuculi	L.	2	1	—	3200	3200	3200	20	4	225
Lychnis flos-cuculi	L.	3	1	—	75	75	75	6.5	5	68
Lychnis flos-cuculi	L.	3	3	—	396	708	519	3	5	208
Lychnis flos-cuculi	L.	3	1	—	688	688	688	13	5	68
Lychnis flos-cuculi	L.	3	2	—	416	2392	1404	6	5	99
Lychnis flos-cuculi	L.	3	5	—	1132	3990	1834	6	5	208
Lychnis flos-cuculi	L.	3	2	—	3172	4472	3822	10	5	27
Lychnis flos-cuculi	L.	4	4	—	163	325	213	10	5	255
Lychnis flos-cuculi	L.	4	5	—	83	676	401	12	5	174
Lychnis flos-cuculi	L.	4	6	—	85	4302	1094	3	5	208
Lychnis flos-cuculi	L.	4	5	—	364	3952	2122	6	5	99
Lychnis flos-cuculi	L.	4	3	—	1600	8400	3867	20	4	225
Lychnis flos-cuculi	L.	4	1	—	9125	9125	9125	5	4	88
Lychnis flos-cuculi	L.	4	1	—	1429	1429	1429	8.5	4	254
Lychnis viscaria	L.	4	1	—	47	47	47	5	5	55
Minuartia recurva	(All.) Schinz & Thell.	4	1	—	92	92	92	5	5	55
Minuartia verna	(L.) Hiern	4	3	—	800	2700	1806	3	5	96
Minuartia verna	(L.) Hiern	1	1	—	0	0	0	20	5	67
Moehringia trinervia	(L.) Clairv.	1	1	—	0	0	0	7.5	5	84
Moehringia trinervia	(L.) Clairv.	1	1	—	0	0	0	5	6	170
Moehringia trinervia	(L.) Clairv.	1	1	—	0	0	0	12	5	171
Moehringia trinervia	(L.) Clairv.	2	2	—	2	9	6	5	6	170
Moehringia trinervia	(L.) Clairv.	2	1	—	10	10	10	6	5	49
Moehringia trinervia	(L.) Clairv.	2	2	—	50	100	75	3	5	96
Moehringia trinervia	(L.) Clairv.	2	3	—	58	116	89	5	5	220
Moehringia trinervia	(L.) Clairv.	2	2	—	240	240	240	15	5	257
Moehringia trinervia	(L.) Clairv.	4	2	—	2	20	11	4	5	176
Moehringia trinervia	(L.) Clairv.	4	1	—	14	14	14	15	5	32
Moehringia trinervia	(L.) Clairv.	4	2	—	17	45	31	6	5	49
Moehringia trinervia	(L.) Clairv.	4	1	—	133	133	133	3	5	96
Moehringia trinervia	(L.) Clairv.	4	2	—	58	470	264	5	5	220
Moehringia trinervia	(L.) Clairv.	4	2	—	398	652	525	7	5	58
Polycarpon tetraphyllum	(L.) L.	1	2	—	0	0	0	3	5	132
Polycarpon tetraphyllum	(L.) L.	4	3	—	88	88	88	3	5	132
Sagina apetala	Ard.	2	1	—	250	250	250	3	5	96
Sagina apetala	Ard.	4	1	—	433	433	433	3	5	96
Sagina apetala	Ard.	4	1	—	750	750	750	1	5	96
Sagina apetala	Ard.	4	4	—	263	1463	797	3	5	132
Sagina apetala	Ard.	3	1	>14	208	208	208	6	5	11
Sagina caespitosa	(J. Vahl) Lange	3		—	5000	5000	5000	20	5	210
Sagina maritima	G. Don	3	1	—	500	500	500	20	5	210
Sagina nodosa	(L.) Fenzl	1	2	—	0	0	0	25	4	36
Sagina procumbens	L.				0	0	0			

Species	Authority	Seed bank type	Number of records	Longevity (y)	Minimum density (seeds m^{-2})	Maximum density (seeds m^{-2})	Mean density (seeds m^{-2})	Depth (cm)	Method	Source code
Sagina procumbens	L	1	1	—	0	0	0	3	5	96
Sagina procumbens	L	1	1	—	0	0	0	6	5	99
Sagina procumbens	L	1	2	—	26	39	33	2	7	99
Sagina procumbens	L	1	4	—	53	53	53	8	7	275
Sagina procumbens	L	1	4	—	156	156	156	6	5	28
Sagina procumbens	L	1	1	—	355	355	355	30	5	45
Sagina procumbens	L	1	1	—	500	500	500	5	5	251
Sagina procumbens	L	2	1	2	0	0	0	25	3	209
Sagina procumbens	L	2	1	—	5	5	5	15	5	32
Sagina procumbens	L	2	1	—	12	12	12	6	5	211
Sagina procumbens	L	2	1	—	16	16	16	5	5	233
Sagina procumbens	L	2	1	—	40	40	40	25	4	36
Sagina procumbens	L	2	1	—	40	40	40	2	5	153
Sagina procumbens	L	2	1	—	113	113	113	25	5	2
Sagina procumbens	L	2	2	—	112	150	131	12	5	162
Sagina procumbens	L	2	1	—	133	133	133	1	5	96
Sagina procumbens	L	2	3	—	5	363	139	3	7	235
Sagina procumbens	L	2	3	—	104	250	166	2	7	99
Sagina procumbens	L	2	1	—	247	247	247	6	5	151
Sagina procumbens	L	2	16	—	38	1263	348	10	5	255
Sagina procumbens	L	2	1	—	435	435	435	5	5	244
Sagina procumbens	L	2	1	—	614	614	614	22	5	162
Sagina procumbens	L	2	3	—	31	1878	942	5	5	222
Sagina procumbens	L	2	19	—	50	4333	949	3	5	96
Sagina procumbens	L	2	1	—	1184	1184	1184	5	5	69
Sagina procumbens	L	2	4	—	312	3120	1716	10	5	152
Sagina procumbens	L	2	4	—	260	6136	1807	6	5	11
Sagina procumbens	L	2	1	—	2188	2188	2188	17	5	162
Sagina procumbens	L	2	5	—	398	6986	2192	30	5	45
Sagina procumbens	L	2	1	—	2808	2808	2808	6	5	99
Sagina procumbens	L	2	1	—	4576	4576	4576	10	5	27
Sagina procumbens	L	3	1	>12	32	32	32	50	5	219
Sagina procumbens	L	3	1	>100	44	44	44	32	5	169
Sagina procumbens	L	3	2	>40	56	100	78	24	5	168
Sagina procumbens	L	3	1	>26	84	84	84	12	4	51
Sagina procumbens	L	3	1	>35	100	100	100	32	5	169
Sagina procumbens	L	3	1	—	117	117	117	10	5	150
Sagina procumbens	L	3	1	—	260	260	260	6	5	11
Sagina procumbens	L	3	2	>18	222	389	306	32	5	169
Sagina procumbens	L	3	1	>30	322	322	322	16	5	168
Sagina procumberis	L	3	1	—	383	383	383	9	5	84
Sagina procumbens	L	3	4	—	156	1976	793	6	5	99

Species	Author									
Sagina procumbens	L.	3	1	—	832	832	832	40	5	11
Sagina procumbens	L.	3	2	—	104	2444	1274	6	5	28
Sagina procumbens	L.	3	4	—	269	5199	2045	30	5	45
Sagina procumbens	L.	3	6	—	156	11284	2531	10	5	27
Sagina procumbens	L.	3	1	—	5400	5400	5400	20	5	69
Sagina procumbens	L.	4	1	—	69	69	69	7	4	58
Sagina procumbens	L.	4	3	—	10	134	77	25	5	36
Sagina procumbens	L.	4	2	—	76	126	101	17.8	5	156
Sagina procumbens	L.	4	1	—	1383	1383	1383	3	5	96
Sagina procumbens	L.	4	1	—	1400	1400	1400	8	5	137
Sagina procumbens	L.	4	1	—	1891	1891	1891	20	4	111
Sagina procumbens	L.	4	3	—	624	5096	2253	6	5	99
Sagina procumbens	L.	4	2	—	4965	10868	7917	17.8	5	155
Sagina procumbens	L.	4	1	—	49408	49408	49408	5	5	69
Scleranthus annuus	L.	1	2	—	0	0	175	10	5	255
Scleranthus annuus	L.	2	1	—	175	175	175	10	5	255
Scleranthus annuus	L.	2	2	—	3444	15333	9389	50	1	260
Scleranthus annuus	L.	3	1	13	0	0	0	25	5	125
Scleranthus annuus	L.	3	1	>18	33	33	33	32	5	169
Scleranthus annuus	L.	3	1	>12	88	88	88	50	5	219
Scleranthus annuus	L.	4	1	—	36	36	36	13.5	5	163
Scleranthus annuus	L.	4	1	—	67	67	67	3	5	96
Scleranthus annuus	L.	4	1	—	500	500	500	15	5	249
Scleranthus annuus	L.	4	1	—	537	537	537	20	4	111
Scleranthus annuus	L.	4	3	—	200	1421	674	25	5	260
Scleranthus annuus	L.	4	1	—	1490	1490	1490	25	4	3
Scleranthus annuus	L.	4	1	—	2943	2943	2943	25	5	2
Silene acaulis	(L.) Jacq.	1	1	—	0	0	0	2	5	73
Silene acaulis	(L.) Jacq.	1	2	—	0	0	0	5	4	91
Silene acaulis	(L.) Jacq.	4	2	—	21	93	57	5	5	55
Silene acaulis	(L.) Jacq.	4	4	—	20	647	263	6	5	39
Silene dioica	(L.) Clairv.	1	2	—	0	0	0	3	5	96
Silene dioica	(L.) Clairv.	1	1	—	0	0	0	6.5	5	128
Silene dioica	(L.) Clairv.	1	4	—	0	0	0	20	4	225
Silene dioica	(L.) Clairv.	1	3	—	0	0	0	10	5	255
Silene dioica	(L.) Clairv.	1	1	—	160	160	160	15	5	257
Silene dioica	(L.) Clairv.	2	2	—	5	8	7	3	7	235
Silene dioica	(L.) Clairv.	2	1	—	44	44	44	15	5	32
Silene dioica	(L.) Clairv.	2	1	—	133	133	133	3	5	96
Silene dioica	(L.) Clairv.	2	2	—	140	500	320	15	5	257
Silene dioica	(L.) Clairv.	2	1	—	1200	1200	1200	20	4	225
Silene dioica	(L.) Clairv.	3	1	>5	0	0	0	7.5	2	195
Silene dioica	(L.) Clairv.	3	1	—	72	72	72	12.5	5	126
Silene dioica	(L.) Clairv.	4	1	—	6	6	6	5	5	213
Silene dioica	(L.) Clairv.	4	2	—	35	46	41	7	5	58
Silene dioica	(L.) Clairv.	4	1	—	50	50	50	5	5	220

Species	Authority	Seed bank type	Number of records	Longevity (y)	Minimum density (seeds m⁻²)	Maximum density (seeds m⁻²)	Mean density (seeds m⁻²)	Depth (cm)	Method	Source code
Silene dioica	(L.) Clairv.	4	1	—		250	250	3	5	96
Silene dioica	(L.) Clairv.	4	3	—	1600	5600	3067	20	4	225
Silene gallica	L.	1	1	—	0	0	0	10	4	22
Silene gallica	L.	2	2	—	186	326	256	5	5	139
Silene gallica	L.	2	1	—	2230	2230	2230	10	5	140
Silene latifolia	Poiret	2	1	1	0	0	0	26	1	133
Silene latifolia	Poiret	3	1	>20	0	0	0	13	1	133
Silene latifolia	Poiret	3	1	>20	0	0	0	26	1	133
Silene latifolia	Poiret	3	1	>20	0	0	0	39	1	133
Silene latifolia	Poiret	3	3	>5	0	0	0	7.5	2	198
Silene latifolia	Poiret	3	1	—	2	2	2	20	5	40
Silene latifolia	Poiret	3	2	>5	60	60	60	15	5	257
Silene noctiflora	L.	3	3	>5	0	0	0	7	2	44
Silene noctiflora	L.	4	1	—	21	21	21	20	4	111
Silene nutans	L.	1	1	—	0	0	0	10	5	152
Silene nutans	L.	2	2	—	6	44	25	3	7	235
Silene nutans	L.	2	1	—	958	958	958	6	5	48
Silene otites	(L.) Wibel	4	1	—	2653	2653	2653	8.5	4	254
Silene rupestris	L.	1	1	—	0	0	0	5	5	55
Silene viscosa	(L.) Pers.	1	1	—	0	0	0	3	5	208
Silene vulgaris	Garcke	1	2	—	0	0	0	10	5	255
Silene vulgaris	Garcke	2	1	4	0	0	0	7	2	44
Silene vulgaris	Garcke	2	1	—	1053	1053	1053	22	5	162
Silene vulgaris	Garcke	3	2	>5	0	0	0	7	2	44
Silene vulgaris	Garcke	3	1	>4	0	0	0	8	1	60
Silene vulgaris	Garcke	3	1	>4	0	0	0	20	1	60
Silene vulgaris	Garcke	3	1	>4	0	0	0	30	1	60
Silene vulgaris	Garcke	3	1	>6	0	0	0	20	1	61
Silene vulgaris	Garcke	3	1	5	0	0	0	8	1	61
Silene vulgaris	Garcke	3	1	5	0	0	0	30	1	61
Spergula arvensis	L.	1	1	—	53	53	53	8	7	275
Spergula arvensis	L.	1	1	—	624	624	624	38	5	186
Spergula arvensis	L.	2	1	>2	0	0	0	2	1	47
Spergula arvensis	L.	2	1	>2	0	0	0	15	1	47
Spergula arvensis	L.	2	1	—	7	7	7	15	5	62
Spergula arvensis	L.	2	1	—	97	97	97	30	5	45
Spergula arvensis	L.	2	10	—	38	1088	273	10	5	255
Spergula arvensis	L.	2	1	—	5738	5738	5738	17.8	5	155
Spergula arvensis	L.	3	3	>5	0	0	0	7	2	44
Spergula arvensis	L.	3	1	>5	0	0	0	2.5	1	203
Spergula arvensis	L.	3	1	>5	0	0	0	2.5	2	203
Spergula arvensis	L.	3	1	>5	0	0	0	7.5	1	203

Species	Authority									
Spergula arvensis	L.	3	1	>5	0	0	0	7.5	2	203
Spergula arvensis	L.	3	1	>5	0	0	0	15	1	203
Spergula arvensis	L.	3	1	>5	0	0	0	15	2	203
Spergula arvensis	L.	3	1	>10	0	0	0	25	3	209
Spergula arvensis	L.	3	1	>4	0	0	0	15	1	258
Spergula arvensis	L.	3	1	>6	7	7	7	15	5	62
Spergula arvensis	L.	3	1	>1700	97	97	97	133	5	161
Spergula arvensis	L.	3	1	>6	183	183	183	30	5	45
Spergula arvensis	L.	3	1	>8	183	183	183	30	5	45
Spergula arvensis	L.	3	1	—	2947	2947	2947	17.8	5	155
Spergula arvensis	L.	4	1	—	100	100	100	3	5	96
Spergula arvensis	L.	4	1	—	175	175	175	8	7	275
Spergula arvensis	L.	4	1	—	380	380	380	25	5	2
Spergula arvensis	L.	4	1	—	1784	1784	1784	20	4	111
Spergula arvensis	L.	4	1	—	2250	2250	2250	20	5	118
Spergula arvensis	L.	4	1	—	4248	4248	4248	15	5	207
Spergula arvensis	L.	4	1	—	6789	6789	6789	25	5	260
Spergula morisonii	Boreau	2	1	>1	10486	19932	15209	15	4	31
Spergula pentandra		1	2	—	0	0	0	10	3	178
Spergularia marina	(L.) Griseb.	1	4	—	0	0	0	3	5	132
Spergularia marina	(L.) Griseb.	1	1	—	0	0	0	10	4	20
Spergularia marina	(L.) Griseb.	2	2	—	300	400	350	1	5	57
Spergularia marina	(L.) Griseb.	2	4	—	600	1200	900	1	5	57
Spergularia marina	(L.) Griseb.	2	1	—	5000	5000	5000	10	4	20
Spergularia marina	(L.) Griseb.	2	1	—	488708	488708	488708	20	5	210
Spergularia marina	(L.) Griseb.	3	1	—	25000	25000	25000	10	5	243
Spergularia marina	(L.) Griseb.	4	3	—	1200	2400	1867	20	5	210
Spergularia marina	(L.) Griseb.	4	3	—	400	8400	3667	10	4	20
Spergularia marina	(L.) Griseb.	4	2	—	67198	310000	188599	10	5	57
Spergularia marina	(L.) Griseb.	1	1	—	376234	376234	376234	5	5	243
Spergularia media	(L.) C. Presl	2	1	—	201	201	201	1	5	242
Spergularia media	(L.) C. Presl	2	1	—	201	201	201	5	5	57
Spergularia rubra	(L.) J.S. Presl & C. Presl	2	1	—	5450	5450	5450	25	4	244
Spergularia rubra	(L.) J.S. Presl & C. Presl	4	1	—	76	76	76	17.8	5	3
Spergularia rupicola	Lebel ex Le Jolis	2	1	—	488	488	488	10	5	156
Stellaria graminea	L.	1	1	—	0	0	0	25	4	255
Stellaria graminea	L.	1	4	—	0	0	0	3	5	36
Stellaria graminea	L.	1	1	—	0	0	0	5	5	59
Stellaria graminea	L.	1	1	—	0	0	0	3	5	69
Stellaria graminea	L.	1	1	—	0	0	0	10	5	96
Stellaria graminea	L.	1	1	—	0	0	0	6	5	150
Stellaria graminea	L.	1	1	—	0	0	0	5	6	151
Stellaria graminea	L.	1	1	—	0	0	0	12	5	170
Stellaria gramifea	L.	1	5	—	0	0	0	3	5	171
Stellaria graminea	L.	1	4	—	0	0	0	20	4	208
Stellaria graminea	L.	1	10	—	0	0	0	10	5	225
Stellaria graminea	L.	1	1	—	0	0	0	10	5	255

Species	Authority	Seed bank type	Number of records	Longevity (y)	Minimum density (seeds m⁻²)	Maximum density (seeds m⁻²)	Mean density (seeds m⁻²)	Depth (cm)	Method	Source code
Stellaria graminea	L.	1	1	—	8	8	8	3	7	235
Stellaria graminea	L.	2	1	—	85	85	85	6	5	208
Stellaria graminea	L.	2	2	—	38	225	132	10	5	255
Stellaria graminea	L.	2	1	—	216	216	216	13	5	68
Stellaria graminea	L.	2	4	—	234	2418	1190	10	5	152
Stellaria graminea	L.	2	3	—	1100	5400	2567	1	5	57
Stellaria graminea	L.	2	2	—	1200	6400	3800	20	4	225
Stellaria graminea	L.	3	1	—	75	75	75	30	5	29
Stellaria graminea	L.	3	1	—	142	142	142	6	5	208
Stellaria graminea	L.	3	1	—	275	275	275	6.5	5	68
Stellaria graminea	L.	3	1	—	5300	5300	5300	20	5	69
Stellaria graminea	L.	4	1	—	32	32	32	3	5	59
Stellaria graminea	L.	4	1	—	67	67	67	3	5	96
Stellaria graminea	L.	4	1	—	113	113	113	3	5	208
Stellaria graminea	L.	4	6	—	38	288	138	10	5	255
Stellaria graminea	L.	4	2	—	143	173	158	5	5	222
Stellaria graminea	L.	4	1	—	390	390	390	10	5	152
Stellaria graminea	L.	4	2	—	144	640	392	5	5	69
Stellaria graminea	L.	4	9	—	88	906	419	10	5	150
Stellaria graminea	L.	4	4	—	350	3140	1740	15	5	94
Stellaria holostea	L.	1	1	—	0	0	0	6	5	49
Stellaria holostea	L.	1	2	—	0	0	0	20	5	67
Stellaria holostea	L.	1	1	—	0	0	0	10	5	126
Stellaria holostea	L.	1	1	—	0	0	0	17.5	5	126
Stellaria holostea	L.	1	1	—	0	0	0	12	5	171
Stellaria holostea	L.	1	1	—	0	0	0	10	5	172
Stellaria holostea	L.	1	2	—	0	0	0	4	5	176
Stellaria holostea	L.	4	2	—	10	55	33	6	5	49
Stellaria holostea	L.	4	1	—	43	43	43	12.5	5	126
Stellaria humifusa	Rottb.	4	1	—	13	13	13	2	5	73
Stellaria media	(L.) Villars	1	1	—	0	0	0	10	4	22
Stellaria media	(L.) Villars	1	2	—	0	0	0	25	4	36
Stellaria media	(L.) Villars	1	5	—	0	0	0	5	5	69
Stellaria media	(L.) Villars	1	1	—	0	0	0	5	5	92
Stellaria media	(L.) Villars	1	11	—	0	0	0	6	5	99
Stellaria media	(L.) Villars	1	1	—	0	0	0	10	5	150
Stellaria media	(L.) Villars	1	1	—	0	0	0	5	5	222
Stellaria media	(L.) Villars	1	1	—	0	0	0	5	5	250
Stellaria media	(L.) Villars	1	5	—	0	120	45	2	7	99
Stellaria media	(L.) Villars	1	1	—	65	65	65	30	5	45
Stellaria media	(L.) Villars	1	1	—	96	96	96	38	5	186
Stellaria media	(L.) Villars	1	3	—	53	342	154	8	7	275

Taxon	Authority									
Stellaria media	(L.) Villars	1	1	—	156	156	156	40	5	11
Stellaria media	(L.) Villars	1	2	—	156	156	156	10	5	27
Stellaria media	(L.) Villars	1	1	>2	378	378	378	50	5	260
Stellaria media	(L.) Villars	2	1	>2	0	0	0	2	1	47
Stellaria media	(L.) Villars	2	1	4	0	0	0	15	1	47
Stellaria media	(L.) Villars	2	1	>2	0	0	0	23	1	64
Stellaria media	(L.) Villars	2	1	>2	0	0	0	0	1	75
Stellaria media	(L.) Villars	2	1	>2	0	0	0	0	2	75
Stellaria media	(L.) Villars	2	1	4	0	0	0	5	1	75
Stellaria media	(L.) Villars	2	1	—	0	0	0	5	2	75
Stellaria media	(L.) Villars	2	3	—	0	0	0	2.5	2	203
Stellaria media	(L.) Villars	3	1	—	6	8	7	25	4	36
Stellaria media	(L.) Villars	2	1	—	8	8	8	3	7	235
Stellaria media	(L.) Villars	2	2	—	13	13	13	8	5	108
Stellaria media	(L.) Villars	2	2	—	43	43	43	15	5	175
Stellaria media	(L.) Villars	2	5	—	31	71	51	5	5	222
Stellaria media	(L.) Villars	2	1	—	50	83	63	3	5	96
Stellaria media	(L.) Villars	2	19	—	112	112	112	5	5	69
Stellaria media	(L.) Villars	2	2	—	38	625	147	10	5	255
Stellaria media	(L.) Villars	2	1	—	220	360	290	10	5	149
Stellaria media	(L.) Villars	2	2	—	300	300	300	12	5	162
Stellaria media	(L.) Villars	2	2	—	300	300	300	40	5	162
Stellaria media	(L.) Villars	2	1	—	313	313	313	17	5	162
Stellaria media	(L.) Villars	2	4	—	342	342	342	20	5	107
Stellaria media	(L.) Villars	2	7	—	352	352	352	100	5	219
Stellaria media	(L.) Villars	2	2	—	75	875	358	10	5	162
Stellaria media	(L.) Villars	2	2	—	105	1210	485	8	7	275
Stellaria media	(L.) Villars	2	1	—	140	882	511	5	5	250
Stellaria media	(L.) Villars	2	1	—	600	600	600	20	5	69
Stellaria media	(L.) Villars	2	1	—	728	728	728	10	5	27
Stellaria media	(L.) Villars	2	2	—	800	800	800	40	5	219
Stellaria media	(L.) Villars	2	1	—	1012	1012	1012	30	5	45
Stellaria media	(L.) Villars	2	1	—	1200	1200	1200	30	5	66
Stellaria media	(L.) Villars	2	1	—	1204	1204	1204	30	4	12
Stellaria media	(L.) Villars	2	1	—	1311	1311	1311	27	5	162
Stellaria media	(L.) Villars	2	1	—	1800	1800	1800	50	5	66
Stellaria media	(L.) Villars	2	1	—	1976	1976	1976	6	5	99
Stellaria media	(L.) Villars	2	1	—	2817	2817	2817	10	5	184
Stellaria media	(L.) Villars	2	3	—	412	8000	3346	50	5	260
Stellaria media	(L.) Villars	2	2	—	4400	31600	18000	20	4	225
Stellaria media	(L.) Villars	2	1	—	35650	35650	35650	20	5	4
Stellaria media	(L.) Villars	3	1	30	0	0	0	—	3	124
Stellaria media	(L.) Villars	3	1	>5	0	0	0	7.5	2	196
Stellaria media	(L.) Villars	3	1	>5	0	0	0	2.5	1	203
Stellaria media	(L.) Villars	3	1	>5	0	0	0	7.5	1	203
Stellaria media	(L.) Villars	3	1	>5	0	0	0	7.5	2	203
Stellaria media	(L.) Villars	3	1	>5	0	0	0	15	1	203

Species	Authority	Seed bank type	Number of records	Longevity (y)	Minimum density (seeds m⁻²)	Maximum density (seeds m⁻²)	Mean density (seeds m⁻²)	Depth (cm)	Method	Source code
Stellaria media	(L.) Villars	3	1	>5	0	0	0	15	2	203
Stellaria media	(L.) Villars	3	1	10	0	0	0	15	1	240
Stellaria media	(L.) Villars	3	1	10	0	0	0	50	1	240
Stellaria media	(L.) Villars	3	1	10	0	0	0	100	1	240
Stellaria media	(L.) Villars	3	1	>4	0	0	0	15	1	258
Stellaria media	(L.) Villars	3	1	>100	33	33	33	32	5	169
Stellaria media	(L.) Villars	3	1	>660	37	37	37	55	5	161
Stellaria media	(L.) Villars	3	1	>660	50	50	50	43	5	161
Stellaria media	(L.) Villars	3	1	>460	53	53	53	62	5	161
Stellaria media	(L.) Villars	3	3	—	43	161	90	30	5	29
Stellaria media	(L.) Villars	3	1	>20	97	97	97	20	5	40
Stellaria media	(L.) Villars	3	1	>660	100	100	100	50	5	161
Stellaria media	(L.) Villars	3	4	—	43	215	102	30	5	45
Stellaria media	(L.) Villars	3	1	—	256	256	256	10	5	252
Stellaria media	(L.) Villars	3	1	>8	592	592	592	30	5	45
Stellaria media	(L.) Villars	3	4	—	550	1150	800	20	5	4
Stellaria media	(L.) Villars	3	1	—	5900	5900	5900	70	5	219
Stellaria media	(L.) Villars	4	1	—	29	29	29	25	5	2
Stellaria media	(L.) Villars	4	2	—	20	50	35	25	4	36
Stellaria media	(L.) Villars	4	6	—	15	243	73	15	4	30
Stellaria media	(L.) Villars	4	3	—	35	167	81	7	5	58
Stellaria media	(L.) Villars	4	5	—	31	214	121	5	5	222
Stellaria media	(L.) Villars	4	2	—	79	166	123	12.5	5	126
Stellaria media	(L.) Villars	4	3	—	138	225	175	10	5	255
Stellaria media	(L.) Villars	4	1	—	260	260	260	6	5	99
Stellaria media	(L.) Villars	4	20	—	20	1418	262	30	4	13
Stellaria media	(L.) Villars	4	12	—	50	847	286	10	5	15
Stellaria media	(L.) Villars	4	1	—	311	311	311	5	5	92
Stellaria media	(L.) Villars	4	2	—	153	611	382	15	5	142
Stellaria media	(L.) Villars	4	1	—	393	393	393	15	5	62
Stellaria media	(L.) Villars	4	2	—	470	670	570	25	4	3
Stellaria media	(L.) Villars	4	1	—	592	592	592	5	5	69
Stellaria media	(L.) Villars	4	1	—	729	729	729	23	5	202
Stellaria media	(L.) Villars	4	2	—	346	1132	739	15	4	31
Stellaria media	(L.) Villars	4	6	—	101	2621	824	17.8	5	155
Stellaria media	(L.) Villars	4	4	—	47	2570	825	10	4	22
Stellaria media	(L.) Villars	4	1	—	1100	1100	1100	15	5	205
Stellaria media	(L.) Villars	4	2	—	310	2180	1245	15	5	94
Stellaria media	(L.) Villars	4	11	—	235	5134	1349	25	5	260
Stellaria media	(L.) Villars	4	4	—	833	2333	1500	20	5	74
Stellaria media	(L.) Villars	4	2	—	350	2715	1533	25	4	185
Stellaria media	(L.) Villars	4	1	—	1725	1725	1725	20	5	118

Species	Authority									
Stellaria media	(L.) Villars	4	1	—	2314	2314	2314	10	5	184
Stellaria media	(L.) Villars	4	1	—	2667	2667	2667	20	4	111
Stellaria media	(L.) Villars	4	1	—	3504	3504	3504	5	5	250
Stellaria media	(L.) Villars	4	1	—	4110	4110	4110	10	5	140
Stellaria media	(L.) Villars	4	1	—	23984	23984	23984	15	5	207
Stellaria nemorum	L.	1	1	—	0	0	0	10	5	126
Stellaria nemorum	L.	1	1	—	0	0	0	12.5	5	126
Stellaria nemorum	L.	1	2	—	0	0	0	4	5	176
Stellaria nemorum	L.	4	5	—	29	1054	556	5	5	220
Stellaria palustris	Retz.	1	3	—	0	0	0	3	5	208
Stellaria palustris	Retz.	1	6	—	85	170	128	6	5	208
Stellaria palustris	Retz.	3	1	—	85	85	85	6	5	208
Stellaria palustris	Retz.	3	1	—	285	285	285	20	5	107
Stellaria palustris	Retz.	4	3	—	283	764	453	3	5	208
Stellaria uliginosa	Murray	1	1	—	0	0	0	10	5	27
Stellaria uliginosa	Murray	1	1	—	0	0	0	5	5	92
Stellaria uliginosa	Murray	1	5	—	0	0	0	6	5	99
Stellaria uliginosa	Murray	1	3	—	26	52	35	2	7	99
Stellaria uliginosa	Murray	2	1	—	463	463	463	30	5	45
Stellaria uliginosa	Murray	2	1	—	15	15	15	3	7	235
Stellaria uliginosa	Murray	2	3	—	41	173	105	5	5	222
Stellaria uliginosa	Murray	2	2	—	26	263	145	2	7	99
Stellaria uliginosa	Murray	2	1	—	990	990	990	30	5	45
Stellaria uliginosa	Murray	3	2	—	2548	2756	2652	6	5	99
Stellaria uliginosa	Murray	3	1	—	171	171	171	20	5	107
Stellaria uliginosa	Murray	3	2	—	208	260	234	6	5	99
Stellaria uliginosa	Murray	4	6	—	312	2496	1300	10	5	27
Stellaria uliginosa	Murray	4	1	—	83	83	83	5	5	213
Stellaria uliginosa	Murray	4	1	—	156	156	156	6	5	99
Stellaria uliginosa	Murray	4	2	—	158	383	271	5	5	98
Stellaria uliginosa	Murray	4	2	—	378	2367	1373	5	5	222
Stellaria uliginosa	Murray	2	1	—	46433	46433	46433	3	5	96
Vaccaria hispanica	(Miller) Rauschert	2	6	1	0	0	0	7	2	44
Vaccaria hispanica	(Miller) Rauschert	2	2	2	0	0	0	7	2	44
Vaccaria hispanica	(Miller) Rauschert	2	1	3	0	0	0	7	2	44
Vaccaria hispanica	(Miller) Rauschert	4	1	—	43	43	43	7.5	5	10
Celastraceae		—	—	—	—	—	—	—	—	—
Euonymus europaeus	L.	1	2	—	0	0	0	15	5	257
Ceratophyllaceae		—	—	—	—	—	—	—	—	—
Ceratophyllum demersum	L.	1	5	—	0	0	0	10	5	87
Ceratophyllum demersum	L.	1	4	—	0	0	0	35	5	247
Chenopodiaceae		—	—	—	—	—	—	—	—	—
Atriplex patula	L.	1	1	—	0	0	0	25	4	36
Atriplex patula	L.	1	1	—	5	5	5	3	7	235

Species	Authority	Seed bank type	Number of records	Longevity (y)	Minimum density (seeds m⁻²)	Maximum density (seeds m⁻²)	Mean density (seeds m⁻²)	Depth (cm)	Method	Source code
Atriplex patula	L.	2	1	—	10	10	10	15	5	62
Atriplex patula	L.	2	1	—	18	18	18	5	5	262
Atriplex patula	L.	2	1	—	240	240	240	10	5	149
Atriplex patula	L.	2	1	—	258	258	258	30	5	29
Atriplex patula	L.	3	3	>5	0	0	0	7.5	2	204
Atriplex patula	L.	3	1	>20	54	54	54	30	5	29
Atriplex patula	L.	3	1	>41	75	75	75	30	5	29
Atriplex patula	L.	3	1	>58	108	108	108	30	5	29
Atriplex patula	L.	3	1	—	646	646	646	30	5	29
Atriplex patula	L.	3	1	>10	1141	1141	1141	30	5	29
Atriplex patula	L.	4	1	—	4	4	4	10	5	33
Atriplex patula	L.	4	1	—	11	11	11	20	4	111
Atriplex patula	L.	4	1	—	25	25	25	5	5	166
Atriplex patula	L.	4	6	4	20	86	50	30	4	13
Atriplex patula	L.	4	1	—	228	228	228	25	5	260
Atriplex patula	L.	4	7	—	6	629	254	15	4	30
Atriplex patula	L.	4	1	—	889	889	889	15	5	207
Atriplex prostrata	Boucher ex DC.	2	1	4	0	0	0	7	2	44
Atriplex prostrata	Boucher ex DC.	2	1	—	4	4	4	2	5	113
Atriplex prostrata	Boucher ex DC.	2	3	—	600	600	600	10	4	20
Atriplex prostrata	Boucher ex DC	2	5	—	600	2700	1580	1	5	57
Atriplex prostrata	Boucher ex DC.	3	2	>5	0	0	0	7	2	44
Atriplex prostrata	Boucher ex DC.	3	3	>5	0	0	0	7.5	2	204
Atriplex prostrata	Boucher ex DC.	4	1	—	30	30	30	15	4	21
Atriplex prostrata	Boucher ex DC.	4	1	—	2655	2655	2655	5	5	242
Bassia scoparia	(L.) Voss	2	1	>3	0	0	0	1	1	274
Bassia scoparia	(L.) Voss	2	1	>3	0	0	0	10	1	274
Bassia scoparia	(L.) Voss	2	1	>3	0	0	0	30	1	274
Bassia scoparia	(L.) Voss	3	1	>10	0	0	0	23	1	38
Bassia scoparia	(L.) Voss	4	1	—	146	146	146	10	5	8
Beta vulgaris subsp. *maritima*	(L.) Arcang.	3	1	5	0	0	0	8	1	61
Beta vulgaris subsp. *maritima*	(L.) Arcang.	3	1	5	0	0	0	20	1	61
Beta vulgaris subsp. *maritima*	(L.) Arcang.	3	1	5	0	0	0	30	1	61
Beta vulgaris subsp. *maritima*	(L.) Arcang.	3	1	10	0	0	0	15	1	240
Beta vulgaris subsp. *maritima*	(L.) Arcang.	3	1	21	0	0	0	50	1	240
Beta vulgaris subsp. *maritima*	(L.) Arcang.	3	1	21	0	0	0	100	1	240
Chenopodium album	L.	1	1	—	0	0	0	10	4	22
Chenopodium album	L.	1	1	—	0	0	0	6.5	5	128
Chenopodium album	L.	1	1	—	0	0	0	11.5	5	134
Chenopodium album	L.	1	1	—	0	0	0	5	5	189
Chenopodium album	L.	1	7	—	53	263	117	8	7	275
Chenopodium album	L.	2	3	>3	0	0	0	7	2	44

Chenopodium album	L.	2	1	>2	0	0	0	2	1	47
Chenopodium album	L.	2	1	>2	0	0	0	15	1	47
Chenopodium album	L.	2	1	—	8	8	8	4	5	217
Chenopodium album	L.	2	1	—	16	16	16	15	5	32
Chenopodium album	L.	2	1	—	24	24	24	13	5	183
Chenopodium album	L.	2	1	—	31	31	31	5	5	222
Chenopodium album	L.	2	1	—	37	37	37	11.5	5	134
Chenopodium album	L.	2	1	—	43	43	43	5	5	220
Chenopodium album	L.	2	2	—	47	47	47	5	5	244
Chenopodium album	L.	2	1	—	50	50	50	10	5	184
Chenopodium album	L.	2	1	—	54	54	54	3	5	59
Chenopodium album	L.	2	1	—	67	67	67	3	5	96
Chenopodium album	L.	2	5	—	75	75	75	62	5	162
Chenopodium album	L.	2	8	—	75	75	75	67	5	162
Chenopodium album	L.	2	1	—	34	232	92	25	4	36
Chenopodium album	L.	2	3	—	48	272	124	5	5	69
Chenopodium album	L.	2	1	—	200	200	200	40	5	162
Chenopodium album	L.	2	6	—	250	500	354	17	5	162
Chenopodium album	L.	2	12	—	423	423	423	15	5	62
Chenopodium album	L.	2	1	—	487	487	487	57	5	162
Chenopodium album	L.	2	5	—	75	875	513	10	5	255
Chenopodium album	L.	2	3	—	38	2863	567	10	5	260
Chenopodium album	L.	2	2	—	944	944	944	50	5	149
Chenopodium album	L.	2	1	—	300	2300	952	10	5	272
Chenopodium album	L.	2	1	—	914	2159	1439	15	5	94
Chenopodium album	L.	2	2	—	1440	1440	1440	15	5	162
Chenopodium album	L.	2	3	—	2188	2188	2188	32	4	3
Chenopodium album	L.	2	3	—	2290	2290	2290	25	4	225
Chenopodium album	L.	2	1	—	1600	7200	4400	20	4	50
Chenopodium album	L.	2	3	—	11100	23300	17767	30	3	14
Chenopodium album	L.	3	3	>5	0	0	0	7	2	44
Chenopodium album	L.	3	1	>5	0	0	0	7	2	44
Chenopodium album	L.	3	1	>6	0	0	0	25	1	125
Chenopodium album	L.	3	2	>26	0	0	0	13	1	133
Chenopodium album	L.	3	1	>20	0	0	0	26	1	133
Chenopodium album	L.	3	1	>20	0	0	0	39	1	133
Chenopodium album	L.	3	1	>20	0	0	0	7.5	1	196
Chenopodium album	L.	3	1	>5	0	0	0	2.5	2	203
Chenopodium album	L.	3	1	>5	0	0	0	2.5	1	203
Chenopodium album	L.	3	1	>5	0	0	0	7.5	2	203
Chenopodium album	L.	3	1	>5	0	0	0	7.5	1	203
Chenopodium album	L.	3	1	>5	0	0	0	15	2	203
Chenopodium album	L.	3	1	>5	0	0	0	15	1	203
Chenopodium album	L.	3	1	>4	0	0	0	20	2	228
Chenopodium album	L.	3	1	>39	0	0	0	50	1	240
Chenopodium album	L.	3	1	>39	0	0	0	100	1	240

Species	Authority	Seed bank type	Number of records	Longevity (y)	Minimum density (seeds m⁻²)	Maximum density (seeds m⁻²)	Mean density (seeds m⁻²)	Depth (cm)	Method	Source code
Chenopodium album	L.	3	1	16	0	0	0	15	1	240
Chenopodium album	L.	3	1	>4	0	0	0	15	1	258
Chenopodium album	L.	3	1	—	8	8	8	8	5	108
Chenopodium album	L.	3	1	>10	9	9	9	10	4	241
Chenopodium album	L.	3	1	—	12	12	12	10	5	252
Chenopodium album	L.	3	1	—	16	16	16	5	5	234
Chenopodium album	L.	3	1	—	28	28	28	70	5	219
Chenopodium album	L.	3	1	>660	30	30	30	50	5	161
Chenopodium album	L.	3	1	>20	33	33	33	24	5	168
Chenopodium album	L.	3	1	>660	37	37	37	55	5	161
Chenopodium album	L.	3	1	>50	65	65	65	30	5	45
Chenopodium album	L.	3	1	>60	65	65	65	30	5	45
Chenopodium album	L.	3	1	>30	75	75	75	40	5	162
Chenopodium album	L.	3	1	—	108	108	108	30	5	29
Chenopodium album	L.	3	1	>6	194	194	194	30	5	45
Chenopodium album	L.	3	1	—	238	238	238	12.5	5	126
Chenopodium album	L.	3	1	>24	377	377	377	30	5	45
Chenopodium album	L.	3	1	>8	1206	1206	1206	30	5	45
Chenopodium album	L.	3	1	—	3700	3700	3700	20	5	69
Chenopodium album	L.	4	1	—	9	9	9	10	5	33
Chenopodium album	L.	4	2	—	11	11	11	10	5	8
Chenopodium album	L.	4	1	—	20	20	20	10	4	22
Chenopodium album	L.	4	3	—	12	248	97	4	5	217
Chenopodium album	L.	4	2	—	9	200	105	15	4	31
Chenopodium album	L.	4	3	—	20	286	135	5	5	189
Chenopodium album	L.	4	1	—	139	139	139	25	5	2
Chenopodium album	L.	4	5	—	76	360	162	25	4	36
Chenopodium album	L.	4	7	—	18	883	187	7.5	5	10
Chenopodium album	L.	4	35	—	20	2277	276	30	4	13
Chenopodium album	L.	4	7	—	171	1600	423	25	5	260
Chenopodium album	L.	4	1	—	438	438	438	10	5	255
Chenopodium album	L.	4	1	—	490	490	490	25	4	136
Chenopodium album	L.	4	2	—	440	590	515	15	5	94
Chenopodium album	L.	4	1	—	1443	1443	1443	7	5	58
Chenopodium album	L.	4	1	—	1528	1528	1528	20	4	111
Chenopodium album	L.	4	1	—	3300	3300	3300	7	4	50
Chenopodium album	L.	4	1	—	9211	9211	9211	23	5	202
Chenopodium album	L.	4	2	—	8951	13440	11196	15	4	46
Chenopodium album	L.	4	4	—	11213	54906	32360	15	5	272
Chenopodium album	L.	4	1	—	54859	54859	54859	15	5	207
Chenopodium ficifolium	Smith	3	1	>5	0	0	0	7.5	2	195
Chenopodium glaucum	L.	2	1	—	3	3	3	5	5	244

Species	Author									
Chenopodium hybridum	L.	3	3	>5	0	0	0	7.5	2	204
Chenopodium hybridum	L.	3	1	>39	0	0	0	100	1	240
Chenopodium hybridum	L.	3	1	10	0	0	0	15	1	240
Chenopodium hybridum	L.	3	1	30	0	0	0	50	1	240
Chenopodium murale	L.	2	3	—	15	190	78	10	4	22
Chenopodium opulifolium	Schrader ex Koch & Ziz	1	1	—	0	0	0	10	4	22
Chenopodium polyspermum	L.	2	1	—	2	2	2	5	5	244
Chenopodium polyspermum	L.	2	1	—	144	144	144	13	5	68
Chenopodium polyspermum	L.	2	7	—	12	776	255	25	4	36
Chenopodium polyspermum	L.	2	1	—	472	472	472	40	5	219
Chenopodium polyspermum	L.	2	1	—	861	861	861	12	5	162
Chenopodium polyspermum	L.	2	1	—	2128	2128	2128	5	5	69
Chenopodium polyspermum	L.	3	3	>5	0	0	0	7.5	2	204
Chenopodium polyspermum	L.	3	1	—	48	48	48	13	5	68
Chenopodium polyspermum	L.	3	1	>35	611	611	611	32	5	169
Chenopodium polyspermum	L.	3	5	—	13400	13400	13400	20	5	69
Chenopodium polyspermum	L.	4	1	—	20	180	68	30	4	13
Chenopodium polyspermum	L.	4	1	—	80	80	80	25	4	36
Chenopodium polyspermum	L.	4	1	—	182	182	182	25	5	260
Chenopodium rubrum	L.	2	1	—	272	272	272	15	5	207
Chenopodium rubrum	L.	2	3	—	5	5	5	5	5	244
Chenopodium rubrum	L.	2	1	—	50	167	95	3	5	96
Chenopodium rubrum	L.	2	3	—	180	180	180	5	5	262
Chenopodium rubrum	L.	2	8	—	40	938	361	4	5	217
Chenopodium rubrum	L.	3	3	—	86	6397	1870	5	5	177
Chenopodium rubrum	L.	4	1	>5	0	0	0	7.5	2	204
Chenopodium rubrum	L.	3	1	>54	75	75	75	65	5	161
Chenopodium rubrum	L.	4	1	—	245	245	245	5	5	166
Chenopodium rubrum	L.	4	1	—	324	324	324	4	5	217
Chenopodium rubrum	L.	4	2	—	989	989	989	5	5	177
Chenopodium rubrum	L.	4	2	—	504	1480	992	12	4	230
Chenopodium rubrum	L.	4	1	—	1877	1877	1877	15	5	207
Salicornia europaea	L.	1	6	—	0	0	0	—	5	109
Salicornia europaea	L.	2	1	—	6	6	6	15	4	21
Salicornia europaea	L.	2	1	—	500	500	500	20	5	210
Salicornia europaea	L.	2	1	—	1000	1000	1000	1	5	57
Salicornia europaea	L.	2	12	—	600	4000	1800	10	4	20
Salicornia europaea	L.	2	1	—	9700	9700	9700	10	5	243
Salicornia europaea	L.	4	1	—	8	8	8	0	6	90
Salicornia europaea	L.	4	1	—	126	126	126	17.8	5	156
Salicornia europaea	L.	4	1	—	379	379	379	5	5	242
Salicornia europaea	L.	4	2	—	3360	18600	10980	10	5	243
Salicornia europaea	L.	4	4	—	2720	23500	12030	—	5	109
Suaeda maritima	(L.) Dumort.	1	9	—	0	0	0	10	4	20
Suaeda maritima	(L.) Dumort.	1	1	—	0	0	0	10	4	103
Suaeda maritima	(L.) Dumort.	2	1	—	3	3	3	15	4	21

Species	Authority	Seed bank type	Number of records	Longevity (y)	Minimum density (seeds m^{-2})	Maximum density (seeds m^{-2})	Mean density (seeds m^{-2})	Depth (cm)	Method	Source code
Cistaceae										
Helianthemum nummularium	(L.) Miller	1	1	—	0	0	0	5	5	55
Helianthemum nummularium	(L.) Miller	1	1	—	0	0	0	1	5	96
Helianthemum nummularium	(L.) Miller	1	1	—	0	0	0	2	5	153
Helianthemum nummularium	(L.) Miller	1	2	—	0	0	0	6	5	181
Helianthemum nummularium	(L.) Miller	1	2	—	0	0	0	6.5	7	182
Helianthemum nummularium	(L.) Miller	2	2	—	5	6	5	3	7	235
Helianthemum nummularium	(L.) Miller	4	2	—	38	123	81	5	5	122
Helianthemum oelandicum	(L.) DC.	4	1	—	35	35	35	5	5	55
Tuberaria guttata	(L.) Fourr.	1	1	—	0	0	0	3	5	132
Tuberaria guttata	(L.) Fourr.	2	1	—	213	213	213	3	5	132
Tuberaria guttata	(L.) Fourr.	4	2	—	275	313	294	3	5	132
Compositae (Asteraceae)										
Achillea collina	J. Becker ex Reichenb.	4	1	—	408	408	408	8.5	4	254
Achillea millefolium	L.	1	3	—	0	0	0	30	5	45
Achillea millefolium	L.	1	1	—	0	0	0	7	5	58
Achillea millefolium	L.	1	3	—	0	0	0	3	5	59
Achillea millefolium	L.	1	1	—	0	0	0	15	5	62
Achillea millefolium	L.	1	1	—	0	0	0	6	5	67
Achillea millefolium	L.	1	1	—	0	0	0	13	5	68
Achillea millefolium	L.	1	5	—	0	0	0	5	5	69
Achillea millefolium	L.	1	1	—	0	0	0	9	5	84
Achillea millefolium	L.	1	3	—	0	0	0	10	4	88
Achillea millefolium	L.	1	6	—	0	0	0	3	5	96
Achillea millefolium	L.	1	5	—	0	0	0	20	5	97
Achillea millefolium	L.	1	2	—	0	0	0	10	5	110
Achillea millefolium	L.	1	3	—	0	0	0	6.5	5	128
Achillea millefolium	L.	1	1	—	0	0	0	11.5	5	134
Achillea millefolium	L.	1	2	—	0	0	0	10	5	149
Achillea millefolium	L.	1	8	—	0	0	0	10	5	150
Achillea millefolium	L.	1	12	—	0	0	0	10	5	152
Achillea millefolium	L.	1	1	—	0	0	0	2	5	153
Achillea millefolium	L.	1	1	—	0	0	0	6	5	181
Achillea millefolium	L.	1	1	—	0	0	0	6.5	7	182
Achillea millefolium	L.	1	1	—	0	0	0	7	5	232
Achillea millefolium	L.	1	1	—	0	0	0	5	5	233
Achillea millefolium	L.	1	2	—	0	0	0	5	5	250
Achillea millefolium	L.	1	6	—	0	0	0	5	5	251
Achillea millefolium	L.	1	19	—	0	0	0	10	5	255
Achillea millefolium	L.	1	4	—	0	3	1	3	7	235
Achillea millefolium	L.	1	4	—	53	105	66	8	7	275

Achillea millefolium	L.	2	5	>1	0	0	0	—	1	275
Achillea millefolium	L.	2	1	—	8	8	8	8	5	108
Achillea millefolium	L.	2	1	—	20	20	20	30	5	219
Achillea millefolium	L.	2	1	—	75	75	75	6.5	5	68
Achillea millefolium	L.	2	1	—	75	75	75	10	5	255
Achillea millefolium	L.	2	1	—	1716	1716	1716	10	5	152
Achillea millefolium	L.	3	1	>5	0	0	0	7.5	2	195
Achillea millefolium	L.	3	1	>20	32	32	32	13	5	183
Achillea millefolium	L.	4	1	—	4	4	4	20	4	111
Achillea millefolium	L.	4	1	—	13	13	13	10	5	110
Achillea millefolium	L.	4	1	—	32	32	32	6	5	181
Achillea millefolium	L.	4	2	—	50	100	75	3	5	96
Achillea millefolium	L.	4	2	—	67	83	75	10	5	184
Achillea millefolium	L.	4	1	—	80	80	80	30	4	13
Achillea millefolium	L.	4	2	—	88	88	88	10	5	150
Achillea millefolium	L.	4	1	—	92	92	92	5	5	222
Achillea millefolium	L.	4	4	—	50	200	116	10	5	255
Achillea millefolium	L.	4	1	—	151	151	151	17.8	5	155
Achillea millefolium	L.	4	3	—	234	468	312	10	5	152
Achillea ptarmica	L.	1	1	—	0	0	0	20	5	67
Achillea ptarmica	L.	1	1	—	0	0	0	5	5	69
Achillea ptarmica	L.	1	1	—	0	0	0	5	4	88
Achillea ptarmica	L.	1	1	—	0	0	0	3	5	96
Achillea ptarmica	L.	1	8	—	0	0	0	3	5	208
Achillea ptarmica	L.	1	3	—	0	0	0	6	5	208
Achillea ptarmica	L.	1	5	—	0	0	0	20	4	225
Achillea ptarmica	L.	1	1	—	0	0	0	3	7	235
Achillea ptarmica	L.	1	8	—	0	0	0	10	5	255
Achillea ptarmica	L.	3	1	—	54	54	54	30	5	45
Achillea ptarmica	L.	4	1	—	85	85	85	3	5	208
Achillea ptarmica	L.	4	1	—	113	113	113	10	5	255
Achillea setacea	Waldst. & Kit.	3	1	—	538	538	538	6	5	208
Ambrosia artemisiifolia	L.	1	1	—	0	0	0	11.5	5	134
Ambrosia artemisiifolia	L.	1	2	—	0	0	0	5	5	189
Ambrosia artemisiifolia	L.	3	1	40	0	0	0	—	3	124
Ambrosia artemisiifolia	L.	3	1	>30	0	0	0	15	1	240
Ambrosia artemisiifolia	L.	3	1	>39	0	0	0	50	1	240
Ambrosia artemisiifolia	L.	3	1	>39	0	0	0	100	1	240
Ambrosia artemisiifolia	L.	3	1	—	49	49	49	11.5	5	134
Ambrosia artemisiifolia	L.	3	1	—	94	94	94	16	5	23
Ambrosia artemisiifolia	L.	4	1	—	3	3	3	10	5	33
Ambrosia artemisiifolia	L.	4	2	—	5	65	35	5	5	189
Anaphalis margaritacea	(L.) Benth.	1	1	—	0	0	0	2	5	95
Anaphalis margaritacea	(L.) Benth.	4	1	—	90	90	90	10	5	120
Antennaria alpina	(L.) Gaertner	1	1	—	0	0	0	5	5	105
Antennaria carpatica	(Wahlenb.) Bluff & Fingerh.	1	1	—	0	0	0	5	5	55
Antennaria dioica	(L.) Gaertner	1	1	—	0	0	0	2	5	153

Species	Authority	Seed bank type	Number of records	Longevity (y)	Minimum density (seeds m⁻²)	Maximum density (seeds m⁻²)	Mean density (seeds m⁻²)	Depth (cm)	Method	Source code
Antennaria dioica	(L.) Gaertner	1	1	—	0	0	0	10	5	255
Anthemis arvensis	L.	1	1	—	0	0	0	10	4	22
Anthemis arvensis	L.	1	1	—	0	0	0	25	4	36
Anthemis arvensis	L.	1	1	—	0	0	0	20	4	111
Anthemis arvensis	L.	1	1	—	0	0	0	20	5	111
Anthemis arvensis	L.	1	1	—	0	0	0	3	5	132
Anthemis arvensis	L.	1	2	—	0	0	0	10	5	255
Anthemis arvensis	L.	2	7	—	63	4025	697	10	5	255
Anthemis arvensis	L.	3	1	>11	0	0	0	25	3	209
Anthemis arvensis	L.	4	2	—	153	213	183	30	4	13
Anthemis arvensis	L.	4	1	—	655	655	655	25	5	255
Anthemis arvensis	L.	4	3	—	538	6875	2863	10	5	36
Anthemis cotula	L.	1	2	—	0	0	0	25	4	36
Anthemis cotula	L.	1	1	—	0	0	0	7	4	50
Anthemis cotula	L.	1	1	—	0	0	0	10	5	255
Anthemis cotula	L.	2	1	—	112	112	112	16	5	23
Anthemis cotula	L.	2	1	—	200	200	200	7	4	50
Anthemis cotula	L.	3	1	30	0	0	0	—	3	124
Anthemis cotula	L.	3	3	>5	0	0	0	7.5	2	206
Anthemis cotula	L.	3	1	>11	0	0	0	25	3	209
Anthemis cotula	L.	3	1	>6	65	65	65	30	5	45
Anthemis cotula	L.	4	2	—	27	27	27	13.5	5	163
Anthemis tinctoria	L.	3	1	>11	0	0	0	25	3	209
Arctium lappa	L.	1	1	—	0	0	0	6	5	49
Arctium lappa	L.	1	1	—	0	0	0	5	5	69
Arctium lappa	L.	3	3	>5	0	0	0	7.5	2	206
Arctium lappa	L.	3	1	>39	0	0	0	100	1	240
Arctium lappa	L.	3	1	16	0	0	0	15	1	240
Arctium lappa	L.	3	1	21	0	0	0	50	1	240
Arctium minus	(Hill) Bernh.	1	1	—	0	0	0	15	5	257
Arctium minus	(Hill) Bernh.	2	1	2	0	0	0	7.5	2	206
Arctium minus	(Hill) Bernh.	2	1	—	30	30	30	5	5	177
Arctium minus	(Hill) Bernh.	3	2	>5	0	0	0	7.5	2	206
Arnica montana	L.	1	4	—	0	0	0	10	5	255
Arnoseris minima	(L.) Schweigger & Koerte	4	1	—	420	420	420	25	4	3
Artemisia absinthium	L.	2	1	—	125	125	125	10	5	162
Artemisia absinthium	L.	2	1	—	225	225	225	27	5	162
Artemisia absinthium	L.	2	1	—	317	317	317	3	5	96
Artemisia campestris	L.	1	3	—	0	0	0	6	5	39
Artemisia norvegica	Fries	1	1	—	0	0	0	5	5	71
Artemisia vulgaris	L.	1	1	—	0	0	0	10	5	255
Artemisia vulgaris	L.	1	3	—	88	526	345	8	7	275

Species										
Artemisia vulgaris	L	2	1	—	38	38	38	10	5	255
Artemisia vulgaris	L	2	1	—	125	125	125	10	5	162
Artemisia vulgaris	L	2	1	—	262	262	262	12	5	162
Artemisia vulgaris	L	2	4	—	74	789	535	8	7	275
Artemisia vulgaris	L	2	1	—	702	702	702	22	5	162
Artemisia vulgaris	L	2	2	—	467	1150	809	3	5	96
Artemisia vulgaris	L	2	2	—	688	1625	1157	17	5	162
Artemisia vulgaris	L	3	1	>5	0	0	0	7.5	2	195
Artemisia vulgaris	L	4	1	—	4	4	4	20	4	111
Artemisia vulgaris	L	4	1	—	83	83	83	3	5	96
Artemisia vulgaris	L	4	1	—	138	138	138	10	5	255
Aster alpinus	L	1	1	—	0	0	0	5	5	55
Aster amellus	L	1	2	—	0	0	0	6.5	7	182
Aster tripolium	L	2	1	—	0	0	0	10	4	103
Aster tripolium	L	2	5	—	3	3	3	2	5	113
Aster tripolium	L	2	14	—	600	6000	2586	10	4	20
Bellis perennis	L	1	2	—	0	0	0	6	5	28
Bellis perennis	L	1	1	—	0	0	0	5	5	55
Bellis perennis	L	1	1	—	0	0	0	6.5	5	68
Bellis perennis	L	1	1	—	0	0	0	13	5	68
Bellis perennis	L	1	7	—	0	0	0	5	5	69
Bellis perennis	L	1	1	—	0	0	0	5	4	88
Bellis perennis	L	1	1	—	0	0	0	6	5	99
Bellis perennis	L	1	2	—	0	0	0	6.5	5	128
Bellis perennis	L	1	1	—	0	0	0	6	5	181
Bellis perennis	L	1	1	—	0	0	0	6.5	7	182
Bellis perennis	L	1	1	—	0	0	0	5	5	250
Bellis perennis	L	2	4	—	26	68	45	2	7	99
Bellis perennis	L	2	1	—	54	54	54	30	5	29
Bellis perennis	L	1	1	—	75	75	75	90	5	162
Bellis perennis	L	2	2	—	208	208	208	10	5	27
Bellis perennis	L	2	1	—	400	400	400	5	5	251
Bellis perennis	L	2	1	4	0	0	0	7.5	2	195
Bellis perennis	L	2	2	—	21	21	21	6.5	7	182
Bellis perennis	L	2	1	—	99	99	99	15	5	62
Bellis perennis	L	2	1	>1	156	156	156	6	5	28
Bellis perennis	L	2	1	—	183	183	183	30	5	29
Bellis perennis	L	2	2	>2	104	286	195	6	5	28
Bellis perennis	L	2	4	—	156	260	208	6	5	99
Bellis perennis	L	2	1	—	234	234	234	5	5	244
Bellis perennis	L	2	2	—	300	300	300	12	5	162
Bellis perennis	L	2	2	—	130	910	520	6	5	28
Bellis perennis	L	2	1	—	642	642	642	2	7	99
Bellis perennis	L	2	2	—	800	800	800	5	5	251
Bellis perennis	L	2	2	—	863	863	863	10	5	255
Bellis perennis	L	2	7	—	161	3520	1602	30	5	45
Bellis perennis	L	2	2	—	833	3750	2292	20	5	97

Species	Authority	Seed bank type	Number of records	Longevity (y)	Minimum density (seeds m^{-2})	Maximum density (seeds m^{-2})	Mean density (seeds m^{-2})	Depth (cm)	Method	Source code
Bellis perennis	L	3	1	—	97	97	97	30	5	29
Bellis perennis	L	3	1	—	125	125	125	8	5	108
Bellis perennis	L	3	1	—	208	208	208	6	5	99
Bellis perennis	L	3	1	—	300	300	300	5	5	251
Bellis perennis	L	3	1	—	334	334	334	30	5	45
Bellis perennis	L	3	3	—	417	2708	1250	20	5	97
Bellis perennis	L	4	2	—	7	35	21	15	5	62
Bellis perennis	L	4	1	—	42	42	42	20	4	111
Bellis perennis	L	4	1	—	48	48	48	6	5	181
Bellis perennis	L	4	3	—	9	132	61	10	5	110
Bellis perennis	L	4	4	—	15	315	108	12	5	174
Bellis perennis	L	4	4	—	234	234	234	5	5	244
Bellis perennis	L	4	4	—	80	880	400	5	5	69
Bellis perennis	L	4	1	—	2275	2275	2275	5	5	250
Bellis perennis	L	4	2	—	2392	3224	2808	6	5	99
Bellis perennis	L	4	2	—	4483	6002	5243	17.8	5	155
Bidens cernua	L	2	2	—	1397	10648	6023	35	5	247
Bidens frondosa	L	3	1	10	0	0	0	50	1	240
Bidens frondosa	L	3	1	10	0	0	0	100	1	240
Bidens frondosa	L	3	1	16	0	0	0	15	1	240
Bidens tripartita	L	1	1	—	0	0	0	5	5	117
Bidens tripartita	L	1	3	—	0	0	0	3	5	208
Bidens tripartita	L	2	4	—	780	9096	3964	50	5	216
Bidens tripartita	L	3	1	>4	0	0	0	7.5	2	195
Bidens tripartita	L	3	1	—	156	156	156	10	5	27
Bidens tripartita	L	4	1	—	3	3	3	3	7	235
Bidens tripartita	L	4	3	—	20	215	97	15	5	272
Buphthalmum salicifolium	L	1	1	—	0	0	0	6	5	181
Buphthalmum salicifolium	L	1	1	—	18	18	18	6.5	7	182
Buphthalmum salicifolium	L	2	1	—	59	59	59	6.5	7	182
Buphthalmum salicifolium	L	3	1	>30	32	32	32	10	5	181
Calendula arvensis	L	1	1	—	0	0	0	10	4	22
Calendula arvensis	L	2	1	—	20	20	20	10	4	22
Carduus acanthoides	L	4	1	—	185	185	185	10	4	22
Carduus acanthoides	L	1	4	—	0	0	0	7	4	50
Carduus crispus	L	3	1	>40	299	299	299	18	5	162
Carduus nutans	L	1	1	—	0	0	0	10	5	110
Carduus nutans	L	2	1	3	0	0	0	7.5	2	201
Carduus nutans	L	3	1	>10	0	0	0	23	1	38
Carduus nutans	L	4	1	—	49	49	49	15	5	62
Carduus pycnocephalus	L	1	3	—	0	0	0	3	5	132
Carduus pycnocephalus	L	2	3	—	300	600	433	7	4	50

Species	Authority									
Carlina acaulis	L.	1	1	—	0	0	0	6	5	181
Carlina acaulis	L.	1	1	—	0	0	0	6.5	7	182
Carlina acaulis	L.	1	4	—	0	0	0	10	5	255
Carlina vulgaris	L.	1	1	—	0	0	0	5	5	69
Carlina vulgaris	L.	1	1	—	0	0	0	3	5	132
Carlina vulgaris	L.	1	2	—	0	0	0	6	5	181
Carlina vulgaris	L.	1	1	—	0	0	0	8.5	4	254
Carlina vulgaris	L.	1	2	—	0	0	0	10	5	255
Carlina vulgaris	L.	1	1	—	16	16	16	6.5	7	182
Carlina vulgaris	L.	2	1	—	0	0	0	6	3	179
Carlina vulgaris	L.	4	1	>3	22	22	22	10	5	110
Centaurea calcitrapa	L.	2	1	>3	0	0	0	15	1	34
Centaurea calcitrapa	L.	2	3	>3	0	0	0	45	1	34
Centaurea cyanus	L.	1	1	—	0	0	0	25	4	36
Centaurea cyanus	L.	1	1	—	0	0	0	10	5	255
Centaurea cyanus	L.	1	1	—	0	0	0	25	5	260
Centaurea cyanus	L.	2	2	2	0	0	0	50	5	260
Centaurea cyanus	L.	2	1	3	0	0	0	30	3	14
Centaurea cyanus	L.	2	1	3	0	0	0	8	1	61
Centaurea cyanus	L.	2	2	3	0	0	0	20	1	61
Centaurea cyanus	L.	3	1	>2	0	0	0	30	1	219
Centaurea cyanus	L.	4	3	8	0	0	0	20	1	125
Centaurea cyanus	L.	4	1	—	25	25	25	25	4	111
Centaurea cyanus	L.	4	1	—	55	55	55	20	5	2
Centaurea jacea	L.	1	2	—	0	0	0	25	5	68
Centaurea jacea	L.	1	2	—	0	0	0	6.5	5	68
Centaurea jacea	L.	1	3	—	0	0	0	13	5	69
Centaurea jacea	L.	1	4	—	0	0	0	5	4	88
Centaurea jacea	L.	1	3	—	0	0	0	10	5	128
Centaurea jacea	L.	1	1	—	0	0	0	6.5	5	150
Centaurea jacea	L.	1	7	—	0	0	0	10	5	152
Centaurea jacea	L.	1	3	—	0	0	0	10	5	174
Centaurea jacea	L.	1	4	—	0	0	0	12	5	181
Centaurea jacea	L.	1	4	—	0	0	0	6	4	225
Centaurea jacea	L.	1	2	—	0	0	0	20	5	250
Centaurea jacea	L.	1	14	—	0	0	0	5	5	251
Centaurea jacea	L.	1	6	—	0	0	0	5	5	255
Centaurea jacea	L.	1	1	—	22	22	22	10	7	182
Centaurea jacea	L.	1	5	>1	521	521	521	6.5	5	97
Centaurea jacea	L.	2	5	—	0	0	0	20	1	275
Centaurea jacea	L.	2	2	—	14	14	14	25	4	36
Centaurea jacea	L.	2	1	—	40	40	40	6.5	7	182
Centaurea jacea	L.	2	1	—	400	400	400	6.5	5	251
Centaurea jacea	L.	4	1	—	24	24	24	5	5	181
Centaurea jacea	L.	4	1	—	563	563	563	6	4	88
Centaurea melitensis	L.	1	3	—	0	0	0	3	5	132

Species	Authority	Seed bank type	Number of records	Longevity (y)	Minimum density (seeds m⁻²)	Maximum density (seeds m⁻²)	Mean density (seeds m⁻²)	Depth (cm)	Method	Source code
Centaurea montana	L.	1	1	—	0	0	0	10	5	255
Centaurea nigra	L.	1	1	—	0	0	0	30	5	45
Centaurea nigra	L.	1	1	—	0	0	0	6	5	48
Centaurea nigra	L.	1	2	—	0	0	0	7	5	58
Centaurea nigra	L.	1	2	—	0	0	0	1	5	96
Centaurea nigra	L.	1	6	—	0	0	0	3	5	96
Centaurea nigra	L.	1	1	—	0	0	0	10	5	110
Centaurea nigra	L.	1	1	—	0	0	0	10	5	255
Centaurea nigra	L.	1	2	—	3	7	5	3	7	235
Centaurea nigra	L.	2	1	4	0	0	0	7.5	2	195
Centaurea nigra	L.	2	1	—	8	8	8	3	7	235
Centaurea nigra	L.	2	2	—	27	34	31	10	5	252
Centaurea nigra	L.	2	2	—	54	344	199	30	5	29
Centaurea nigra	L.	3	1	—	479	479	479	6	5	48
Centaurea nigra	L.	4	1	—	67	67	67	3	5	96
Centaurea nigra	L.	4	1	—	479	479	479	17.8	5	155
Centaurea nigra	L.	4	3	—	13	1719	942	10	5	110
Centaurea scabiosa	L.	1	2	—	0	0	0	7	5	58
Centaurea scabiosa	L.	1	2	—	0	0	0	6	5	67
Centaurea scabiosa	L.	1	1	—	0	0	0	20	5	97
Centaurea scabiosa	L.	1	2	—	0	0	0	3	7	235
Centaurea scabiosa	L.	1	3	—	0	0	0	10	5	255
Centaurea scabiosa	L.	2	1	>3	0	0	0	6	3	179
Centaurea scabiosa	L.	2	1	4	0	0	0	7.5	2	195
Centaurea scabiosa	L.	4	1	—	50	50	50	3	5	96
Centaurea solstitialis	L.	2	1	>3	0	0	0	15	1	34
Centaurea solstitialis	L.	2	1	>3	0	0	0	45	1	34
Centaurea solstitialis	L.	2	1	4	0	0	0	30	1	35
Chrysanthemum segetum	L.	1	1	—	118	118	118	30	5	45
Chrysanthemum segetum	L.	2	1	—	580	580	580	17.8	5	155
Chrysanthemum segetum	L.	3	1	9	0	0	0	25	1	125
Chrysanthemum segetum	L.	3	3	>5	0	0	0	7.5	2	206
Chrysanthemum segetum	L.	3	1	>8	54	54	54	30	5	45
Chrysanthemum segetum	L.	3	1	>20	258	258	258	20	5	40
Chrysanthemum segetum	L.	3	1	>6	312	312	312	30	5	45
Chrysanthemum segetum	L.	4	1	—	275	275	275	20	4	111
Cichorium intybus	L.	1	1	—	0	0	0	10	4	22
Cichorium intybus	L.	2	1	4	0	0	0	7.5	2	195
Cirsium acaule	(L.) Scop.	1	1	—	0	0	0	5	5	55
Cirsium acaule	(L.) Scop.	1	1	—	0	0	0	6	5	67
Cirsium acaule	(L.) Scop.	1	1	—	0	0	0	5	5	122
Cirsium acaule	(L.) Scop.	1	2	—	0	0	0	6	5	181

Cirsium acaule	(L.) Scop.	1	8	—	0	0	0	10	5	255
Cirsium acaule	(L.) Scop.	1	2	—	13	21	17	6.5	7	182
Cirsium arvense	(L.) Scop.	1	4	—	0	0	0	25	4	36
Cirsium arvense	(L.) Scop.	1	1	—	0	0	0	30	5	66
Cirsium arvense	(L.) Scop.	1	2	—	0	0	0	50	5	66
Cirsium arvense	(L.) Scop.	1	1	—	0	0	0	5	5	69
Cirsium arvense	(L.) Scop.	1	1	—	0	0	0	10	4	88
Cirsium arvense	(L.) Scop.	1	3	—	0	0	0	15	5	94
Cirsium arvense	(L.) Scop.	1	3	—	0	0	0	2	5	95
Cirsium arvense	(L.) Scop.	1	1	—	0	0	0	3	5	96
Cirsium arvense	(L.) Scop.	1	10	—	0	0	0	10	5	150
Cirsium arvense	(L.) Scop.	1	1	—	0	0	0	13	5	183
Cirsium arvense	(L.) Scop.	1	2	—	0	0	0	10	5	184
Cirsium arvense	(L.) Scop.	1	2	—	0	0	0	6	5	208
Cirsium arvense	(L.) Scop.	1	10	—	0	0	0	5	5	222
Cirsium arvense	(L.) Scop.	1	1	—	0	0	0	20	4	225
Cirsium arvense	(L.) Scop.	1	4	—	0	0	0	5	5	250
Cirsium arvense	(L.) Scop.	1	1	—	0	0	0	5	5	251
Cirsium arvense	(L.) Scop.	1	6	—	0	3	2	10	5	252
Cirsium arvense	(L.) Scop.	1	3	—	5	5	5	10	5	255
Cirsium arvense	(L.) Scop.	1	1	—	43	43	43	3	7	235
Cirsium arvense	(L.) Scop.	1	1	—	53	210	99	10	4	127
Cirsium arvense	(L.) Scop.	1	4	—	106	106	106	30	5	45
Cirsium arvense	(L.) Scop.	2	1	1	0	0	0	8	7	275
Cirsium arvense	(L.) Scop.	2	1	2	0	0	0	12	4	51
Cirsium arvense	(L.) Scop.	2	2	>3	0	0	0	7	2	44
Cirsium arvense	(L.) Scop.	2	2	4	0	0	0	7	2	44
Cirsium arvense	(L.) Scop.	2	1	>1	17	17	17	20	1	61
Cirsium arvense	(L.) Scop.	2	5	—	50	67	59	7.5	2	201
Cirsium arvense	(L.) Scop.	2	1	—	60	60	60	—	1	275
Cirsium arvense	(L.) Scop.	2	2	—	66	66	66	6	5	49
Cirsium arvense	(L.) Scop.	2	2	—	70	70	70	3	5	96
Cirsium arvense	(L.) Scop.	2	1	—	38	125	79	15	5	257
Cirsium arvense	(L.) Scop.	2	1	—	80	80	80	5	5	177
Cirsium arvense	(L.) Scop.	2	3	—	688	688	688	8	5	108
Cirsium arvense	(L.) Scop.	3	2	—	0	0	0	10	5	255
Cirsium arvense	(L.) Scop.	3	1	—	0	0	0	10	5	149
Cirsium arvense	(L.) Scop.	3	1	—	0	0	0	38	5	186
Cirsium arvense	(L.) Scop.	3	1	>25	0	0	0	25	1	125
Cirsium arvense	(L.) Scop.	3	1	>5	0	0	0	7.5	2	201
Cirsium arvense	(L.) Scop.	3	1	21	0	0	0	15	1	240
Cirsium arvense	(L.) Scop.	3	1	21	0	0	0	50	1	240
Cirsium arvense	(L.) Scop.	3	1	21	0	0	0	100	1	240
Cirsium arvense	(L.) Scop.	3	1	—	10	10	10	15	4	21
Cirsium arvense	(L.) Scop.	3	1	—	16	16	16	70	5	219
Cirsium arvense	(L.) Scop.	3	1	—	36	36	36	12.5	5	126

Species	Authority	Seed bank type	Number of records	Longevity (y)	Minimum density (seeds m⁻²)	Maximum density (seeds m⁻²)	Mean density (seeds m⁻²)	Depth (cm)	Method	Source code
Cirsium arvense	(L.) Scop.	3	1	—	113	113	113	6	5	208
Cirsium arvense	(L.) Scop.	3	1	—	151	151	151	30	5	45
Cirsium arvense	(L.) Scop.	3	2	—	60	500	280	15	5	257
Cirsium arvense	(L.) Scop.	3	1	—	600	600	600	30	5	66
Cirsium arvense	(L.) Scop.	4	1	—	9	9	9	10	5	110
Cirsium arvense	(L.) Scop.	4	1	—	10	10	10	5	5	166
Cirsium arvense	(L.) Scop.	4	3	—	7	49	27	15	5	62
Cirsium arvense	(L.) Scop.	4	1	—	42	42	42	20	5	111
Cirsium arvense	(L.) Scop.	4	1	—	185	185	185	5	5	177
Cirsium arvense	(L.) Scop.	4	13	—	38	663	193	10	5	255
Cirsium arvense	(L.) Scop.	4	3	—	280	390	330	2	5	95
Cirsium arvense	(L.) Scop.	4	2	—	35	774	405	7	5	58
Cirsium arvense	(L.) Scop.	4	1	—	1200	1200	1200	20	4	225
Cirsium eriophorum	(L.) Scop.	1	2	—	0	0	0	10	5	110
Cirsium eriophorum	(L.) Scop.	1	1	—	0	0	0	13	5	183
Cirsium eriophorum	(L.) Scop.	2	3	1	0	0	0	7.5	2	201
Cirsium heterophyllum	(L.) Hill	1	1	—	8	8	8	6.5	7	182
Cirsium oleraceum	(L.) Scop.	1	3	—	0	0	0	25	4	36
Cirsium oleraceum	(L.) Scop.	1	1	—	0	0	0	6.5	5	68
Cirsium oleraceum	(L.) Scop.	1	1	—	0	0	0	13	5	68
Cirsium oleraceum	(L.) Scop.	1	4	—	0	0	0	6.5	5	128
Cirsium oleraceum	(L.) Scop.	1	5	—	0	0	0	12	5	174
Cirsium oleraceum	(L.) Scop.	1	4	—	0	0	0	10	5	255
Cirsium oleraceum	(L.) Scop.	1	5	—	0	169	104	2	7	99
Cirsium oleraceum	(L.) Scop.	4	1	—	60	60	60	6.5	5	128
Cirsium palustre	(L.) Scop.	1	2	—	0	0	0	6	5	11
Cirsium palustre	(L.) Scop.	1	1	—	0	0	0	40	5	11
Cirsium palustre	(L.) Scop.	1	4	—	0	0	0	10	5	27
Cirsium palustre	(L.) Scop.	1	3	—	0	0	0	6	5	28
Cirsium palustre	(L.) Scop.	1	1	—	0	0	0	20	5	67
Cirsium palustre	(L.) Scop.	1	1	—	0	0	0	5	5	69
Cirsium palustre	(L.) Scop.	1	1	—	0	0	0	5	4	88
Cirsium palustre	(L.) Scop.	1	3	—	0	0	0	3	5	96
Cirsium palustre	(L.) Scop.	1	8	—	0	0	0	6	5	99
Cirsium palustre	(L.) Scop.	1	8	—	0	0	0	10	5	149
Cirsium palustre	(L.) Scop.	1	5	—	0	0	0	12	5	174
Cirsium palustre	(L.) Scop.	1	2	—	0	0	0	5	5	222
Cirsium palustre	(L.) Scop.	1	6	—	0	0	0	10	5	255
Cirsium palustre	(L.) Scop.	1	1	—	6	6	6	3	7	235
Cirsium palustre	(L.) Scop.	1	5	—	0	26	16	2	7	99
Cirsium palustre	(L.) Scop.	1	1	—	118	118	118	30	5	45
Cirsium palustre	(L.) Scop.	2	1	>3	0	0	0	7.5	2	201

Species	Author									
Cirsium palustre	(L.) Scop.	2	5	>1	0	0	0	—	1	275
Cirsium palustre	(L.) Scop.	2	1	—	12	12	12	15	5	32
Cirsium palustre	(L.) Scop.	2	1	—	32	32	32	30	5	45
Cirsium palustre	(L.) Scop.	2	1	—	38	38	38	10	5	255
Cirsium palustre	(L.) Scop.	2	2	—	39	39	39	3	7	235
Cirsium palustre	(L.) Scop.	3	1	—	113	142	128	3	5	208
Cirsium palustre	(L.) Scop.	3	1	—	85	85	85	3	5	208
Cirsium palustre	(L.) Scop.	3	2	—	104	104	104	6	5	28
Cirsium palustre	(L.) Scop.	3	2	—	156	208	182	6	5	99
Cirsium palustre	(L.) Scop.	4	1	—	368	509	439	6	5	208
Cirsium palustre	(L.) Scop.	4	1	—	61	61	61	5	5	222
Cirsium palustre	(L.) Scop.	4	1	—	67	67	67	5	5	98
Cirsium palustre	(L.) Scop.	4	1	—	151	151	151	17.8	5	155
Cirsium palustre	(L.) Scop.	4	1	—	260	260	260	6	5	99
Cirsium palustre	(L.) Scop.	4	12	—	50	750	265	10	5	255
Cirsium palustre	(L.) Scop.	4	6	—	170	821	349	3	5	208
Cirsium palustre	(L.) Scop.	4	1	—	400	400	400	3	5	96
Cirsium rivulare	(Jacq.) All.	1	4	—	0	0	0	12	5	174
Cirsium spinosissimum	(L.) Scop.	1	1	—	0	0	0	5	5	55
Cirsium vulgare	(Savi) Ten.	1	1	—	0	0	0	5	5	19
Cirsium vulgare	(Savi) Ten.	1	4	—	0	0	0	5	5	69
Cirsium vulgare	(Savi) Ten.	1	1	—	0	0	0	3	5	96
Cirsium vulgare	(Savi) Ten.	1	2	—	0	0	0	20	5	97
Cirsium vulgare	(Savi) Ten.	1	2	—	0	0	0	10	5	110
Cirsium vulgare	(Savi) Ten.	1	1	—	0	0	0	10	5	150
Cirsium vulgare	(Savi) Ten.	1	1	—	0	0	0	6	5	151
Cirsium vulgare	(Savi) Ten.	1	1	—	0	0	0	6	5	181
Cirsium vulgare	(Savi) Ten.	1	1	—	0	0	0	3	7	235
Cirsium vulgare	(Savi) Ten.	1	7	—	0	0	0	10	5	255
Cirsium vulgare	(Savi) Ten.	1	1	—	0	0	0	15	5	257
Cirsium vulgare	(Savi) Ten.	2	2	4	0	0	0	7.5	2	201
Cirsium vulgare	(Savi) Ten.	2	1	—	17	17	17	10	4	127
Cirsium vulgare	(Savi) Ten.	2	1	—	500	500	500	10	5	184
Cirsium vulgare	(Savi) Ten.	2	1	—	760	760	760	15	5	257
Cirsium vulgare	(Savi) Ten.	2	1	—	896	896	896	10	5	252
Cirsium vulgare	(Savi) Ten.	2	2	—	1800	1800	1800	12	5	81
Cirsium vulgare	(Savi) Ten.	3	2	>5	0	0	0	7.5	2	201
Cirsium vulgare	(Savi) Ten.	4	1	—	9	9	9	10	5	110
Cirsium vulgare	(Savi) Ten.	4	1	—	23	23	23	7	5	58
Cirsium vulgare	(Savi) Ten.	4	1	—	300	300	300	10	5	184
Conyza canadensis	(L.) Cronq.	1	2	—	0	0	0	3	5	132
Conyza canadensis	(L.) Cronq.	2	1	—	24	24	24	16	5	23
Conyza canadensis	(L.) Cronq.	2	3	—	17	37	28	5	5	189
Conyza canadensis	(L.) Cronq.	2	1	—	30	30	30	5	5	177
Conyza canadensis	(L.) Cronq.	2	1	—	40	40	40	5	5	244
Conyza canadensis	(L.) Cronq.	2	1	—	124	124	124	11.5	5	134
Conyza canadensis	(L.) Cronq.	2	3	—	56	261	181	15	5	272

Species	Authority	Seed bank type	Number of records	Longevity (y)	Minimum density (seeds m⁻²)	Maximum density (seeds m⁻²)	Mean density (seeds m⁻²)	Depth (cm)	Method	Source code
Conyza canadensis	(L.) Cronq.	3	1	>11	0	0	0	25	3	209
Conyza canadensis	(L.) Cronq.	4	1	—	14	14	14	10	5	8
Conyza canadensis	(L.) Cronq.	4	1	—	19	19	19	10	5	33
Conyza canadensis	(L.) Cronq.	4	4	—	39	417	186	15	5	272
Conyza canadensis	(L.) Cronq.	4	1	—	346	346	346	11.5	5	134
Cotula coronopifolia	L.	2	1	1	0	0	0	5	1	246
Crepis biennis	L.	1	1	—	0	0	0	20	5	97
Crepis biennis	L.	1	1	—	0	0	0	6.5	5	128
Crepis biennis	L.	4	1	—	313	313	313	5	4	88
Crepis capillaris	(L.) Wallr.	1	2	—	0	0	0	5	5	69
Crepis capillaris	(L.) Wallr.	1	1	—	0	0	0	3	5	96
Crepis capillaris	(L.) Wallr.	1	1	—	0	0	0	20	4	111
Crepis capillaris	(L.) Wallr.	1	1	—	0	0	0	20	5	111
Crepis capillaris	(L.) Wallr.	1	1	—	210	210	210	8	7	275
Crepis capillaris	(L.) Wallr.	2	1	—	1580	1580	1580	12	5	81
Crepis capillaris	(L.) Wallr.	3	1	>5	0	0	0	7.5	2	195
Crepis capillaris	(L.) Wallr.	4	2	—	5	23	14	5	5	122
Crepis capillaris	(L.) Wallr.	4	3	—	9	119	51	10	5	110
Crepis capillaris	(L.) Wallr.	4	2	—	383	450	417	3	5	96
Crepis foetida	L.	2	1	—	15	15	15	10	4	22
Crepis mollis	(Jacq.) Asch.	1	1	—	0	0	0	10	5	255
Crepis paludosa	(L.) Moench	1	1	—	0	0	0	10	5	126
Crepis paludosa	(L.) Moench	1	3	—	0	0	0	10	5	149
Crepis paludosa	(L.) Moench	1	8	—	0	0	0	10	5	255
Crepis paludosa	(L.) Moench	1	3	—	113	113	113	3	5	208
Crepis paludosa	(L.) Moench	4	3	—	198	1443	698	3	5	208
Crepis praemorsa	(L.) F. Walther	1	2	—	0	0	0	10	5	150
Crepis praemorsa	(L.) F. Walther	1	1	—	0	0	0	10	5	152
Crepis tectorum	L.	4	1	—	14	14	14	20	4	111
Crepis vesicaria	L.	1	1	—	0	0	0	3	5	208
Erigeron acer	L.	1	1	—	0	0	0	5	5	69
Erigeron annuus	(L.) Pers.	2	4	—	8	18	14	5	5	189
Erigeron annuus	(L.) Pers.	2	2	—	37	74	56	11.5	5	134
Erigeron annuus	(L.) Pers.	2	2	—	172	172	172	15	5	175
Erigeron annuus	(L.) Pers.	3	1	—	99	99	99	11.5	5	134
Erigeron annuus	(L.) Pers.	4	1	—	263	263	263	10	5	33
Eupatorium cannabinum	L.	1	1	—	0	0	0	5	5	222
Eupatorium cannabinum	L.	1	2	—	0	0	0	20	4	225
Eupatorium cannabinum	L.	1	2	—	26	26	26	2	7	99
Eupatorium cannabinum	L.	1	7	—	53	263	102	8	7	275
Eupatorium cannabinum	L.	2	1	—	102	102	102	5	5	222
Eupatorium cannabinum	L.	2	2	—	218	2913	1566	8	7	275

Species	Author									
Eupatorium cannabinum	L.	3	1	—	22	22	22	12.5	5	126
Eupatorium cannabinum	L.	3	1	—	170	170	170	6	5	208
Eupatorium cannabinum	L.	3	1	—	3306	3306	3306	25	5	107
Eupatorium cannabinum	L.	4	3	—	25	592	236	5	5	98
Filago gallica	L.	1	1	—	0	0	0	3	5	132
Filago lutescens	Jordan	1	3	—	0	0	0	3	5	132
Filago lutescens	Jordan	4	1	—	100	100	100	20	4	132
Filago minima	(Smith) Pers.	1	1	—	0	0	0	20	5	111
Filago minima	(Smith) Pers.	1	1	—	0	0	0	24	5	111
Filago minima	(Smith) Pers.	3	1	>40	67	67	67	3	5	168
Filago minima	(Smith) Pers.	4	4	—	38	200	106	4	5	132
Galinsoga parviflora	Cav.	2	1	—	50	50	50	67	5	162
Galinsoga parviflora	Cav.	2	1	—	113	113	113	7	5	162
Galinsoga parviflora	Cav.	2	1	>5	250	250	250	7.5	2	162
Galinsoga parviflora	Cav.	3	1	>11	0	0	0	25	3	195
Galinsoga parviflora	Cav.	3	1	—	0	0	0	8	7	209
Galinsoga quadriradiata	Ruiz Lopez & Pavon	1	2	—	53	105	79	8	7	275
Galinsoga quadriradiata	Ruiz Lopez & Pavon	2	1	—	164	164	164	10	5	275
Gnaphalium luteoalbum	L.	4	1	—	32	32	32	5	5	147
Gnaphalium supinum	L.	2	1	—	15	15	15	5	6	55
Gnaphalium sylvaticum	L.	2	2	—	0	1	1	10	5	170
Gnaphalium sylvaticum	L.	4	4	—	38	100	57	2	5	255
Gnaphalium sylvaticum	L.	4	1	—	61	61	61	11.5	5	82
Gnaphalium uliginosum	L.	1	1	—	0	0	0	100	5	134
Gnaphalium uliginosum	L.	1	1	—	12	12	12	2	7	219
Gnaphalium uliginosum	L.	1	2	—	0	26	13	8	7	99
Gnaphalium uliginosum	L.	1	2	—	53	79	66	25	4	275
Gnaphalium uliginosum	L.	2	1	—	8	8	8	5	5	36
Gnaphalium uliginosum	L.	2	1	—	10	10	10	5	5	117
Gnaphalium uliginosum	L.	2	2	—	24	24	24	5	5	244
Gnaphalium uliginosum	L.	2	1	—	41	51	46	11.5	5	222
Gnaphalium uliginosum	L.	2	1	—	49	49	49	5	5	134
Gnaphalium uliginosue	L.	2	1	—	256	256	256	40	5	69
Gnaphalium uliginosum	L.	2	1	—	260	260	260	6	5	11
Gnaphalium uliginosum	L.	2	2	—	285	285	285	6	5	151
Gnaphalium uliginosum	L.	2	2	—	156	468	312	17	5	11
Gnaphalium uliginosum	L.	2	19	—	313	438	376	10	5	162
Gnaphalium uliginosum	L.	2	2	—	63	1963	492	3	5	255
Gnaphalium uliginosum	L.	2	2	—	283	1367	825	12	5	96
Gnaphalium uliginosum	L.	2	2	—	1311	1311	1311	15	5	162
Gnaphalium uliginosum	L.	2	4	—	2880	4650	3765	50	5	94
Gnaphalium uliginosum	L.	3	1	—	7536	35863	18451	5	5	216
Gnaphalium uliginosum	L.	3	1	—	6	6	6	25	5	234
Gnaphalium uliginosum	L.	3	1	>50	28	28	28	16	5	219
Gnaphalium uliginosum	L.	3	2	>40	44	44	44	24	5	168
Gnaphalium uliginosum	L.	3	1	—	56	78	67	15	5	168
Gnaphalium uliginosum	L.	3	1	—	80	80	80	15	5	257

Species	Authority	Seed bank type	Number of records	Longevity (y)	Minimum density (seeds m⁻²)	Maximum density (seeds m⁻²)	Mean density (seeds m⁻²)	Depth (cm)	Method	Source code
Gnaphalium uliginosum	L.	3	2	>18	78	100	89	32	5	169
Gnaphalium uliginosum	L.	3	1	—	130	130	130	8	5	108
Gnaphalium uliginosum	L.	3	1	>100	211	211	211	32	5	169
Gnaphalium uliginosum	L.	3	1	>12	372	372	372	50	5	219
Gnaphalium uliginosum	L.	3	1	>35	378	378	378	32	5	169
Gnaphalium uliginosum	L.	3	1	—	399	399	399	20	5	107
Gnaphalium uliginosum	L.	3	1		1000	1000	1000	20	5	69
Gnaphalium uliginosum	L.	3	1	>50	1119	1119	1119	30	5	45
Gnaphalium uliginosum	L.	3	2	—	1976	2080	2028	10	5	27
Gnaphalium uliginosum	L.	4	1	—	33	33	33	30	4	13
Gnaphalium uliginosum	L.	4	1	—	63	63	63	8	7	275
Gnaphalium uliginosum	L.	4	1	—	150	150	150	3	5	96
Gnaphalium uliginosum	L.	4	1	—	182	182	182	25	5	260
Gnaphalium uliginosum	L.	4	2	—	426	1246	836	15	4	31
Gnaphalium uliginosum	L.	4	1	—	1013	1013	1013	15	5	207
Gnaphalium uliginosum	L.	4	1	—	1253	1253	1253	20	4	111
Gnaphalium uliginosum	L.	4	2	—	1880	4820	3350	15	5	94
Helianthus annuus	L.	2	1	1	0	0	0	15	1	240
Helianthus annuus	L.	2	1	1	0	0	0	50	1	240
Helianthus annuus	L.	2	1	1	0	0	0	100	1	240
Helianthus annuus	L.	3	1	>10	0	0	0	23	1	38
Helichrysum arenarium	(L.) Moench	4	2	—	2	9	6	20	4	226
Hieracium alpinum	L.	1	1	—	0	0	0	5	5	55
Hieracium lactucella	Wallr.	1	1	—	0	0	0	5	5	69
Hieracium lactucella	Wallr.	1	2	—	0	0	0	10	5	150
Hieracium lactucella	Wallr.	1	3	—	0	0	0	10	5	152
Hieracium laevigatum	—	1	5	—	0	0	0	10	5	255
Hieracium umbellatum	L.	1	4	—	0	0	0	10	5	255
Hieracium umbellatum	L.	2	5	>1	0	0	0	—	1	275
Homogyne alpina	(L.) Cass.	1	2	—	0	0	0	5	5	55
Homogyne alpina	(L.) Cass.	1	4	—	0	0	0	5	4	91
Hypochaeris glabra	L.	1	3	—	0	0	0	3	5	132
Hypochaeris glabra	L.	4	1	—	63	63	63	3	5	132
Hypochaeris glabra	L.	4	1	—	4140	4140	4140	10	5	140
Hypochaeris radicata	L.	1	1	—	0	0	0	30	5	45
Hypochaeris radicata	L.	1	2	—	0	0	0	5	5	69
Hypochaeris radicata	L.	1	1	—	0	0	0	10	4	88
Hypochaeris radicata	L.	1	4	—	0	0	0	3	5	96
Hypochaeris radicata	L.	1	3	—	0	0	0	20	5	97
Hypochaeris radicata	L.	1	1	—	0	0	0	8	5	108
Hypochaeris radicata	L.	1	1	—	0	0	0	5	5	234
Hypochaeris radicata	L.	1	2	—	0	0	0	5	5	251

Species											
Hypochaeris radicata	L.		1	1	—	0	0	0	10	5	255
Hypochaeris radicata	L.		1	1	—	5	5	5	3	7	235
Hypochaeris radicata	L.		1	1	—	11	11	11	10	5	119
Hypochaeris radicata	L.		1	1	—	53	53	53	8	7	275
Hypochaeris radicata	L.		1	4	—	182	364	273	6	5	28
Hypochaeris radicata	L.		2	1	4	0	0	0	7.5	2	195
Hypochaeris radicata	L.		2	5	>1	0	0	0	—	1	275
Hypochaeris radicata	L.		2	1		188	188	188	17	5	162
Hypochaeris radicata	L.		2	1	>1	208	208	208	6	5	28
Hypochaeris radicata	L.		2	2	—	390	442	416	6	5	28
Hypochaeris radicata	L.		3	1	—	32	32	32	30	5	29
Hypochaeris radicata	L.		4	1	—	64	64	64	10	5	120
Hypochaeris radicata	L.		4	2	—	144	160	152	5	5	69
Hypochaeris radicata	L.		4	1	—	475	475	475	5	5	250
Inula conyzae	(Griess.) Meikle		3	1	—	22	22	22	6	5	211
Inula conyzae	(Griess.) Meikle		4	2	—	24	80	52	6	5	181
Inula salicina	L.		1	16	—	0	0	0	10	5	152
Inula salicina	L.		1	1	—	0	0	0	6	5	181
Lactuca serriola	L.		1	1	—	6	6	6	10	4	127
Lactuca serriola	L.		2	1	4	0	0	0	7	2	44
Lactuca serriola	L.		2	1	4	0	0	0	7.5	2	195
Lactuca serriola	L.		2	1	3	0	0	0	15	1	240
Lactuca serriola	L.		2	1	3	0	0	0	50	1	240
Lactuca serriola	L.		2	1	3	0	0	0	100	1	240
Lactuca serriola	L.		2	2	—	34	340	187	5	5	189
Lactuca serriola	L.		3	2	>5	0	0	0	7	2	44
Lactuca serriola	L.		4	1	—	6	6	6	10	5	33
Lactuca serriola	L.		4	2	—	100	250	175	10	5	184
Lactuca virosa	L.		2	3	>3	0	0	0	7	2	44
Lactuca virosa	L.		2	3	2	0	0	0	7	2	44
Lactuca virosa	L.		2	1	3	0	0	0	7	2	44
Lactuca virosa	L.		3	2	>6	0	0	0	7	2	44
Lapsana communis	L.		1	1	—	0	0	0	10	5	255
Lapsana communis	L.		2	1	2	0	0	0	30	3	14
Lapsana communis	L.		2	1	—	6	6	6	25	4	36
Lapsana communis	L.		2	2	—	163	163	163	10	5	255
Lapsana communis	L.		2	1	—	167	167	167	3	5	96
Lapsana communis	L.		2	1	—	180	180	180	11	5	162
Lapsana communis	L.		2	1	—	188	188	188	32	5	162
Lapsana communis	L.		2	4	—	100	875	316	10	5	162
Lapsana communis	L.		3	3	>5	0	0	0	7.5	2	206
Lapsana communis	L.		3	1	—	6	6	6	5	5	234
Lapsana communis	L.		3	1	—	54	54	54	30	5	45
Lapsana communis	L.		3	1	—	60	60	60	15	5	257
Lapsana communis	L.		3	1	>6	474	474	474	30	5	45
Lapsana communis	L.		3	1	—	800	800	800	30	5	66
Lapsana communis	L.		4	1	—	36	36	36	25	5	2

Species	Authority	Seed bank type	Number of records	Longevity (y)	Minimum density (seeds m⁻²)	Maximum density (seeds m⁻²)	Mean density (seeds m⁻²)	Depth (cm)	Method	Source code
Lapsana communis	L.	4	1	—	102	102	102	15	5	142
Lapsana communis	L.	4	7	—	20	286	125	30	4	13
Lapsana communis	L.	4	4	—	88	888	304	10	5	255
Lapsana communis	L.	4	1	—	333	333	333	3	5	96
Leontodon autumnalis	L.	1	4	—	0	0	0	6	5	11
Leontodon autumnalis	L.	1	3	—	0	0	0	10	5	27
Leontodon autumnalis	L.	1	3	—	0	0	0	3	5	59
Leontodon autumnalis	L.	1	5	—	0	0	0	5	5	69
Leontodon autumnalis	L.	1	6	—	0	0	0	6	5	99
Leontodon autumnalis	L.	1	2	—	0	0	0	7	5	112
Leontodon autumnalis	L.	1	4	—	0	0	0	10	5	150
Leontodon autumnalis	L.	1	1	—	0	0	0	5	5	250
Leontodon autumnalis	L.	1	1	—	3	3	3	3	7	235
Leontodon autumnalis	L.	1	1	—	20	20	20	10	5	252
Leontodon autumnalis	L.	1	5	—	0	390	78	2	7	99
Leontodon autumnalis	L.	1	3	—	113	113	113	6	5	208
Leontodon autumnalis	L.	1	4	—	130	468	312	6	5	28
Leontodon autumnalis	L.	2	1	—	7	7	7	3	7	235
Leontodon autumnalis	L.	2	1	—	400	400	400	3	5	96
Leontodon autumnalis	L.	2	4	—	286	4238	1911	6	5	28
Leontodon autumnalis	L.	3	1	>5	0	0	0	7.5	2	195
Leontodon autumnalis	L.	3	1	—	183	183	183	8	5	108
Leontodon autumnalis	L.	4	1	—	76	76	76	6	5	151
Leontodon autumnalis	L.	4	2	—	176	294	235	7	5	112
Leontodon autumnalis	L.	4	2	—	175	350	263	5	5	250
Leontodon hispidus	L.	1	4	—	0	0	0	5	4	91
Leontodon hispidus	L.	1	1	—	0	0	0	3	5	96
Leontodon hispidus	L.	1	4	—	0	0	0	20	5	97
Leontodon hispidus	L.	1	1	—	0	0	0	10	5	110
Leontodon hispidus	L.	1	5	—	0	0	0	10	5	149
Leontodon hispidus	L.	1	2	—	0	0	0	10	5	152
Leontodon hispidus	L.	1	1	—	0	0	0	12	5	174
Leontodon hispidus	L.	1	2	—	0	0	0	6	5	181
Leontodon hispidus	L.	1	1	—	0	0	0	3	7	235
Leontodon hispidus	L.	1	4	—	0	0	0	10	5	255
Leontodon hispidus	L.	1	1	—	205	205	205	30	5	45
Leontodon hispidus	L.	2	1	>3	0	0	0	6	3	179
Leontodon hispidus	L.	2	2	—	6	7	7	3	7	235
Leontodon hispidus	L.	2	3	—	9	34	22	12	5	174
Leontodon hispidus	L.	2	2	—	20	25	23	6.5	7	182
Leontodon hispidus	L.	2	2	—	32	54	43	30	5	45
Leontodon hispidus	L.	2	1	—	180	180	180	6	5	48

Species	Authority									
Leontodon hispidus	L.	3	1	>5	0	0	0	7.5	2	195
Leontodon hispidus	L.	4	1	—	11	11	11	5	5	122
Leontodon hispidus	L.	4	3	—	9	22	16	12	5	174
Leontodon hispidus	L.	4	2	—	29	46	38	7	5	58
Leontodon hispidus	L.	4	3	—	25	75	45	10	5	110
Leontodon hispidus	L.	4	1	—	50	50	50	10	5	255
Leontodon hispidus	L.	4	1	—	151	151	151	17.8	5	155
Leontodon hispidus	L.	4	1	—	1735	1735	1735	8.5	4	254
Leontodon pyrenaicus	Gouan	4	1	—	51	51	51	5	5	55
Leontodon saxatilis	Lam.	1	2	—	0	0	0	3	5	132
Leontodon saxatilis	Lam.	1	1	—	0	0	0	10	5	252
Leontodon saxatilis	Lam.	2	1	4	0	0	0	7.5	2	195
Leontodon saxatilis	Lam.	2	1	—	75	75	75	5	5	250
Leontodon saxatilis	Lam.	4	1	—	128	128	128	5	5	69
Leontodon saxatilis	Lam.	4	1	—	910	910	910	5	5	250
Leucanthemum vulgare	Lam.	1	1	—	0	0	0	5	5	19
Leucanthemum vulgare	Lam.	1	1	—	0	0	0	25	4	36
Leucanthemum vulgare	Lam.	1	1	—	0	0	0	6	5	49
Leucanthemum vulgare	Lam.	1	1	—	0	0	0	3	5	59
Leucanthemum vulgare	Lam.	1	1	—	0	0	0	20	5	67
Leucanthemum vulgare	Lam.	1	3	—	0	0	0	5	5	69
Leucanthemum vulgare	Lam.	1	1	—	0	0	0	10	4	88
Leucanthemum vulgare	Lam.	1	2	—	0	0	0	3	5	96
Leucanthemum vulgare	Lam.	1	1	—	0	0	0	6.5	5	128
Leucanthemum vulgare	Lam.	1	1	—	0	0	0	11.5	5	134
Leucanthemum vulgare	Lam.	1	2	—	0	0	0	10	5	149
Leucanthemum vulgare	Lam.	1	2	—	0	0	0	10	5	152
Leucanthemum vulgare	Lam.	1	2	—	0	0	0	12	5	174
Leucanthemum vulgare	Lam.	1	1	—	0	0	0	6	5	181
Leucanthemum vulgare	Lam.	1	1	—	0	0	0	5	5	244
Leucanthemum vulgare	Lam.	1	10	—	0	0	0	10	5	250
Leucanthemum vulgare	Lam.	1	1	—	43	43	43	30	5	255
Leucanthemum vulgare	Lam.	1	3	—	53	105	70	8	7	45
Leucanthemum vulgare	Lam.	1	11	—	400	400	400	5	7	275
Leucanthemum vulgare	Lam.	1	2	—	2292	2292	2292	20	5	251
Leucanthemum vulgare	Lam.	2	1	—	10	10	10	15	5	97
Leucanthemum vulgare	Lam.	2	1	—	16	16	16	16	5	62
Leucanthemum vulgare	Lam.	2	1	—	37	37	37	11.5	5	23
Leucanthemum vulgare	Lam.	2	2	—	22	93	58	12	5	134
Leucanthemum vulgare	Lam.	2	5	—	38	150	65	10	5	174
Leucanthemum vulgare	Lam.	2	2	—	77	230	154	6.5	7	255
Leucanthemum vulgare	Lam.	2	1	—	270	270	270	13	5	182
Leucanthemum vulgare	Lam.	2	4	—	234	390	312	10	5	68
Leucanthemum vulgare	Lam.	2	2	—	122	515	319	10	5	152
Leucanthemum vulgare	Lam.	2	1	—	600	600	600	5	5	251

Species	Authority	Seed bank type	Number of records	Longevity (y)	Minimum density (seeds m⁻²)	Maximum density (seeds m⁻²)	Mean density (seeds m⁻²)	Depth (cm)	Method	Source code
Leucanthemum vulgare	Lam.	2	4	—	313	1563	678	20	5	97
Leucanthemum vulgare	Lam.	3	1	>5	0	0	0	7.5	2	195
Leucanthemum vulgare	Lam.	3	1	>30	0	0	0	15	1	240
Leucanthemum vulgare	Lam.	3	1	>39	0	0	0	50	1	240
Leucanthemum vulgare	Lam.	3	1	30	0	0	0	100	1	240
Leucanthemum vulgare	Lam.	3	1	>15	33	33	33	24	5	168
Leucanthemum vulgare	Lam.	3	1	>35	33	33	33	24	5	168
Leucanthemum vulgare	Lam.	3	1	>20	78	78	78	24	5	168
Leucanthemum vulgare	Lam.	3	1	—	313	313	313	20	5	97
Leucanthemum vulgare	Lam.	3	3		65	538	330	30	5	45
Leucanthemum vulgare	Lam.	3	1	—	375	375	375	10	4	88
Leucanthemum vulgare	Lam.	3	1	—	600	600	600	6.5	5	68
Leucanthemum vulgare	Lam.	3	1	—	1000	1000	1000	20	5	69
Leucanthemum vulgare	Lam.	4	1	—	9	9	9	12	5	174
Leucanthemum vulgare	Lam.	4	1	—	14	14	14	10	5	33
Leucanthemum vulgare	Lam.	4	2	—	23	40	32	7	5	58
Leucanthemum vulgare	Lam.	4	3	—	50	100	67	3	5	96
Leucanthemum vulgare	Lam.	4	1	—	99	99	99	11.5	5	134
Leucanthemum vulgare	Lam.	4	7	—	38	150	100	10	5	255
Leucanthemum vulgare	Lam.	4	2	—	88	263	176	10	5	150
Leucanthemum vulgare	Lam.	4	2	—	151	427	289	17.8	5	155
Leucanthemum vulgare	Lam.	4	5	—	20	780	347	10	5	152
Leucanthemum vulgare	Lam.	4	2	—	320	688	504	5	5	69
Leucanthemum vulgare	Lam.	4	1	—	1238	1238	1238	3	5	59
Leucanthemum vulgare	Lam.	4	1	—	1438	1438	1438	5	4	88
Matricaria discoidea	DC.	1	1	—	0	0	0	3	5	96
Matricaria discoidea	DC.	1	2	—	53	53	53	8	7	275
Matricaria discoidea	DC.	2	1	>2	0	0	0	2	1	47
Matricaria discoidea	DC.	2	1	>2	0	0	0	15	1	47
Matricaria discoidea	DC.	2	2	—	20	23	22	3	7	235
Matricaria discoidea	DC.	2	1	—	33	33	33	8	5	108
Matricaria discoidea	DC.	2	1	—	38	38	38	4	5	162
Matricaria discoidea	DC.	2	1	—	88	88	88	5	5	244
Matricaria discoidea	DC.	2	1	—	138	138	138	10	5	255
Matricaria discoidea	DC.	2	1	—	208	208	208	5	5	69
Matricaria discoidea	DC.	2	1	—	233	233	233	8	7	275
Matricaria discoidea	DC.	2	8	—	50	2517	565	3	5	96
Matricaria discoidea	DC.	3	1	>4	0	0	0	100	1	54
Matricaria discoidea	DC.	3	1	>5	0	0	0	2.5	1	203
Matricaria discoidea	DC.	3	1	>5	0	0	0	2.5	2	203
Matricaria discoidea	DC.	3	1	>5	0	0	0	7.5	1	203
Matricaria discoidea	DC.	3	1	>5	0	0	0	7.5	2	203

Species	Authority									
Matricaria discoidea	DC.	3	1	>5	0	0	0	15	1	203
Matricaria discoidea	DC.	3	1	>5	0	0	0	15	2	203
Matricaria discoidea	DC.	3	3	>5	0	0	0	7.5	2	206
Matricaria discoidea	DC.	4	1	—	182	182	182	20	4	111
Matricaria discoidea	DC.	4	1	—	425	425	425	20	5	118
Matricaria discoidea	DC.	4	18	—	40	5188	1304	30	4	13
Matricaria discoidea	DC.	4	2	—	867	2300	1584	3	5	96
Matricaria discoidea	DC.	4	1	—	51747	51747	51747	15	5	207
Matricaria recutita	L.	1	2	—	0	0	0	25	4	36
Matricaria recutita	L.	1	8	—	53	158	105	8	7	275
Matricaria recutita	L.	2	1	—	16	16	16	100	5	219
Matricaria recutita	L.	2	1	—	192	192	192	5	5	69
Matricaria recutita	L.	2	4	—	52	562	274	15	5	272
Matricaria recutita	L.	2	1	—	492	492	492	25	4	36
Matricaria recutita	L.	3	3	>5	0	0	0	7.5	2	206
Matricaria recutita	L.	3	1	>11	0	0	0	25	3	209
Matricaria recutita	L.	3	1	>20	1780	1780	1780	20	5	40
Matricaria recutita	L.	4	2	—	53	88	71	8	7	275
Matricaria recutita	L.	4	3	—	6	258	101	25	4	36
Matricaria recutita	L.	4	1	—	333	333	333	25	5	260
Matricaria recutita	L.	4	8	—	55	882	561	10	5	15
Matricaria recutita	L.	4	1	—	778	778	778	50	5	260
Matricaria recutita	L.	4	2	—	1238	4148	2693	25	4	185
Matricaria recutita	L.	4	1	—	3112	3112	3112	15	5	207
Matricaria recutita	L.	4	4	—	402	6542	3134	15	5	272
Mycelis muralis	(L.) Dumort.	1	1	—	0	0	0	12	5	171
Mycelis muralis	(L.) Dumort.	2	1	—	0	0	0	5	6	170
Mycelis muralis	(L.) Dumort.	4	1	—	6	6	6	10	5	120
Onopordum acanthium	L.	2	1	—	750	750	750	17	5	162
Onopordum acanthium	L.	3	3	>5	0	750	0	7.5	2	201
Onopordum acanthium	L.	3	1	>30	0	0	0	15	1	240
Onopordum acanthium	L.	3	1	>39	0	0	0	50	1	240
Onopordum acanthium	L.	3	1	>39	0	0	0	100	1	240
Petasites albus	(L.) Gaertner	1	1	—	0	0	0	5	5	213
Petasites hybridus	(L.) P. Gaertner B. Meyer & Scherb.	1	2	—	0	0	0	20	4	225
Petasites hybridus	(L.) P. Gaertner B. Meyer & Scherb.	1	2	—	0	0	0	10	5	255
Picris echioides	L.	1	2	—	0	0	0	10	4	22
Picris echioides	L.	4	1	—	78	78	78	25	4	136
Picris hieracioides	L.	1	1	—	0	0	0	25	4	36
Picris hieracioides	L.	2	1	—	9	9	9	6	5	211
Picris hieracioides	L.	3	3	>5	0	0	0	7.5	2	206
Picris hieracioides	L.	3	1	>20	44	44	44	24	5	168
Pilosella aurantiaca	(L.) F. Schultz & Schultz-Bip.	1	2	—	0	0	0	3	5	59
Pilosella aurantiaca	(L.) F. Schultz & Schultz-Bip.	1	1	—	0	0	0	11.5	5	134
Pilosella aurantiaca	(L.) F. Schultz & Schultz-Bip.	2	2	—	49	74	62	11.5	5	134
Pilosella aurantiaca	(L.) F. Schultz & Schultz-Bip.	4	1	—	65	65	65	3	5	59
Pilosella officinarum	(Pugsley) Sell & C. West	1	1	—	0	0	0	6	5	67

Species	Authority	Seed bank type	Number of records	Longevity (y)	Minimum density (seeds m⁻²)	Maximum density (seeds m⁻²)	Mean density (seeds m⁻²)	Depth (cm)	Method	Source code
Pilosella officinarum	(Pugsley) Sell & C. West	1	1	—	0	0	0	5	5	69
Pilosella officinarum	(Pugsley) Sell & C. West	1	1	—	0	0	0	9	5	84
Pilosella officinarum	(Pugsley) Sell & C. West	1	5	—	0	0	0	3	5	96
Pilosella officinarum	(Pugsley) Sell & C. West	1	2	—	0	0	0	10	5	110
Pilosella officinarum	(Pugsley) Sell & C. West	1	2	—	0	0	0	5	5	122
Pilosella officinarum	(Pugsley) Sell & C. West	1	2	—	0	0	0	10	5	150
Pilosella officinarum	(Pugsley) Sell & C. West	1	1	—	0	0	0	2	5	153
Pilosella officinarum	(Pugsley) Sell & C. West	1	1	—	0	0	0	6	5	181
Pilosella officinarum	(Pugsley) Sell & C. West	1	1	—	0	0	0	8.5	4	254
Pilosella officinarum	(Pugsley) Sell & C. West	1	4	—	0	0	0	10	5	255
Pilosella officinarum	(Pugsley) Sell & C. West	1	2	—	0	3	2	3	7	235
Pilosella officinarum	(Pugsley) Sell & C. West	1	1	>1	26	26	26	2	7	99
Pilosella officinarum	(Pugsley) Sell & C. West	2	5	—	0	0	0	—	1	275
Pilosella officinarum	(Pugsley) Sell & C. West	2	1	>5	898	898	898	6	5	48
Pilosella officinarum	(Pugsley) Sell & C. West	3	1		0	0	0	7.5	2	195
Pilosella officinarum	(Pugsley) Sell & C. West	4	2	—	25	28	27	10	5	110
Pilosella officinarum	(Pugsley) Sell & C. West	4	1	—	50	50	50	3	5	96
Pilosella officinarum	(Pugsley) Sell & C. West	4	1	—	112	112	112	5	5	69
Pulicaria dysenterica	(L.) Bernh.	1	1	—	0	0	0	5	5	250
Pulicaria dysenterica	(L.) Bernh.	2	5	>1	0	0	0	—	1	275
Scolymus hispanicus	L.	1	1	—	0	0	0	3	5	132
Scorzonera humilis	L.	1	2	—	0	0	0	10	5	149
Scorzonera humilis	L.	1	1	—	0	0	0	10	5	152
Senecio aquaticus	Hill	1	2	—	0	0	0	6.5	5	128
Senecio aquaticus	Hill	1	3	—	0	0	0	6	5	208
Senecio aquaticus	Hill	4	2	—	200	540	370	6.5	5	128
Senecio erucifolius	L.	1	1	—	0	0	0	6	5	181
Senecio erucifolius	L.	1	1	—	0	0	0	10	5	255
Senecio erucifolius	L.	3	1	>5	0	0	0	7.5	2	195
Senecio helenitis	(L.) Schinz & Thell.	4	1	—	20	20	20	30	4	13
Senecio jacobaea	L.	1	2	—	0	0	0	5	5	69
Senecio jacobaea	L.	1	6	—	0	0	0	3	5	96
Senecio jacobaea	L.	1	2	—	0	0	0	6.5	7	182
Senecio jacobaea	L.	1	1	—	3	3	3	3	7	235
Senecio jacobaea	L.	1	1	—	24	24	24	20	5	219
Senecio jacobaea	L.	2	1	4	0	0	0	7.5	2	195
Senecio jacobaea	L.	2	1	—	2583	2583	2583	30	5	45
Senecio jacobaea	L.	3	1	>4	0	0	0	200	1	54
Senecio jacobaea	L.	3	1	>6	0	0	0	1	3	231
Senecio jacobaea	L.	3	1	>6	0	0	0	5	3	231
Senecio jacobaea	L.	3	1	>6	0	0	0	10	3	231
Senecio jacobaea	L.	3	1	>6	0	0	0	20	3	231

Species										
Senecio jacobaea	L	4	3	—	13	44	31	10	5	110
Senecio nemorensis	L	1	1	—	0	0	0	10	5	181
Senecio nemorensis	L	1	3	—	0	0	0	10	5	255
Senecio nemorensis	L	1	1	—	56	56	56	50	5	219
Senecio squalidus	L	3	1	>5	0	0	0	7.5	2	195
Senecio sylvaticus	L	3	1	>5	0	0	0	7.5	2	195
Senecio sylvaticus	L	4	2	—	67	167	117	3	5	96
Senecio sylvaticus	L	2	1	—	2224	2224	2224	10	5	120
Senecio viscosus	L	3	1	—	40	40	40	6	5	211
Senecio viscosus	L	3	1	>5	0	0	0	7.5	2	195
Senecio viscosus	L	1	1	—	36	36	36	6	5	211
Senecio vulgaris	L	1	4	—	0	0	0	10	4	22
Senecio vulgaris	L	1	1	—	0	0	0	25	4	36
Senecio vulgaris	L	1	1	—	0	0	0	50	5	66
Senecio vulgaris	L	1	1	—	0	0	0	15	5	142
Senecio vulgaris	L	1	1	—	0	0	0	10	5	255
Senecio vulgaris	L	1	1	—	6	6	6	6	5	211
Senecio vulgaris	L	1	1	—	53	53	53	8	7	275
Senecio vulgaris	L	2	1	4	0	0	0	2.5	1	203
Senecio vulgaris	L	2	1	4	0	0	0	2.5	2	203
Senecio vulgaris	L	2	1	4	0	0	0	7.5	2	203
Senecio vulgaris	L	2	1	4	0	0	0	15	2	203
Senecio vulgaris	L	2	1	3	0	0	0	25	3	209
Senecio vulgaris	L	2	1	>2	0	0	0	20	1	219
Senecio vulgaris	L	2	1	—	4	4	4	5	5	244
Senecio vulgaris	L	2	1	—	14	14	14	25	4	36
Senecio vulgaris	L	2	1	—	22	22	22	10	5	110
Senecio vulgaris	L	2	1	—	76	76	76	40	5	219
Senecio vulgaris	L	2	2	—	83	100	92	3	5	96
Senecio vulgaris	L	2	1	—	112	112	112	38	5	186
Senecio vulgaris	L	2	1	—	166	166	166	15	5	62
Senecio vulgaris	L	2	1	—	2242	2242	2242	7.5	5	10
Senecio vulgaris	L	2	2	—	1800	6400	4100	30	5	66
Senecio vulgaris	L	3	1	>5	0	0	0	7.5	2	196
Senecio vulgaris	L	3	1	>5	0	0	0	7.5	1	203
Senecio vulgaris	L	3	1	>5	0	0	0	15	1	203
Senecio vulgaris	L	3	1	—	8	8	8	5	5	234
Senecio vulgaris	L	3	1	—	16	16	16	70	5	219
Senecio vulgaris	L	3	1	—	150	150	150	20	5	96
Senecio vulgaris	L	4	2	—	8	12	10	25	5	4
Senecio vulgaris	L	4	2	—	26	73	50	30	4	36
Senecio vulgaris	L	4	1	—	50	50	50	15	4	13
Senecio vulgaris	L	4	2	—	48	104	76	15	5	205
Senecio vulgaris	L	4	7	—	84	84	84	20	4	31
Senecio vulgaris	L	4	7	—	28	411	196	15	5	111
Senecio vulgaris	L	4	1	—	400	400	400	20	4	30
Senecio vulgaris	L	4	1	—	457	457	457	20	5	118
Senecio vulgaris	L	4	1	—	457	457	457	23	5	202

Species	Authority	Seed bank type	Number of records	Longevity (y)	Minimum density (seeds m^{-2})	Maximum density (seeds m^{-2})	Mean density (seeds m^{-2})	Depth (cm)	Method	Source code
Senecio vulgaris	L.	4	1	—	4001	4001	4001	15	5	207
Seriphidium maritimum	(L.) Polj.	1	2	—	0	0	0	1	5	57
Seriphidium maritimum	(L.) Polj.	2	1	—	1000	1000	1000	1	5	57
Seriphidium maritimum	(L.) Polj.	4	2	—	800	1200	1000	1	5	57
Serratula tinctoria	L.	1	1	—	4	4	4	3	7	235
Solidago canadensis	L.	1	2	—	0	0	0	11.5	5	134
Solidago canadensis	L.	4	1	—	1	1	1	5	5	166
Solidago canadensis	L.	4	3	—	37	421	202	11.5	5	134
Solidago gigantea	Aiton	1	1	—	0	0	0	10	5	255
Solidago gigantea	Aiton	4	1	—	14	14	14	10	5	33
Solidago graminifolia	(L.) Salisb.	1	1	—	0	0	0	5	5	19
Solidago graminifolia	(L.) Salisb.	2	1	—	590	590	590	15	5	94
Solidago graminifolia	(L.) Salisb.	4	1	—	19	19	19	5	5	117
Solidago graminifolia	(L.) Salisb.	4	2	—	1180	1310	1245	15	5	94
Solidago virgaurea	L.	1	1	—	0	0	0	2	5	82
Solidago virgaurea	L.	1	2	—	0	0	0	5	5	82
Solidago virgaurea	L.	1	3	—	0	0	0	10	5	150
Solidago virgaurea	L.	1	1	—	0	0	0	5	6	170
Solidago virgaurea	L.	1	1	—	0	0	0	12	5	171
Solidago virgaurea	L.	1	1	—	0	0	0	3	7	235
Solidago virgaurea	L.	1	3	—	0	0	0	10	5	255
Solidago virgaurea	L.	1	4	—	0	0	0	15	5	257
Solidago virgaurea	L.	2	1	—	0	0	0	5	6	170
Sonchus arvensis	L.	1	1	—	160	160	160	38	5	186
Sonchus arvensis	L.	2	3	>3	0	0	0	7	2	44
Sonchus arvensis	L.	2	2	1	0	0	0	7	2	44
Sonchus arvensis	L.	2	1	3	0	0	0	7	2	44
Sonchus arvensis	L.	2	2	4	0	0	0	7	2	44
Sonchus arvensis	L.	2	1	—	10	10	10	6	5	49
Sonchus arvensis	L.	2	3	—	37	133	92	5	5	177
Sonchus arvensis	L.	2	3	—	60	220	140	10	5	149
Sonchus arvensis	L.	2	1	—	150	150	150	20	5	4
Sonchus arvensis	L.	2	1	—	193	193	193	15	5	62
Sonchus arvensis	L.	2	2	—	600	600	600	10	4	20
Sonchus arvensis	L.	3	1	>6	0	0	0	7	2	44
Sonchus arvensis	L.	3	3	>5	0	0	0	7.5	2	206
Sonchus arvensis	L.	4	1	—	5	5	5	5	5	166
Sonchus arvensis	L.	4	1	—	25	25	25	20	4	111
Sonchus arvensis	L.	4	3	—	14	58	31	10	5	8
Sonchus arvensis	L.	4	7	—	6	149	48	15	4	30
Sonchus arvensis	L.	4	1	—	74	74	74	5	5	177
Sonchus arvensis	L.	4	7	—	20	346	109	30	4	13

Species	Authority									
Sonchus arvensis	L.	4	2	—	158	667	413	25	5	260
Sonchus arvensis	L.	4	1	—	775	775	775	20	5	118
Sonchus asper	(L.) Hill	1	4	—	0	0	0	10	4	22
Sonchus asper	(L.) Hill	1	1	—	0	0	0	25	4	36
Sonchus asper	(L.) Hill	1	1	—	0	0	0	6.5	5	128
Sonchus asper	(L.) Hill	1	5	—	53	158	84	10	5	255
Sonchus asper	(L.) Hill	1	1	—	240	240	240	8	7	275
Sonchus asper	(L.) Hill	1	1	—	800	800	800	6	5	48
Sonchus asper	(L.) Hill	1	1	—	6	6	6	30	5	66
Sonchus asper	(L.) Hill	2	1	—	14	14	14	5	5	244
Sonchus asper	(L.) Hill	2	3	—	50	83	67	25	4	36
Sonchus asper	(L.) Hill	2	1	—	100	100	100	3	5	96
Sonchus asper	(L.) Hill	2	1	—	108	108	108	6.5	5	128
Sonchus asper	(L.) Hill	2	4	—	112	112	112	10	5	252
Sonchus asper	(L.) Hill	2	1	—	114	822	310	5	5	69
Sonchus asper	(L.) Hill	2	3	—	1200	1200	1200	8	7	275
Sonchus asper	(L.) Hill	2	1	—	0	0	0	50	5	66
Sonchus asper	(L.) Hill	3	3	>5	0	0	0	7.5	2	206
Sonchus asper	(L.) Hill	3	1	>11	0	0	0	25	3	209
Sonchus asper	(L.) Hill	3	1	—	431	431	431	30	5	29
Sonchus asper	(L.) Hill	3	3	>10	1820	1820	1820	12	5	81
Sonchus asper	(L.) Hill	4	3	—	6	10	7	25	4	36
Sonchus asper	(L.) Hill	4	3	—	28	60	40	10	5	110
Sonchus asper	(L.) Hill	4	1	—	17	98	56	7	5	58
Sonchus asper	(L.) Hill	4	20	—	79	79	79	20	4	111
Sonchus asper	(L.) Hill	4	1	—	20	1465	147	30	4	13
Sonchus asper	(L.) Hill	4	1	—	176	176	176	25	5	260
Sonchus asper	(L.) Hill	4	1	—	963	963	963	15	5	207
Sonchus oleraceus	L.	1	1	—	0	0	0	10	4	22
Sonchus oleraceus	L.	1	3	—	0	0	0	25	4	36
Sonchus oleraceus	L.	1	1	—	30	30	30	30	5	66
Sonchus oleraceus	L.	2	3	—	20	52	31	10	5	255
Sonchus oleraceus	L.	2	2	—	42	46	44	15	5	22
Sonchus oleraceus	L.	2	2	—	32	75	54	6	5	272
Sonchus oleraceus	L.	2	2	—	144	144	144	30	5	211
Sonchus oleraceus	L.	2	1	—	50	317	150	70	5	45
Sonchus oleraceus	L.	2	3	—	180	180	180	3	5	219
Sonchus oleraceus	L.	2	1	—	203	203	203	100	5	96
Sonchus oleraceus	L.	2	1	—	460	460	460	17.8	5	219
Sonchus oleraceus	L.	3	3	—	0	0	0	15	5	155
Sonchus oleraceus	L.	3	3	—	0	0	0	7.5	5	94
Sonchus oleraceus	L.	3	1	>5				25	2	206
Sonchus oleraceus	L.	3	1	>11				8	3	209
Sonchus oleraceus	L.	3	1	—	8	8	8	30	5	108
Sonchus oleraceus	L.	3	1	—	43	43	43	16	5	45
Sonchus oleraceus	L.	3	1	>20	67	67	67	16	5	168
Sonchus oleraceus	L.	3	1	—	252	252	252	17.8	5	155

Species	Authority	Seed bank type	Number of records	Longevity (y)	Minimum density (seeds m^{-2})	Maximum density (seeds m^{-2})	Mean density (seeds m^{-2})	Depth (cm)	Method	Source code
Sonchus oleraceus	L.	4	1	—	13	13	13	10	5	33
Sonchus oleraceus	L.	4	1	—	37	37	37	10	4	22
Sonchus oleraceus	L.	4	2	—	38	100	69	10	5	255
Sonchus oleraceus	L.	4	1	—	100	100	100	3	5	96
Sonchus oleraceus	L.	4	4	—	37	487	218	7.5	5	10
Sonchus oleraceus	L.	4	2	—	99	424	262	15	5	272
Sonchus oleraceus	L.	4	2	—	170	1290	730	15	5	94
Sonchus oleraceus	L.	4	1	—	1581	1581	1581	15	5	207
Tanacetum vulgare	L.	1	1	—	0	0	0	5	5	69
Tanacetum vulgare	L.	1	1	—	0	0	0	3	7	235
Tanacetum vulgare	L.	1	2	—	0	0	0	10	5	255
Tanacetum vulgare	L.	1	1	—	12	12	12	50	5	219
Tanacetum vulgare	L.	1	3	—	53	79	62	8	7	275
Tanacetum vulgare	L.	2	1	—	338	338	338	10	5	255
Tanacetum vulgare	L.	4	3	—	113	263	192	10	5	255
Taraxacum officinale	—	1	1	—	0	0	0	25	5	2
Taraxacum officinale	—	1	7	—	0	0	0	10	5	27
Taraxacum officinale	—	1	6	—	0	0	0	6	5	28
Taraxacum officinale	—	1	3	—	0	0	0	5	4	53
Taraxacum officinale	—	1	3	—	0	0	0	3	5	59
Taraxacum officinale	—	1	1	—	0	0	0	30	5	66
Taraxacum officinale	—	1	1	—	0	0	0	6	5	67
Taraxacum officinale	—	1	1	—	0	0	0	20	5	67
Taraxacum officinale	—	1	1	—	0	0	0	6.5	5	68
Taraxacum officinale	—	1	1	—	0	0	0	13	5	68
Taraxacum officinale	—	1	8	—	0	0	0	5	5	69
Taraxacum officinale	—	1	1	—	0	0	0	5	4	88
Taraxacum officinale	—	1	1	—	0	0	0	15	5	94
Taraxacum officinale	—	1	8	—	0	0	0	1	5	96
Taraxacum officinale	—	1	6	—	0	0	0	3	5	96
Taraxacum officinale	—	1	9	—	0	0	0	20	5	97
Taraxacum officinale	—	1	2	—	0	0	0	6	5	99
Taraxacum officinale	—	1	1	—	0	0	0	10	5	110
Taraxacum officinale	—	1	4	—	0	0	0	6.5	5	128
Taraxacum officinale	—	1	2	—	0	0	0	10	5	149
Taraxacum officinale	—	1	1	—	0	0	0	10	5	150
Taraxacum officinale	—	1	1	—	0	0	0	6	5	151
Taraxacum officinale	—	1	4	—	0	0	0	12	5	171
Taraxacum officinale	—	1	2	—	0	0	0	12	5	174
Taraxacum officinale	—	1	6	—	0	0	0	6	5	181
Taraxacum officinale	—	1	6	—	0	0	0	3	5	208
Taraxacum officinale	—	1	3	—	0	0	0	6	5	208

Species										
Taraxacum officinale	—	1	2	—	0	0	0	20	4	225
Taraxacum officinale	—	1	1	—	0	0	0	5	5	244
Taraxacum officinale	—	1	2	—	0	0	0	5	5	250
Taraxacum officinale	—	1	1	—	0	0	0	8.5	4	254
Taraxacum officinale	—	1	12	—	0	0	0	10	5	255
Taraxacum officinale	—	1	1	—	0	0	0	15	5	257
Taraxacum officinale	—	1	2	—	0	7	3	3	7	235
Taraxacum officinale	—	1	1	—	18	18	18	6.5	7	182
Taraxacum officinale	—	1	2	—	0	43	22	30	5	45
Taraxacum officinale	—	1	6	—	53	105	72	8	7	275
Taraxacum officinale	—	2	3	—	0	130	74	2	7	99
Taraxacum officinale	—	2	1	—	0	0	0	7	2	44
Taraxacum officinale	—	2	1	—	0	0	0	5	6	170
Taraxacum officinale	—	2	1	—	0	3	3	4	5	176
Taraxacum officinale	—	2	4	—	3	6	6	16	5	23
Taraxacum officinale	—	2	1	—	8	22	14	25	4	36
Taraxacum officinale	—	2	1	—	14	14	14	6.5	7	182
Taraxacum officinale	—	2	1	—	28	28	28	8	5	108
Taraxacum officinale	—	2	1	—	40	40	40	10	5	252
Taraxacum officinale	—	2	1	—	44	44	44	30	5	219
Taraxacum officinale	—	2	2	—	49	49	49	15	5	62
Taraxacum officinale	—	2	2	—	50	50	50	10	5	255
Taraxacum officinale	—	2	1	—	47	104	76	2.5	5	114
Taraxacum officinale	—	2	1	—	102	102	102	5	4	53
Taraxacum officinale	—	2	1	—	107	107	107	2	7	99
Taraxacum officinale	—	2	1	—	150	150	150	15	5	162
Taraxacum officinale	—	2	2	—	179	179	179	8	7	275
Taraxacum officinale	—	2	2	—	156	208	182	6	5	11
Taraxacum officinale	—	2	1	—	275	275	275	10	5	162
Taraxacum officinale	—	2	1	—	313	313	313	20	5	97
Taraxacum officinale	—	2	5	—	161	646	353	30	5	45
Taraxacum officinale	—	2	2	—	364	416	390	6	5	99
Taraxacum officinale	—	2	2	—	188	688	438	17	5	162
Taraxacum officinale	—	2	1	—	700	700	700	6	5	162
Taraxacum officinale	—	2	1	—	1248	1248	1248	10	5	27
Taraxacum officinale	—	2	2	—	1311	1311	1311	12	5	162
Taraxacum officinale	—	2	2	—	350	23550	11950	20	5	4
Taraxacum officinale	—	3	3	>5	0	0	0	7.5	2	206
Taraxacum officinale	—	3	1	—	10	10	10	5	5	234
Taraxacum officinale	—	3	1	>12	24	24	24	50	5	219
Taraxacum officinale	—	3	3	—	32	32	32	30	5	45
Taraxacum officinale	—	3	1	>660	40	40	40	55	5	161
Taraxacum officinale	—	3	1	>660	75	75	75	50	5	161
Taraxacum officinale	—	3	1	>660	260	260	260	40	5	11
Taraxacum officinale	—	3	1	—	800	800	800	50	5	66
Taraxacum officinale	—	4	1	—	16	16	16	10	5	33

Species	Authority	Seed bank type	Number of records	Longevity (y)	Minimum density (seeds m⁻²)	Maximum density (seeds m⁻²)	Mean density (seeds m⁻²)	Depth (cm)	Method	Source code
Taraxacum officinale	—	4	3	—	10	26	17	25	4	36
Taraxacum officinale	—	4	3	—	9	34	17	12	5	174
Taraxacum officinale	—	4	3	—	5	30	22	15	5	62
Taraxacum officinale	—	4	1	—	25	25	25	15	5	267
Taraxacum officinale	—	4	1	—	28	28	28	20	5	111
Taraxacum officinale	—	4	1	—	57	57	57	6	5	151
Taraxacum officinale	—	4	4	—	50	75	60	10	5	255
Taraxacum officinale	—	4	1	—	79	79	79	8	7	275
Taraxacum officinale	—	4	2	—	80	80	80	6.5	5	128
Taraxacum officinale	—	4	1	—	88	88	88	5	5	244
Taraxacum officinale	—	4	2	—	50	133	92	10	5	184
Taraxacum officinale	—	4	1	—	101	101	101	17.8	5	156
Taraxacum officinale	—	4	2	—	100	105	103	5	5	250
Taraxacum officinale	—	4	1	—	122	122	122	5	4	53
Taraxacum officinale	—	4	2	—	130	130	130	15	5	94
Taraxacum officinale	—	4	2	—	67	280	174	7.5	5	10
Taraxacum officinale	—	4	4	—	88	263	176	10	5	150
Taraxacum officinale	—	4	4	—	20	579	186	30	4	13
Taraxacum officinale	—	4	3	—	37	370	191	2.5	5	114
Taraxacum officinale	—	4	4	—	96	352	212	5	5	69
Taraxacum officinale	—	4	1	—	213	213	213	2	7	99
Taraxacum officinale	—	4	4	—	260	364	286	6	5	99
Taraxacum officinale	—	4	1	—	3372	3372	3372	17.8	5	155
Tephroseris palustris	(L.) Fourr.	2	1	—	0	2	2	5	5	244
Tephroseris palustris	(L.) Fourr.	4	1	—	0	0	0	5	5	166
Tephroseris palustris	(L.) Fourr.	4	2	1	272	47200	23736	12	4	230
Tragopogon dubius	Scop.	2	6	—	0	0	0	7	2	44
Tragopogon dubius	Scop.	4	2	—	40	93	67	30	4	13
Tragopogon pratensis	L.	1	2	—	0	0	0	10	5	150
Tragopogon pratensis	L.	1	1	—	0	0	0	6.5	7	182
Tragopogon pratensis	L.	2	1	2	0	0	0	10	5	252
Tragopogon pratensis	L.	2	1	2	0	0	0	7.5	2	195
Tragopogon pratensis	L.	2	5	1	0	0	0	—	1	275
Tripleurospermum inodorum	(L.) Schultz-Bip.	1	1	0	0	0	0	0	1	75
Tripleurospermum inodorum	(L.) Schultz-Bip.	2	1	2	0	0	0	30	3	14
Tripleurospermum inodorum	(L.) Schultz-Bip.	2	1	>2	0	0	0	0	2	75
Tripleurospermum inodorum	(L.) Schultz-Bip.	2	1	>2	0	0	0	5	1	75
Tripleurospermum inodorum	(L.) Schultz-Bip.	2	1	>2	0	0	0	5	2	75
Tripleurospermum inodorum	(L.) Schultz-Bip.	2	1	4	0	0	0	13	1	133
Tripleurospermum inodorum	(L.) Schultz-Bip.	2	1	4	0	0	0	26	1	133
Tripleurospermum inodorum	(L.) Schultz-Bip.	2	1		0	0	0	39	1	133
Tripleurospermum inodorum	(L.) Schultz-Bip.	2	1	>3	0	0	0	10	3	135

Species	Authority	(1)	(2)	(3)	(4)	(5)	(6)	(7)	(8)	(9)
Tripleurospermum inodorum	(L.) Schultz-Bip.	2	1	>3	0	0	0	25	3	135
Tripleurospermum inodorum	(L.) Schultz-Bip.	2	1	—	18	18	18	3	7	235
Tripleurospermum inodorum	(L.) Schultz-Bip.	2	1	—	38	38	38	67	5	162
Tripleurospermum inodorum	(L.) Schultz-Bip.	2	7	—	67	226	617	3	5	96
Tripleurospermum inodorum	(L.) Schultz-Bip.	2	1	—	412	412	412	12	5	162
Tripleurospermum inodorum	(L.) Schultz-Bip.	3	1	>4	1000	1000	1000	17	5	162
Tripleurospermum inodorum	(L.) Schultz-Bip.	3	1	>6	0	0	0	100	1	54
Tripleurospermum inodorum	(L.) Schultz-Bip.	3	1	5	0	0	0	20	1	61
Tripleurospermum inodorum	(L.) Schultz-Bip.	3	1	5	0	0	0	8	1	61
Tripleurospermum inodorum	(L.) Schultz-Bip.	3	1	>26	0	0	0	30	1	61
Tripleurospermum inodorum	(L.) Schultz-Bip.	3	1	>20	0	0	0	25	2	125
Tripleurospermum inodorum	(L.) Schultz-Bip.	3	1	>5	0	0	0	26	1	133
Tripleurospermum inodorum	(L.) Schultz-Bip.	3	1	>5	0	0	0	7.5	2	196
Tripleurospermum inodorum	(L.) Schultz-Bip.	3	1	>5	0	0	0	2.5	1	203
Tripleurospermum inodorum	(L.) Schultz-Bip.	3	1	>5	0	0	0	2.5	2	203
Tripleurospermum inodorum	(L.) Schultz-Bip.	3	1	>5	0	0	0	7.5	1	203
Tripleurospermum inodorum	(L.) Schultz-Bip.	3	1	>5	0	0	0	7.5	2	203
Tripleurospermum inodorum	(L.) Schultz-Bip.	3	1	>5	0	0	0	15	1	203
Tripleurospermum inodorum	(L.) Schultz-Bip.	4	1	—	0	0	0	15	5	203
Tripleurospermum inodorum	(L.) Schultz-Bip.	4	2	—	28	28	28	10	5	252
Tripleurospermum inodorum	(L.) Schultz-Bip.	4	2	—	108	1077	2045	30	4	29
Tripleurospermum inodorum	(L.) Schultz-Bip.	4	6	—	6	52	114	15	5	30
Tripleurospermum inodorum	(L.) Schultz-Bip.	4	1	—	100	100	100	1	5	96
Tripleurospermum inodorum	(L.) Schultz-Bip.	4	1	—	367	367	367	3	4	96
Tripleurospermum inodorum	(L.) Schultz-Bip.	4	1	—	1284	1284	1284	15	5	207
Tripleurospermum inodorum	(L.) Schultz-Bip.	4	2	—	290	2114	3938	15	5	31
Tripleurospermum inodorum	(L.) Schultz-Bip.	4	6	—	1833	3861	7667	20	3	74
Tripleurospermum maritimum	(L.) Koch	2	6	—	50	311	1013	10	5	255
Tripleurospermum maritimum	(L.) Koch	3	1	>11	0	0	0	25	3	209
Tripleurospermum maritimum	(L.) Koch	4	1	—	149	149	149	20	4	111
Tripleurospermum maritimum	(L.) Koch	3	3	—	550	638	800	10	5	255
Tussilago farfara	L.	1	1	0	0	0	0	15	1	24
Tussilago farfara	L.	1	4	—	0	0	0	10	5	110
Tussilago farfara	L.	1	1	—	0	0	0	20	4	111
Tussilago farfara	L.	1	1	—	0	0	0	20	5	111
Tussilago farfara	L.	1	2	—	0	0	0	6.5	5	128
Tussilago farfara	L.	1	1	—	0	0	0	20	4	225
Tussilago farfara	L.	1	1	—	0	0	0	3	7	235
Tussilago farfara	L.	1	5	—	0	0	0	10	5	255
Tussilago farfara	L.	1	1	—	53	53	53	8	7	275
Tussilago farfara	L.	2	2	—	60	60	60	6.5	5	128
Convolvulaceae										
Calystegia sepium	(L.) R. Br.	1	2	—	0	0	0	30	5	66
Calystegia sepium	(L.) R. Br.	1	1	—	0	0	0	3	5	96
Calystegia sepium	(L.) R. Br.	1	4	—	0	0	0	6.5	5	128
Calystegia sepium	(L.) R. Br.	1	1	—	53	53	53	8	7	275

Species	Authority	Seed bank type	Number of records	Longevity (y)	Minimum density (seeds m⁻²)	Maximum density (seeds m⁻²)	Mean density (seeds m⁻²)	Depth (cm)	Method	Source code
Calystegia sepium	(L.) R. Br.	2	1	—	108	108	108	15	5	175
Calystegia sepium	(L.) R. Br.	3	1	>30	0	0	0	15	1	240
Calystegia sepium	(L.) R. Br.	3	1	>39	0	0	0	50	1	240
Calystegia sepium	(L.) R. Br.	3	1	>39	0	0	0	100	1	240
Calystegia sepium	(L.) R. Br.	3	2	—	140	280	210	15	5	175
Calystegia silvatica	(Kit.) Griseb.	1	1	—	0	0	0	3	7	235
Calystegia soldanella	(L.) R. Br.	1	1	—	97	97	97	15	5	45
Convolvulus arvensis	L.	1	2	—	0	0	0	25	5	2
Convolvulus arvensis	L.	1	2	—	0	0	0	10	4	22
Convolvulus arvensis	L.	1	6	—	0	0	0	25	4	36
Convolvulus arvensis	L.	1	1	—	0	0	0	30	5	66
Convolvulus arvensis	L.	1	1	—	0	0	0	50	5	66
Convolvulus arvensis	L.	1	2	—	0	0	0	6	5	67
Convolvulus arvensis	L.	1	3	—	0	0	0	3	5	96
Convolvulus arvensis	L.	1	1	—	0	0	0	15	5	142
Convolvulus arvensis	L.	1	1	—	0	0	0	20	4	225
Convolvulus arvensis	L.	1	10	—	0	0	0	10	5	255
Convolvulus arvensis	L.	1	1	—	32	32	32	30	5	29
Convolvulus arvensis	L.	2	1	—	188	188	188	17	5	162
Convolvulus arvensis	L.	3	1	>4	0	0	0	15	1	34
Convolvulus arvensis	L.	3	1	>4	0	0	0	45	1	34
Convolvulus arvensis	L.	3	1	>26	0	0	0	30	1	35
Convolvulus arvensis	L.	3	1	>22	0	0	0	—	6	238
Convolvulus arvensis	L.	3	1	>8	5	5	5	15	5	62
Convolvulus arvensis	L.	4	1	—	714	714	714	8.5	4	254
Cornaceae	—	—	—	—	—	—	—	—	—	
Cornus sanguinea	L.	1	1	—	0	0	0	6	5	181
Cornus sanguinea	L.	1	1	—	0	0	0	10	5	181
Cornus sanguinea	L.	1	1	—	0	0	0	10	5	255
Corylaceae	—	—	—	—	—	—	—	—	—	
Carpinus betulus	L.	1	2	—	0	0	0	6	5	49
Carpinus betulus	L.	1	1	—	0	0	0	10	5	255
Carpinus betulus	L.	1	1	—	0	0	0	15	5	257
Carpinus betulus	L.	4	2	—	2	37	20	4	5	176
Corylus avellana	L.	1	2	—	0	0	0	4	5	176
Corylus avellana	L.	1	1	—	0	0	0	10	5	255
Corylus avellana	L.	1	5	—	0	0	0	15	5	257
Crassulaceae	—	—	—	—	—	—	—	—	—	
Crassula tillaea	Lester-Garl.	2	2	—	1375	2475	1925	3	5	132

Species	Authority									
Crassula tillaea	Lester-Garl.	4	2	—	2125	3313	2719	3	5	132
Sedum acre	L.	1	1	—	0	0	0	1	5	57
Sedum acre	L.	4	1	—	4	4	4	3	7	235
Sedum acre	L.	4	1	—	400	400	400	1	5	57
Sedum album	L.	4	1	—	150	150	150	5	5	96
Sedum alpestre	Vill.	4	1	—	20	20	20	5	5	55
Sedum annuum	L.	4	2	—	76	107	92	5	4	91
Sedum rupestre	L.	1	1	—	0	0	0	30	5	66
Sedum telephium	L.	1	5	—	0	0	0	10	5	255

Cruciferae (Brassicaceae)

Species	Authority									
Alliaria petiolata	(M. Bieb.) Cavara & Grande	1	3	—	0	0	0	6	5	49
Alliaria petiolata	(M. Bieb.) Cavara & Grande	1	1	—	0	0	0	10	5	126
Alliaria petiolata	(M. Bieb.) Cavara & Grande	2	1	4	0	0	0	7.5	2	196
Alliaria petiolata	(M. Bieb.) Cavara & Grande	3	2	>5	0	0	0	7.5	2	196
Arabidopsis thaliana	(L.) Heynh.	1	1	—	0	0	0	1	5	96
Arabidopsis thaliana	(L.) Heynh.	1	1	—	0	0	0	3	5	96
Arabidopsis thaliana	(L.) Heynh.	2	7	—	6	70	32	25	4	36
Arabidopsis thaliana	(L.) Heynh.	2	2	—	5	92	48	3	7	235
Arabidopsis thaliana	(L.) Heynh.	2	1	—	85	85	85	16	5	23
Arabidopsis thaliana	(L.) Heynh.	2	1	—	91	91	91	15	5	272
Arabidopsis thaliana	(L.) Heynh.	2	18	—	38	16550	1814	10	5	255
Arabidopsis thaliana	(L.) Heynh.	3	3	>5	0	0	0	7.5	2	196
Arabidopsis thaliana	(L.) Heynh.	3	1	>18	56	56	56	32	5	169
Arabidopsis thaliana	(L.) Heynh.	3	1	—	240	240	240	6	5	48
Arabidopsis thaliana	(L.) Heynh.	4	3	—	20	124	70	15	5	272
Arabidopsis thaliana	(L.) Heynh.	4	5	—	20	120	73	30	4	13
Arabidopsis thaliana	(L.) Heynh.	4	5	—	33	168	87	10	5	15
Arabidopsis thaliana	(L.) Heynh.	4	3	—	138	750	392	3	5	132
Arabidopsis thaliana	(L.) Heynh.	4	1	—	449	449	449	20	5	111
Arabidopsis thaliana	(L.) Heynh.	4	1	—	2483	2483	2483	25	5	2
Arabidopsis thaliana	(L.) Heynh.	4	2	—	538	5088	2813	10	5	255
Arabis alpina	L.	1	1	—	0	0	0	5	5	55
Arabis hirsuta	(L.) Scop.	1	1	—	0	0	0	3	5	96
Arabis hirsuta	(L.) Scop.	1	2	—	0	0	0	10	5	152
Arabis hirsuta	(L.) Scop.	2	1	—	31	31	31	10	5	110
Arabis hirsuta	(L.) Scop.	2	2	—	83	117	100	10	5	184
Arabis hirsuta	(L.) Scop.	2	1	—	479	479	479	6	5	48
Arabis hirsuta	(L.) Scop.	2	9	—	468	2808	1187	10	5	152
Arabis hirsuta	(L.) Scop.	4	1	—	67	67	67	3	5	96
Arabis hirsuta	(L.) Scop.	4	3	—	546	3120	1534	10	5	152
Barbarea vulgaris	R. Br.	2	1	>3	0	0	0	7	1	17
Barbarea vulgaris	R. Br.	2	2	—	2	11	7	5	5	19
Barbarea vulgaris	R. Br.	3	1	>5	0	0	0	7.5	2	195
Berteroa incana	(L.) DC.	2	1	—	460	460	460	8	7	275
Biscutella laevigata	L.	1	2	—	0	0	0	5	4	91
Brassica juncea	(L.) Czernj.	2	7	1	0	0	0	7	2	44

Species	Authority	Seed bank type	Number of records	Longevity (y)	Minimum density (seeds m⁻²)	Maximum density (seeds m⁻²)	Mean density (seeds m⁻²)	Depth (cm)	Method	Source code
Brassica juncea	(L.) Czernj.	2	1	2			0	7	2	44
Brassica juncea	(L.) Czernj.	2	1	4			0	7	2	44
Brassica napus	L.	3	1	16			0	25	1	125
Brassica nigra	(L.) Koch	2	1	—	12	12	12	6	5	211
Brassica nigra	(L.) Koch	3	1	50			0	—	3	124
Brassica nigra	(L.) Koch	3	1	>5			0	7.5	2	195
Brassica nigra	(L.) Koch	3	1	10			0	15	1	240
Brassica nigra	(L.) Koch	3	1	21			0	50	1	240
Brassica nigra	(L.) Koch	3	1	21			0	100	1	240
Brassica oleracea	L.	1	1	0			0	15	1	240
Brassica oleracea	L.	1	1	0			0	50	1	240
Brassica oleracea	L.	1	1	0			0	100	1	240
Brassica rapa	L.	1	1	0			0	15	1	240
Brassica rapa	L.	2	1	—	101	101	101	17.8	5	155
Brassica rapa	L.	3	1	>4			0	8	1	60
Brassica rapa	L.	3	1	>4			0	20	1	60
Brassica rapa	L.	3	1	>4			0	30	1	60
Brassica rapa	L.	3	1	>6			0	20	1	61
Brassica rapa	L.	3	1	5			0	8	1	61
Brassica rapa	L.	3	1	5			0	30	1	61
Brassica rapa	L.	3	1	23			0	25	1	125
Brassica rapa	L.	3	1	10			0	100	1	240
Brassica rapa	L.	3	1	6			0	50	1	240
Brassica rapa	L.	3	1	>40	75	75	75	30	5	45
Brassica rapa	L.	3	1	>660	80	80	80	43	5	161
Brassica rapa	L.	3	1	>6	86	86	86	30	5	45
Brassica rapa	L.	3	1	>8	269	269	269	30	5	45
Brassica rapa	L.	4	1	—	61	61	61	7.5	5	10
Camelina microcarpa	Andrz. ex DC.	2	3	1	0	0	0	7	2	44
Camelina sativa	(L.) Crantz	2	3	1	0	0	0	7	2	44
Capsella bursa-pastoris	(L.) Medikus	1	3	—	0	0	0	25	4	36
Capsella bursa-pastoris	(L.) Medikus	1	2	—	0	0	0	5	5	69
Capsella bursa-pastoris	(L.) Medikus	1	2	—	0	0	0	3	5	132
Capsella bursa-pastoris	(L.) Medikus	1	7	—	53	79	65	8	7	275
Capsella bursa-pastoris	(L.) Medikus	1	1	—	595	595	595	50	5	260
Capsella bursa-pastoris	(L.) Medikus	2	1	3	0	0	0	7	2	44
Capsella bursa-pastoris	(L.) Medikus	2	1	4	0	0	0	7	2	44
Capsella bursa-pastoris	(L.) Medikus	2	1	>2	0	0	0	2	1	47
Capsella bursa-pastoris	(L.) Medikus	2	1	>2	0	0	0	15	1	47
Capsella bursa-pastoris	(L.) Medikus	2	1	>2	0	0	0	0	1	75
Capsella bursa-pastoris	(L.) Medikus	2	1	>2	0	0	0	0	2	75
Capsella bursa-pastoris	(L.) Medikus	2	1	>2	0	0	0	5	1	75

Species	Authority									
Capsella bursa-pastoris	(L.) Medikus	2	1	>2	0	0	0	5	2	75
Capsella bursa-pastoris	(L.) Medikus	2	1	>2	0	0	0	20	1	219
Capsella bursa-pastoris	(L.) Medikus	2	1	2	0	0	0	2.5	1	256
Capsella bursa-pastoris	(L.) Medikus	2	1	2	0	0	0	5	1	256
Capsella bursa-pastoris	(L.) Medikus	2	1	2	0	0	0	7.5	1	256
Capsella bursa-pastoris	(L.) Medikus	2	1	—	0	0	0	17.5	5	256
Capsella bursa-pastoris	(L.) Medikus	2	1	—	10	10	10	15	4	62
Capsella bursa-pastoris	(L.) Medikus	2	1	—	12	12	12	25	5	36
Capsella bursa-pastoris	(L.) Medikus	2	1	—	67	67	67	3	5	96
Capsella bursa-pastoris	(L.) Medikus	2	1	—	97	97	97	10	5	162
Capsella bursa-pastoris	(L.) Medikus	2	1	—	164	164	164	5	5	244
Capsella bursa-pastoris	(L.) Medikus	2	4	—	92	325	196	15	5	272
Capsella bursa-pastoris	(L.) Medikus	2	5	—	80	300	196	10	5	149
Capsella bursa-pastoris	(L.) Medikus	2	1	—	208	208	208	5	5	69
Capsella bursa-pastoris	(L.) Medikus	2	1	—	244	244	244	100	5	219
Capsella bursa-pastoris	(L.) Medikus	2	8	—	50	1025	383	10	5	255
Capsella bursa-pastoris	(L.) Medikus	2	1	—	500	500	500	20	5	69
Capsella bursa-pastoris	(L.) Medikus	2	1	—	544	544	544	38	5	186
Capsella bursa-pastoris	(L.) Medikus	2	1	—	2000	2000	2000	50	5	260
Capsella bursa-pastoris	(L.) Medikus	3	1	>5	0	0	0	30	3	14
Capsella bursa-pastoris	(L.) Medikus	3	1	>5	0	0	0	7	2	44
Capsella bursa-pastoris	(L.) Medikus	3	1	35	0	0	0	—	3	124
Capsella bursa-pastoris	(L.) Medikus	3	1	>5	0	0	0	7.5	2	196
Capsella bursa-pastoris	(L.) Medikus	3	1	>5	0	0	0	2.5	1	203
Capsella bursa-pastoris	(L.) Medikus	3	1	>5	0	0	0	2.5	2	203
Capsella bursa-pastoris	(L.) Medikus	3	1	>5	0	0	0	7.5	1	203
Capsella bursa-pastoris	(L.) Medikus	3	1	>5	0	0	0	7.5	2	203
Capsella bursa-pastoris	(L.) Medikus	3	1	>5	0	0	0	15	1	203
Capsella bursa-pastoris	(L.) Medikus	3	1	>11	0	0	0	15	2	203
Capsella bursa-pastoris	(L.) Medikus	3	1	10	0	0	0	25	3	209
Capsella bursa-pastoris	(L.) Medikus	3	1	10	0	0	0	15	1	240
Capsella bursa-pastoris	(L.) Medikus	3	1	16	0	0	0	50	1	240
Capsella bursa-pastoris	(L.) Medikus	3	1	>5	0	0	0	100	1	240
Capsella bursa-pastoris	(L.) Medikus	3	1	>5	0	0	0	12.5	1	256
Capsella bursa-pastoris	(L.) Medikus	3	1	>20	0	0	0	25	1	256
Capsella bursa-pastoris	(L.) Medikus	3	1	—	39	39	39	20	5	40
Capsella bursa-pastoris	(L.) Medikus	3	1	—	67	67	67	10	5	252
Capsella bursa-pastoris	(L.) Medikus	3	3	—	16	689	249	15	5	175
Capsella bursa-pastoris	(L.) Medikus	3	2	—	250	350	300	20	5	4
Capsella bursa-pastoris	(L.) Medikus	3	1	—	354	354	354	30	4	12
Capsella bursa-pastoris	(L.) Medikus	3	1	—	500	500	500	20	5	69
Capsella bursa-pastoris	(L.) Medikus	3	1	—	1667	1667	1667	50	5	260
Capsella bursa-pastoris	(L.) Medikus	4	1	—	50	50	50	10	5	255
Capsella bursa-pastoris	(L.) Medikus	4	1	—	81	81	81	25	5	2
Capsella bursa-pastoris	(L.) Medikus	4	6	—	39	381	154	15	4	30
Capsella bursa-pastoris	(L.) Medikus	4	2	—	50	400	225	15	5	205
Capsella bursa-pastoris	(L.) Medikus	4	3	—	64	576	304	5	5	69

Species	Authority	Seed bank type	Number of records	Longevity (y)	Minimum density (seeds m⁻²)	Maximum density (seeds m⁻²)	Mean density (seeds m⁻²)	Depth (cm)	Method	Source code
Capsella bursa-pastoris	(L.) Medikus	4	1	—	332	332	332	20	4	111
Capsella bursa-pastoris	(L.) Medikus	4	35	—	20	5088	524	30	4	13
Capsella bursa-pastoris	(L.) Medikus	4	4	—	176	1028	527	25	5	260
Capsella bursa-pastoris	(L.) Medikus	4	4	—	85	1760	681	15	5	272
Capsella bursa-pastoris	(L.) Medikus	4	2	—	315	1881	1098	15	4	31
Capsella bursa-pastoris	(L.) Medikus	4	1	—	1100	1100	1100	20	5	118
Capsella bursa-pastoris	(L.) Medikus	4	2	—	5	2264	1135	15	5	62
Capsella bursa-pastoris	(L.) Medikus	4	1	—	1606	1606	1606	5	5	250
Capsella bursa-pastoris	(L.) Medikus	4	1	—	4775	4775	4775	23	5	202
Capsella bursa-pastoris	(L.) Medikus	4	1	—	11041	11041	11041	15	5	207
Cardamine amara	L.	3	1	—	570	570	570	20	5	107
Cardamine bellidifolia	L.	1	1	—	0	0	0	5	5	105
Cardamine bellidifolia	L.	4	1	—	6	6	6	2	5	73
Cardamine bellidifolia	L.	4	2	—	16	32	24	5	5	105
Cardamine bulbifera	(L.) Crantz	1	3	—	0	0	0	6	5	49
Cardamine flexuosa	With.	1	1	—	60	60	60	15	5	257
Cardamine flexuosa	With.	2	7	—	380	380	380	15	5	257
Cardamine flexuosa	With.	2	7	—	50	2867	495	3	5	96
Cardamine flexuosa	With.	4	1	—	11	11	11	3	7	235
Cardamine flexuosa	With.	4	1	—	533	533	533	5	5	98
Cardamine flexuosa	With.	4	1	—	1150	1150	1150	3	5	96
Cardamine hirsuta	L.	1	1	—	0	0	0	1	5	96
Cardamine hirsuta	L.	1	1	—	0	0	0	3	5	96
Cardamine hirsuta	L.	1	1	—	0	0	0	3	5	132
Cardamine hirsuta	L.	1	1	—	118	118	118	30	5	45
Cardamine hirsuta	L.	1	7	—	53	316	139	8	7	275
Cardamine hirsuta	L.	2	1	—	16	16	16	10	5	110
Cardamine hirsuta	L.	2	1	—	108	108	108	30	5	45
Cardamine hirsuta	L.	2	10	—	50	317	192	3	5	96
Cardamine hirsuta	L.	2	2	—	561	622	592	8	7	275
Cardamine hirsuta	L.	2	1	—	599	599	599	6	5	48
Cardamine hirsuta	L.	2	3	—	150	1700	717	20	5	4
Cardamine hirsuta	L.	2	1	—	800	800	800	30	5	66
Cardamine hirsuta	L.	2	1	—	1608	1608	1608	16	5	23
Cardamine hirsuta	L.	2	1	—	1714	1714	1714	17.8	5	155
Cardamine hirsuta	L.	3	3	>5	0	0	0	7.5	2	196
Cardamine hirsuta	L.	3	1	—	100	100	100	15	5	257
Cardamine hirsuta	L.	3	1	—	403	403	403	17.8	5	155
Cardamine hirsuta	L.	4	1	—	75	75	75	3	5	132
Cardamine hirsuta	L.	4	1	—	317	317	317	3	5	96
Cardamine impatiens	L.	1	1	—	0	0	0	12	5	171
Cardamine impatiens	L.	2	3	—	27	1176	419	6	5	49

Species	Authority									
Cardamine pratensis	L.	1	5	—	0	0	0	6	5	11
Cardamine pratensis	L.	1	2	—	0	0	0	10	5	27
Cardamine pratensis	L.	1	2	—	0	0	0	6	5	28
Cardamine pratensis	L.	1	3	—	0	0	0	25	4	36
Cardamine pratensis	L.	1	1	—	0	0	0	20	5	67
Cardamine pratensis	L.	1	1	—	0	0	0	5	5	69
Cardamine pratensis	L.	1	1	—	0	0	0	5	4	88
Cardamine pratensis	L.	1	1	—	0	0	0	3	5	96
Cardamine pratensis	L.	1	2	—	0	0	0	20	5	97
Cardamine pratensis	L.	1	3	—	0	0	0	6	5	99
Cardamine pratensis	L.	1	2	—	0	0	0	6.5	5	128
Cardamine pratensis	L.	1	3	—	0	0	0	10	5	149
Cardamine pratensis	L.	1	2	—	0	0	0	6	5	151
Cardamine pratensis	L.	1	1	—	0	0	0	3	5	208
Cardamine pratensis	L.	1	1	—	0	0	0	5	5	213
Cardamine pratensis	L.	1	1	—	0	0	0	20	4	225
Cardamine pratensis	L.	1	5	—	0	0	0	10	5	255
Cardamine pratensis	L.	1	2	—	4	21	12	3	7	235
Cardamine pratensis	L.	1	1	—	26	26	26	2	7	99
Cardamine pratensis	L.	1	1	—	300	300	300	5	5	251
Cardamine pratensis	L.	2	1	—	28	28	28	12	5	174
Cardamine pratensis	L.	2	1	>1	130	130	130	6	5	28
Cardamine pratensis	L.	2	7	—	38	838	229	10	5	255
Cardamine pratensis	L.	2	4	—	177	343	236	2	7	99
Cardamine pratensis	L.	2	2	—	204	348	276	13	5	68
Cardamine pratensis	L.	2	4	—	113	594	333	3	5	208
Cardamine pratensis	L.	2	2	—	225	800	513	6.5	5	68
Cardamine pratensis	L.	2	4	—	85	877	516	6	5	208
Cardamine pratensis	L.	2	2	—	416	936	676	6	5	11
Cardamine pratensis	L.	3	3	—	468	32448	11249	10	5	27
Cardamine pratensis	L.	3	2	—	85	113	99	6	5	208
Cardamine pratensis	L.	3	1	—	566	566	566	3	5	208
Cardamine pratensis	L.	3	3	—	468	1872	1127	6	5	99
Cardamine pratensis	L.	3	3	—	1144	2652	1872	10	5	27
Cardamine pratensis	L.	4	4	—	46	120	81	12	5	174
Cardamine pratensis	L.	4	6	—	38	525	213	10	5	255
Cardamine pratensis	L.	4	5	—	311	651	453	3	5	208
Cardamine pratensis	L.	4	12	—	312	3484	1092	6	5	99
Cardamine raphanifolia	Pourret	2	1	—	26	26	26	15	5	272
Cardamine raphanifolia	Pourret	4	1	—	85	85	85	15	5	272
Cochlearia aestuaria	(Lloyd) Heywood	1	1	—	0	0	0	25	4	36
Cochlearia officinalis		4	5	—	752	752	752	2	5	73
Conringia orientalis	(L.) Dumort.	2	5	1	0	0	0	7	2	44
Conringia orientalis	(L.) Dumort.	2	4	>5	0	0	0	7	2	44
Coronopus didymus	(L.) Smith	3	1	—	0	0	0	7.5	2	195
Coronopus didymus	(L.) Smith	3	1	—	2090	2090	2090	5	5	234
Coronopus didymus	(L.) Smith	4	1	—	1235	1235	1235	15	5	207

Species	Authority	Seed bank type	Number of records	Longevity (y)	Minimum density (seeds m^{-2})	Maximum density (seeds m^{-2})	Mean density (seeds m^{-2})	Depth (cm)	Method	Source code
Coronopus squamatus	(Forsskal) Asch.	3	1	>5	0	0	0	7.5	2	195
Descurainia sophia	(L.) Webb ex Prantl	2	1	>2	0	0	0	2	1	47
Descurainia sophia	(L.) Webb ex Prantl	2	1	>2	0	0	0	15	1	47
Descurainia sophia	(L.) Webb ex Prantl	4	2	—	24	140	82	7.5	5	10
Descurainia sophia	(L.) Webb ex Prantl	4	2	—	105	141	123	10	5	8
Diplotaxis tenuifolia	(L.) DC.	1	1	—	0	0	0	10	4	22
Draba aizoides	L.	1	2	—	0	0	0	5	4	91
Erophila verna	(L.) DC.	1	2	—	0	0	0	6	5	67
Erophila verna	(L.) DC.	1	1	—	0	0	0	5	5	69
Erophila verna	(L.) DC.	1	2	—	0	0	0	1	5	96
Erophila verna	(L.) DC.	1	1	—	53	53	53	8	7	275
Erophila verna	(L.) DC.	2	4	—	10	32	21	25	4	36
Erophila verna	(L.) DC.	2	1	—	117	117	117	10	5	184
Erophila verna	(L.) DC.	2	1	—	300	300	300	3	5	132
Erophila verna	(L.) DC.	2	3	—	234	936	468	10	5	152
Erophila verna	(L.) DC.	3	1	—	162	162	162	50	5	260
Erophila verna	(L.) DC.	4	1	—	151	151	151	20	5	111
Erophila verna	(L.) DC.	4	2	—	125	413	269	3	5	132
Erophila verna	(L.) DC.	4	1	—	300	300	300	15	5	249
Erucastrum gallicum	(Willd.) O. Schulz	2	1	>2	0	0	0	7	2	44
Erucastrum gallicum	(Willd.) O. Schulz	2	1	1	0	0	0	7	2	44
Erucastrum gallicum	(Willd.) O. Schulz	3	1	>5	0	0	0	7	2	44
Erysimum cheiranthoides	L.	2	2	3	0	0	0	50	1	240
Erysimum cheiranthoides	L.	2	2	—	38	75	57	10	5	255
Erysimum cheiranthoides	L.	3	2	>5	0	0	0	7.5	2	196
Erysimum cheiranthoides	L.	3	1	6	0	0	0	15	2	240
Erysimum cheiranthoides	L.	3	1	6	0	0	0	100	1	240
Hirschfeldia incana	(L.) Lagr.-Fossat	1	4	—	0	0	0	7	4	50
Lepidium campestre	(L.) R. Br.	2	1	—	50	50	50	10	5	255
Lepidium campestre	(L.) R. Br.	3	3	>5	0	0	0	7.5	2	196
Lepidium draba	L.	1	1	—	0	0	0	30	5	66
Lepidium draba	L.	2	1	1	0	0	0	15	1	34
Lepidium draba	L.	2	1	1	0	0	0	45	1	34
Lepidium draba	L.	2	1	2	0	0	0	30	1	35
Lepidium perfoliatum	L.	2	3	>3	0	0	0	7	2	44
Lepidium perfoliatum	L.	3	3	>5	0	0	0	7	2	44
Lepidium perfoliatum	L.	3	3	>6	0	0	0	7	2	44
Lepidium virginicum	L.	1	1	—	32	32	32	15	5	44
Lepidium virginicum	L.	2	3	>3	0	0	0	7	2	175
Lepidium virginicum	L.	2	1	—	5	5	5	7	5	44
Lepidium virginicum	L.	2	1	—	6	6	6	7	5	1
Lepidium virginicum	L.	2	2	—	0	0	0	5	2	189
Lepidium virginicum	L.	2	2	—	7	16	12	5	5	1

Lepidium virginicum	L.	2	3	—	43	2013	836	15	5	175
Lepidium virginicum	L.	3	3	>5	0	0	0	7	2	44
Lepidium virginicum	L.	3	3	>6	0	0	0	7	2	44
Lepidium virginicum	L.	3	1	40	0	0	0	—	3	124
Lepidium virginicum	L.	3	1	—	43	43	43	15	5	175
Lepidium virginicum	L.	4	1	—	17	17	17	10	5	8
Lepidium virginicum	(L.) Desv.	2	1	—	91	91	91	10	4	22
Lobularia maritima	(L.) Desv.	4	1	—	160	160	160	10	4	22
Lobularia maritima	(L.) Desv.	1	1	—	0	0	0	10	4	22
Neslia paniculata	(L.) Desv.	3	1	10	0	0	0	15	1	240
Neslia paniculata	(L.) Desv.	3	1	10	0	0	0	50	1	240
Neslia paniculata	(L.) Desv.	3	1	10	0	0	0	100	1	240
Raphanus raphanistrum	L.	1	4	—	0	0	0	10	4	22
Raphanus raphanistrum	L.	1	4	—	0	0	0	25	4	36
Raphanus raphanistrum	L.	1	6	—	0	0	0	10	5	255
Raphanus raphanistrum	L.	1	1	—	0	0	0	25	5	260
Raphanus raphanistrum	L.	1	1	—	53	53	53	8	7	275
Raphanus raphanistrum	L.	1	2	—	176	176	176	50	5	260
Raphanus raphanistrum	L.	2	1	>2	0	0	0	0	1	190
Raphanus raphanistrum	L.	2	1	—	48	48	48	5	5	69
Raphanus raphanistrum	L.	2	3	—	38	150	79	10	5	255
Raphanus raphanistrum	L.	2	1	—	1310	1310	1310	15	5	94
Raphanus raphanistrum	L.	3	1	>5	0	0	0	7.5	2	196
Raphanus raphanistrum	L.	3	1	>18	33	33	33	32	5	169
Raphanus raphanistrum	L.	3	1	>20	80	80	80	20	5	40
Raphanus raphanistrum	L.	4	1	—	7	7	7	20	4	111
Raphanus raphanistrum	L.	4	1	—	84	84	84	25	5	2
Raphanus raphanistrum	L.	4	1	—	88	88	88	10	5	255
Raphanus raphanistrum	L.	4	3	—	980	2030	1330	15	5	94
Raphanus raphanistrum	L.	4	3	—	429	3000	1880	25	5	260
Raphanus raphanistrum subsp. *maritimus*	(Smith) Thell.	1	1	—	0	0	0	10	5	255
Rorippa amphibia	(L.) Besser	1	2	—	0	0	0	3	5	208
Rorippa amphibia	(L.) Besser	1	3	—	85	85	85	6	5	208
Rorippa amphibia	(L.) Besser	4	1	—	85	85	85	3	5	208
Rorippa islandica	(Oeder ex Murray) Borbas	2	1	>2	0	0	0	2	1	47
Rorippa islandica	(Oeder ex Murray) Borbas	2	1	>2	0	0	0	15	1	47
Rorippa islandica	(Oeder ex Murray) Borbas	2	1	—	12	12	12	3	7	235
Rorippa islandica	(Oeder ex Murray) Borbas	2	1	—	16	16	16	5	5	244
Rorippa islandica	(Oeder ex Murray) Borbas	2	1	>50	33	33	33	16	5	168
Rorippa islandica	(Oeder ex Murray) Borbas	2	1	—	60	60	60	20	5	111
Rorippa islandica	(Oeder ex Murray) Borbas	2	2	—	217	367	292	3	5	96
Rorippa islandica	(Oeder ex Murray) Borbas	2	1	—	438	438	438	5	4	88
Rorippa islandica	(Oeder ex Murray) Borbas	2	2	—	250	700	475	20	5	4
Rorippa islandica	(Oeder ex Murray) Borbas	2	4	—	625	1750	1125	10	4	88
Rorippa islandica	(Oeder ex Murray) Borbas	2	1	—	2185	2185	2185	35	5	247
Rorippa islandica	(Oeder ex Murray) Borbas	3	1	>5	0	0	0	7.5	2	195
Rorippa islandica	(Oeder ex Murray) Borbas	3	1	—	300	300	300	20	5	4

Species	Authority	Seed bank type	Number of records	Longevity (y)	Minimum density (seeds m^{-2})	Maximum density (seeds m^{-2})	Mean density (seeds m^{-2})	Depth (cm)	Method	Source code
Rorippa islandica	(Oeder ex Murray) Borbas	3	1	—	1092	1092	1092	10	5	27
Rorippa islandica	(Oeder ex Murray) Borbas	3	5	—	455	6793	2815	35	5	247
Rorippa islandica	(Oeder ex Murray) Borbas	4	2	—	67	83	75	3	5	96
Rorippa microphylla	(Boenn.) N. Hylander ex A. Love & D. Löve	2	1	—	67	67	67	3	5	96
Rorippa nasturtium-aquaticum	(L.) Hayek	2	1	—	2	2	2	5	5	244
Rorippa palustris	(L.) Besser	2	1	—	654	654	654	50	5	216
Rorippa palustris	(L.) Besser	3	1	—	10	10	10	10	5	252
Rorippa palustris	(L.) Besser	4	2	—	39	760	400	12	4	230
Rorippa palustris	(L.) Besser	4	2	—	1299	1299	1299	50	5	216
Rorippa sylvestris	(L.) Besser	1	1	—	0	0	0	3	5	96
Sinapis alba	L.	1	1	>5	0	0	0	10	5	149
Sinapis alba	L.	3	1	>41	0	0	0	7.5	2	195
Sinapis alba	L.	3	1	—	32	32	32	30	5	29
Sinapis arvensis	L.	1	2	—	0	0	0	10	4	22
Sinapis arvensis	L.	1	1	—	170	170	170	6	5	208
Sinapis arvensis	L.	1	1	—	216	216	216	50	5	260
Sinapis arvensis	L.	2	1	>3	0	0	0	7	2	44
Sinapis arvensis	L.	2	2	3	0	0	0	7	2	44
Sinapis arvensis	L.	2	1	2	0	0	0	2.5	1	256
Sinapis arvensis	L.	2	1	2	0	0	0	5	1	256
Sinapis arvensis	L.	2	1	—	25	25	25	10	4	22
Sinapis arvensis	L.	2	1	—	52	52	52	15	5	62
Sinapis arvensis	L.	2	1	—	104	104	104	7.5	5	10
Sinapis arvensis	L.	2	1	—	311	311	311	3	5	208
Sinapis arvensis	L.	2	1	—	2845	2845	2845	17.8	5	155
Sinapis arvensis	L.	3	1	>5	0	0	0	30	3	14
Sinapis arvensis	L.	3	1	>4	0	0	0	7	2	44
Sinapis arvensis	L.	3	3	>5	0	0	0	7	2	44
Sinapis arvensis	L.	3	2	6	0	0	0	7	2	60
Sinapis arvensis	L.	3	1	>9	0	0	0	30	1	61
Sinapis arvensis	L.	3	1	>18	0	0	0	30	1	61
Sinapis arvensis	L.	3	1	5	0	0	0	8	1	61
Sinapis arvensis	L.	3	1	6	0	0	0	20	1	125
Sinapis arvensis	L.	3	1	>26	0	0	0	25	1	196
Sinapis arvensis	L.	3	3	>5	0	0	0	7.5	2	256
Sinapis arvensis	L.	3	1	>5	0	0	0	7.5	1	256
Sinapis arvensis	L.	3	1	>5	0	0	0	12.5	1	256
Sinapis arvensis	L.	3	1	>5	0	0	0	17.5	1	256
Sinapis arvensis	L.	3	1	>5	9	9	9	25	1	237
Sinapis arvensis	L.	3	1	>24	43	43	43	—	6	45
Sinapis arvensis	L.	3	1	>6	74	74	74	15	5	62

Species										
Sinapis arvensis	L.	3	1	>80	200	200	200	90	5	162
Sinapis arvensis	L.	3	1	>20	211	211	211	16	5	168
Sinapis arvensis	L.	3	1	>40	506	506	506	30	5	45
Sinapis arvensis	L.	3	1	>6	980	980	980	30	5	45
Sinapis arvensis	L.	3	1	>8	1098	1098	1098	30	5	45
Sinapis arvensis	L.	3	1	—	1300	1300	1300	20	5	69
Sinapis arvensis	L.	4	1	—	9	9	9	10	5	110
Sinapis arvensis	L.	4	1	—	11	11	11	10	5	8
Sinapis arvensis	L.	4	1	—	11	11	11	20	5	111
Sinapis arvensis	L.	4	1	—	29	29	29	7	5	58
Sinapis arvensis	L.	4	13	—	20	219	85	30	4	13
Sinapis arvensis	L.	4	2	—	16	254	135	10	4	22
Sinapis arvensis	L.	4	1	—	200	200	200	10	5	86
Sinapis arvensis	L.	4	3	—	84	737	302	10	5	15
Sinapis arvensis	L.	4	4	—	171	2353	1096	25	5	260
Sisymbrium altissimum	L.	2	3	>3	0	0	0	7	2	44
Sisymbrium altissimum	L.	2	1	3	0	0	0	7	2	44
Sisymbrium altissimum	L.	2	2	4	0	0	0	7	2	44
Sisymbrium altissimum	L.	3	1	—	67	67	67	10	5	184
Sisymbrium altissimum	L.	3	3	>5	0	0	0	7	2	44
Sisymbrium altissimum	L.	3	3	>6	0	0	0	7	2	44
Sisymbrium altissimum	L.	3	1	10	0	0	0	50	1	240
Sisymbrium altissimum	L.	4	1	10	0	0	0	100	1	240
Sisymbrium altissimum	L.	1	35	6	0	0	0	15	1	240
Sisymbrium altissimum	L.	1	1	—	54	2519	386	15	4	37
Sisymbrium officinale	(L.) Scop.	1	1	—	0	0	0	5	5	69
Sisymbrium officinale	(L.) Scop.	1	3	—	0	0	0	10	5	255
Sisymbrium officinale	(L.) Scop.	2	1	—	53	158	95	8	7	275
Sisymbrium officinale	(L.) Scop.	2	2	—	4	4	4	5	5	244
Sisymbrium officinale	(L.) Scop.	2	1	—	170	170	170	10	5	140
Sisymbrium officinale	(L.) Scop.	2	4	—	188	188	188	17	5	162
Sisymbrium officinale	(L.) Scop.	3	1	—	412	412	412	12	5	162
Thlaspi arvense	L.	4	1	>5	0	0	0	7.5	2	196
Thlaspi arvense	L.	1	3	—	0	0	0	25	5	2
Thlaspi arvense	L.	1	4	—	0	0	0	25	5	260
Thlaspi arvense	L.	1	4	—	158	158	158	8	7	275
Thlaspi arvense	L.	2	2	>3	0	0	0	7	2	44
Thlaspi arvense	L.	2	1	4	0	0	0	7	2	44
Thlaspi arvense	L.	2	1	4	0	0	0	2.5	2	203
Thlaspi arvense	L.	2	1	2	0	0	0	2.5	1	256
Thlaspi arvense	L.	2	1	2	0	0	0	5	1	256
Thlaspi arvense	L.	2	4	—	1	1	1	20	5	40
Thlaspi arvense	L.	2	2	—	37	37	37	15	5	62
Thlaspi arvense	L.	2		—	48	48	48	5	5	69
Thlaspi arvense	L.	2		—	80	120	90	10	5	149
Thlaspi arvense	L.	2		—	100	100	100	10	5	184

Species	Authority	Seed bank type	Number of records	Longevity (y)	Minimum density (seeds m⁻²)	Maximum density (seeds m⁻²)	Mean density (seeds m⁻²)	Depth (cm)	Method	Source code
Thlaspi arvense	L.	2	1	—	195	195	195	7.5	5	10
Thlaspi arvense	L.	2	13	—	50	3413	552	10	5	255
Thlaspi arvense	L.	3	1	>5	0	0	0	7	2	44
Thlaspi arvense	L.	3	1	>6	0	0	0	7	2	44
Thlaspi arvense	L.	3	1	>25	0	0	0	25	1	125
Thlaspi arvense	L.	3	1	>5	0	0	0	7.5	2	196
Thlaspi arvense	L.	3	1	>5	0	0	0	2.5	1	203
Thlaspi arvense	L.	3	1	>5	0	0	0	7.5	1	203
Thlaspi arvense	L.	3	1	>5	0	0	0	7.5	2	203
Thlaspi arvense	L.	3	1	>5	0	0	0	15	1	203
Thlaspi arvense	L.	3	1	>5	0	0	0	15	2	203
Thlaspi arvense	L.	3	1	>11	0	0	0	25	3	209
Thlaspi arvense	L.	3	1	30	0	0	0	50	1	240
Thlaspi arvense	L.	3	1	30	0	0	0	100	1	240
Thlaspi arvense	L.	3	1	6	0	0	0	15	1	240
Thlaspi arvense	L.	3	1	>5	0	0	0	7.5	1	256
Thlaspi arvense	L.	3	1	>5	0	0	0	12.5	1	256
Thlaspi arvense	L.	3	1	>5	0	0	0	17.5	1	256
Thlaspi arvense	L.	3	1	>5	0	0	0	25	1	256
Thlaspi arvense	L.	3	1	>20	133	133	133	24	5	168
Thlaspi arvense	L.	4	3	—	20	40	29	30	4	13
Thlaspi arvense	L.	4	1	—	137	137	137	25	4	185
Thlaspi arvense	L.	4	2	—	225	475	350	10	5	255
Thlaspi arvense	L.	4	2	—	28	1020	524	10	5	8
Thlaspi arvense	L.	4	4	—	110	2504	1080	7.5	5	10
Thlaspi arvense	L.	4	1	—	4750	4750	4750	20	5	118
Thlaspi arvense	L.	4	36	—	990	17631	5029	15	4	37
Cupressaceae	—		—	—	—	—	—	—	—	
Juniperus communis	L.	1	1	—	0	0	0	5	5	55
Juniperus communis	L.	1	1	—	0	0	0	2	5	82
Juniperus communis	L.	1	2	—	0	0	0	10	5	150
Juniperus communis	L.	1	1	—	0	0	0	10	5	152
Juniperus communis	L.	1	1	—	0	0	0	2	5	153
Juniperus communis	L.	1	1	—	0	0	0	6.5	7	182
Juniperus communis	L.	1	4	—	0	0	0	10	5	255
Juniperus communis	L.	1	1	—	0	0	0	5	5	264
Cyperaceae	—		—	—	—	—	—	—	—	
Bolboschoenus maritimus	(L.) Palla	1	2	—	0	0	0	7	5	112
Bolboschoenus maritimus	(L.) Palla	2	1	—	5	5	5	2	5	113
Bolboschoenus maritimus	(L.) Palla	2	6	—	6	158	76	4	5	217

Species	Authority									
Bolboschoenus maritimus	(L.) Palla	2	5	—	37	177	118	5	5	177
Bolboschoenus maritimus	(L.) Palla	2	1	—	156	156	156	5	5	116
Bolboschoenus maritimus	(L.) Palla	4	1	—	37	37	37	5	5	177
Bolboschoenus maritimus	(L.) Palla	4	6	—	10	174	61	4	5	217
Carex acuta	L.	1	1	—	0	0	0	5	4	88
Carex acuta	L.	1	6	—	0	0	0	3	5	208
Carex acuta	L.	1	3	—	0	0	0	6	5	208
Carex acuta	L.	2	1	—	0	0	0	5	5	248
Carex acuta	L.	2	3	—	63	300	213	10	5	255
Carex acuta	L.	2	2	—	176	537	357	16	5	216
Carex acuta	L.	2	1	—	764	764	764	6	5	208
Carex acuta	L.	3	3	—	1200	1200	1200	20	4	225
Carex acuta	L.	4	4	—	113	340	255	6	5	208
Carex acuta	L.	4	3	—	85	198	120	3	5	208
Carex acuta	L.	4	2	—	25	575	336	5	5	98
Carex acuta	L.	4	2	—	200	638	419	10	5	255
Carex acuta	L.	4	5	—	780	7536	4158	50	5	216
Carex acutiformis	Ehrh.	1	3	—	0	0	0	6.5	5	128
Carex acutiformis	Ehrh.	1	2	—	0	0	0	12	5	174
Carex acutiformis	Ehrh.	1	1	—	0	0	0	10	5	252
Carex acutiformis	Ehrh.	2	1	—	9	9	9	12	5	174
Carex acutiformis	Ehrh.	2	1	—	720	720	720	6.5	5	128
Carex acutiformis	Ehrh.	3	1	—	456	456	456	25	5	107
Carex acutiformis	Ehrh.	4	2	—	12	28	20	12	5	174
Carex acutiformis	Ehrh.	4	4	—	80	1160	380	6.5	5	128
Carex aquatilis	Wahlenb.	1	1	—	0	0	0	16	5	130
Carex aquatilis	Wahlenb.	1	1	—	208	208	208	10	5	27
Carex aquatilis	Wahlenb.	2	2	—	55	267	161	10	5	63
Carex aquatilis	Wahlenb.	2	1	—	2648	2648	2648	16	5	216
Carex aquatilis	Wahlenb.	3	1	—	1840	1840	1840	16	5	216
Carex arenaria	L.	4	2	—	500	800	650	15	5	249
Carex bicolor	All.	1	1	—	0	0	0	6.5	7	182
Carex bigelowii	Torrey ex Schwein.	1	1	—	0	0	0	7	5	232
Carex bigelowii	Torrey ex Schwein.	2	1	—	144	144	144	10	5	63
Carex bigelowii	Torrey ex Schwein.	3	1	>200	220	220	220	15	5	146
Carex bigelowii	Torrey ex Schwein.	3	1	—	444	444	444	10	5	63
Carex bigelowii	Torrey ex Schwein.	3	1	—	683	683	683	30	5	78
Carex bigelowii	Torrey ex Schwein.	3	1	—	1665	1665	1665	25	5	144
Carex bigelowii	Torrey ex Schwein.	4	2	—	34	128	81	3	4	158
Carex bigelowii	Torrey ex Schwein.	4	5	—	113	584	314	20	5	191
Carex binervis	Smith	1	2	—	0	0	0	2	5	153
Carex binervis	Smith	2	3	—	158	341	265	5	5	92
Carex binervis	Smith	4	1	—	7	7	7	3	7	235
Carex binervis	Smith	4	1	—	33	33	33	2	5	153
Carex binervis	Smith	4	4	—	285	423	343	5	5	92
Carex brunnescens	(Pers.) Poiret	1	1	—	0	0	0	3	4	158
Carex brunnescens	(Pers.) Poiret	1	1	—	403	403	403	12	5	171

Species	Authority	Seed bank type	Number of records	Longevity (y)	Minimum density (seeds m⁻²)	Maximum density (seeds m⁻²)	Mean density (seeds m⁻²)	Depth (cm)	Method	Source code
Carex brunnescens	(Pers.) Poiret	4	1	—	25	25	25	3	4	158
Carex buxbaumii	Wahlenb.	4	1	—	41	41	41	16	5	216
Carex caryophyllea	Latour.	1	1	—	0	0	0	6	5	48
Carex caryophyllea	Latour.	1	1	—	0	0	0	3	5	96
Carex caryophyllea	Latour.	1	1	—	0	0	0	10	5	152
Carex caryophyllea	Latour.	1	1	—	0	0	0	6	5	181
Carex caryophyllea	Latour.	1	1	—	0	0	0	6.5	7	182
Carex caryophyllea	Latour.	1	3	—	0	0	0	3	7	235
Carex caryophyllea	Latour.	1	3	—	0	0	0	10	5	255
Carex caryophyllea	Latour.	3	1	—	240	240	240	6	5	48
Carex caryophyllea	Latour.	4	1	—	112	112	112	5	5	69
Carex caryophyllea	Latour.	4	1	—	2551	2551	2551	8.5	4	254
Carex chordorrhiza	L.f.	2	7	—	144	144	144	10	5	63
Carex chordorrhiza	L.f.	4	2	—	134	227	181	16	5	216
Carex curta	Gooden.	1	1	—	0	0	0	6	5	11
Carex curta	Gooden.	1	1	—	0	0	0	40	5	11
Carex curta	Gooden.	1	1	—	0	0	0	30	5	180
Carex davalliana	Smith	2	1	—	12	12	12	12	5	174
Carex davalliana	Smith	4	1	—	22	22	22	12	5	174
Carex davalliana	Smith	4	1	—	925	925	925	10	5	255
Carex digitata	L.	1	2	—	0	0	0	5	6	170
Carex digitata	L.	1	1	—	0	0	0	4	5	176
Carex digitata	L.	1	2	—	0	83	42	12	5	171
Carex digitata	L.	2	1	—	0	0	0	5	6	170
Carex digitata	L.	4	1	—	2	2	2	4	5	176
Carex dioica	L.	4	2	—	32	232	132	6	5	181
Carex disticha	Hudson	1	1	—	0	0	0	5	4	88
Carex disticha	Hudson	1	1	—	0	0	0	7	5	112
Carex disticha	Hudson	1	9	—	0	0	0	3	5	208
Carex disticha	Hudson	1	5	—	0	0	0	6	5	208
Carex disticha	Hudson	1	5	—	0	0	0	20	4	225
Carex disticha	Hudson	2	1	—	80	80	80	6.5	5	128
Carex disticha	Hudson	2	4	—	75	188	141	10	5	255
Carex disticha	Hudson	3	1	—	85	85	85	6	5	208
Carex disticha	Hudson	4	2	—	38	125	82	10	5	255
Carex disticha	Hudson	4	3	—	120	220	153	6.5	5	128
Carex disticha	Hudson	4	4	—	125	608	296	5	5	98
Carex echinata	Murray	1	3	—	0	0	0	3	5	208
Carex echinata	Murray	1	2	—	0	0	0	10	5	255
Carex echinata	Murray	2	1	—	195	195	195	7	5	232
Carex echinata	Murray	4	1	—	35	35	35	7	5	232
Carex echinata	Murray	4	1	—	1200	1200	1200	3	5	96

Species	Authority									
Carex elata	All.	1	1	—	0	0	0	12	5	174
Carex elata	All.	2	3	—	22	43	9	12	5	174
Carex elata	All.	4	2	—	101	187	15	12	5	174
Carex elongata	L.	1	1	—	0	0	0	5	5	248
Carex filiformis	L.	3	1	—	336	336	336	20	5	67
Carex flacca	Schreber	1	2	—	0	0	0	3	5	96
Carex flacca	Schreber	1	2	—	0	0	0	10	5	110
Carex flacca	Schreber	1	1	—	0	0	0	5	5	122
Carex flacca	Schreber	1	2	—	0	0	0	6.5	5	128
Carex flacca	Schreber	1	4	—	0	0	0	3	7	235
Carex flacca	Schreber	1	1	—	0	0	0	10	5	252
Carex flacca	Schreber	1	2	—	0	0	0	10	5	255
Carex flacca	Schreber	1	1	—	521	521	521	20	5	97
Carex flacca	Schreber	2	2	—	23	29	17	7	5	58
Carex flacca	Schreber	2	1	—	112	112	112	13	5	183
Carex flacca	Schreber	2	3	—	160	240	60	6.5	5	128
Carex flacca	Schreber	2	2	—	365	417	313	20	5	97
Carex flacca	Schreber	2	1	—	778	778	778	6	5	48
Carex flacca	Schreber	2	2	—	2546	2565	2527	6.5	7	182
Carex flacca	Schreber	3	1	>20	24	24	24	13	5	183
Carex flacca	Schreber	3	1	—	1304	1304	1304	10	5	181
Carex flacca	Schreber	4	2	—	18	22	13	10	5	110
Carex flacca	Schreber	4	1	—	60	60	60	6.5	5	128
Carex flacca	Schreber	4	1	—	70	70	70	5	5	122
Carex flacca	Schreber	4	3	—	129	219	35	7	5	58
Carex flacca	Schreber	4	2	—	388	575	200	10	5	255
Carex flacca	Schreber	4	2	—	3088	5128	1048	6	5	181
Carex flacca	Schreber	1	1	—	0	0	0	6.5	5	128
Carex flava	L.	1	1	—	0	0	0	12	5	174
Carex flava	L.	1	1	—	0	0	0	10	5	255
Carex flava	L.	2	3	—	81	162	9	12	5	174
Carex flava	L.	3	1	>6	0	0	0	40	1	212
Carex flava	L.	4	5	—	354	1071	28	12	5	174
Carex globularis	L.	1	1	—	0	0	0	5	5	82
Carex hirta	L.	1	1	—	0	0	0	20	5	67
Carex hirta	L.	1	1	—	0	0	0	3	5	96
Carex hirta	L.	1	2	—	0	0	0	20	4	225
Carex hirta	L.	1	1	—	0	0	0	3	7	235
Carex hirta	L.	1	1	—	0	0	0	10	5	252
Carex hirta	L.	1	2	—	0	0	0	10	5	255
Carex hirta	L.	2	1	—	87	87	87	10	5	126
Carex hirta	L.	4	1	—	42	42	42	5	5	98
Carex hostiana	DC.	2	3	—	17	34	9	12	5	174
Carex hostiana	DC.	4	4	—	73	96	40	12	5	174
Carex lasiocarpa	Ehrh.	1	1	—	0	0	0	5	5	248
Carex lasiocarpa	Ehrh.	3	1	—	765	765	765	16	5	216
Carex lasiocarpa	Ehrh.	4	1	—	25	25	25	5	5	117

Species	Authority	Seed bank type	Number of records	Longevity (y)	Minimum density (seeds m⁻²)	Maximum density (seeds m⁻²)	Mean density (seeds m⁻²)	Depth (cm)	Method	Source code
Carex lasiocarpa	Ehrh.	4	1	—	203	203	203	16	5	216
Carex montana	L.	1	6	—	0	0	0	10	5	152
Carex montana	L.	1	1	—	0	0	0	6	5	181
Carex montana	L.	1	2	—	0	0	0	6.5	7	182
Carex montana	L.	2	1	—	160	168	168	10	5	181
Carex montana	L.	4	1	—	336	336	336	6	5	181
Carex montana	L.	4	1	—	624	624	624	10	5	152
Carex muricata	L.	2	1	—	128	128	128	13	5	162
Carex muricata	L.	3	1	—	167	167	167	32	5	169
Carex nardina	Fries	4	1	—	32	32	32	2	5	73
Carex nigra	(L.) Reichard	1	1	—	0	0	0	40	5	11
Carex nigra	(L.) Reichard	1	4	—	0	0	0	6	5	28
Carex nigra	(L.) Reichard	1	2	—	0	0	0	3	5	96
Carex nigra	(L.) Reichard	1	8	—	0	0	0	6	5	99
Carex nigra	(L.) Reichard	1	2	—	0	0	0	7	5	112
Carex nigra	(L.) Reichard	1	5	—	0	0	0	12	5	174
Carex nigra	(L.) Reichard	1	3	—	0	0	0	3	5	208
Carex nigra	(L.) Reichard	1	2	—	0	0	0	6	5	208
Carex nigra	(L.) Reichard	1	2	—	0	0	0	10	5	255
Carex nigra	(L.) Reichard	1	5	—	0	39	13	2	7	99
Carex nigra	(L.) Reichard	1	1	—	53	53	53	8	7	275
Carex nigra	(L.) Reichard	2	1	—	23	23	23	3	7	235
Carex nigra	(L.) Reichard	2	1	—	238	238	238	10	5	255
Carex nigra	(L.) Reichard	2	1	—	416	416	416	6	5	11
Carex nigra	(L.) Reichard	2	1	—	1002	1002	1002	16	5	216
Carex nigra	(L.) Reichard	3	1	—	113	113	113	6	5	208
Carex nigra	(L.) Reichard	3	1	—	364	364	364	6	5	99
Carex nigra	(L.) Reichard	3	2	—	171	570	371	30	5	107
Carex nigra	(L.) Reichard	4	3	—	9	68	30	12	5	174
Carex nigra	(L.) Reichard	4	1	—	139	139	139	7	5	232
Carex nigra	(L.) Reichard	4	2	—	88	200	144	10	5	255
Carex nigra	(L.) Reichard	4	1	—	156	156	156	6	5	99
Carex ornithopoda	Willd.	1	2	—	0	0	0	6.5	7	182
Carex ornithopoda	Willd.	4	1	—	24	24	24	6	5	181
Carex otrubae	Podp.	4	1	—	32	32	32	3	7	235
Carex ovalis	Gooden.	1	1	—	0	0	0	3	5	96
Carex ovalis	Gooden.	1	4	—	0	0	0	6	5	99
Carex ovalis	Gooden.	1	3	—	0	0	0	6	5	208
Carex ovalis	Gooden.	1	1	—	0	0	0	20	4	225
Carex ovalis	Gooden.	1	4	—	0	0	0	10	5	255
Carex ovalis	Gooden.	1	4	—	26	156	68	2	7	99
Carex ovalis	Gooden.	1	1	—	69	69	69	12	5	171

Species	Authority									
Carex ovalis	Gooden.	1	1	—	924	924	924	13	5	68
Carex ovalis	Gooden.	2	7	—	50	263	163	10	5	255
Carex ovalis	Gooden.	2	1	—	1625	1625	1625	6.5	5	68
Carex ovalis	Gooden.	2	1	—	1768	1768	1768	40	5	11
Carex ovalis	Gooden.	2	1	—	4836	4836	4836	6	5	11
Carex ovalis	Gooden.	3	1	—	104	104	104	20	5	67
Carex ovalis	Gooden.	3	1	—	208	208	208	6	5	11
Carex ovalis	Gooden.	3	3	—	156	1196	641	10	5	27
Carex ovalis	Gooden.	3	3	—	468	1144	919	6	5	99
Carex ovalis	Gooden.	3	1	>14	2548	2548	2548	6	5	11
Carex ovalis	Gooden.	4	1	—	91	91	91	2	7	99
Carex ovalis	Gooden.	4	1	—	128	128	128	5	5	69
Carex ovalis	Gooden.	4	2	—	188	400	294	10	5	255
Carex ovalis	Gooden.	4	1	—	2800	2800	2800	20	4	225
Carex pallescens	L.	2	1	—	29	29	29	15	5	32
Carex pallescens	L.	2	1	—	36	36	36	16	5	216
Carex pallescens	L.	2	5	—	2	94	40	5	6	170
Carex pallescens	L.	2	1	—	108	108	108	20	5	67
Carex pallescens	L.	2	1	—	153	153	153	12	5	171
Carex pallescens	L.	3	1	—	139	139	139	12	5	171
Carex panicea	L.	1	1	—	0	0	0	6	5	11
Carex panicea	L.	1	1	—	0	0	0	3	5	96
Carex panicea	L.	1	3	—	0	0	0	6.5	5	128
Carex panicea	L.	1	2	—	0	0	0	2	5	153
Carex panicea	L.	1	1	—	0	0	0	12	5	174
Carex panicea	L.	1	6	—	0	0	0	3	5	208
Carex panicea	L.	1	1	—	0	0	0	10	5	255
Carex panicea	L.	2	2	—	12	19	16	12	5	174
Carex panicea	L.	2	2	—	38	213	85	10	5	255
Carex panicea	L.	2	4	—	100	100	100	3	5	96
Carex panicea	L.	3	1	—	3090	3090	3090	16	5	216
Carex panicea	L.	4	4	—	75	125	103	10	5	255
Carex panicea	L.	4	7	—	15	454	194	12	5	174
Carex pediformis	C.A. Meyer	1	1	—	0	0	0	12	5	171
Carex pendula	Hudson	2	1	—	4	4	4	15	5	32
Carex pilosa	Scop.	1	2	—	0	0	0	5	6	170
Carex pilosa	Scop.	1	1	—	0	0	0	30	5	173
Carex pilosa	Scop.	2	1	—	82	82	82	6	5	49
Carex pilosa	Scop.	3	1	—	34	34	34	10	5	172
Carex pilosa	Scop.	4	2	—	1	17	9	4	5	176
Carex pilulifera	L.	1	2	—	0	0	0	10	5	255
Carex pilulifera	L.	2	1	—	21	21	21	15	5	32
Carex pilulifera	L.	2	1	—	199	199	199	7	5	232
Carex pilulifera	L.	2	1	—	224	224	224	20	5	67
Carex pilulifera	L.	2	5	—	22	2362	545	5	5	220
Carex pilulifera	L.	2	4	—	60	5160	1350	15	5	257
Carex pilulifera	L.	3	1	—	195	195	195	17.5	5	126

Species	Authority	Seed bank type	Number of records	Longevity (y)	Minimum density (seeds m⁻²)	Maximum density (seeds m⁻²)	Mean density (seeds m⁻²)	Depth (cm)	Method	Source code
Carex pilulifera	L.	3	1	>59	213	213	213	7.5	5	84
Carex pilulifera	L.	3	1	—	266	266	266	9	5	84
Carex pilulifera	L.	3	1	>25	979	979	979	9.5	5	84
Carex pilulifera	L.	3	1	>68	1043	1043	1043	12	5	84
Carex pilulifera	L.	3	8	—	60	5180	1273	15	5	257
Carex pilulifera	L.	3	1	>30	2404	2404	2404	10	5	84
Carex pilulifera	L.	4	1	—	39	39	39	5	5	123
Carex pilulifera	L.	4	1	—	250	250	250	16	5	216
Carex pilulifera	L.	4	1	—	658	658	658	7	5	232
Carex pilulifera	L.	4	1	—	708	708	708	5	5	220
Carex pulicaris	L.	1	1	—	0	0	0	10	5	255
Carex remota	L.	1	1	—	0	0	0	20	4	225
Carex remota	L.	2	2	—	1	3	2	4	5	176
Carex remota	L.	2	1	—	12	12	12	5	5	213
Carex remota	L.	3	1	—	111	111	111	16	5	168
Carex remota	L.	4	1	—	6	6	6	15	5	32
Carex remota	L.	4	2	—	43	58	51	5	5	220
Carex remota	L.	4	1	—	1262	1262	1262	10	5	126
Carex riparia	Curtis	1	1	—	0	0	0	5	4	88
Carex riparia	Curtis	1	3	—	53	79	62	8	7	275
Carex riparia	Curtis	2	1	—	179	179	179	8	7	275
Carex riparia	Curtis	4	2	—	33	100	67	5	5	98
Carex rostrata	Stokes	1	2	—	0	0	0	12	5	174
Carex rostrata	Stokes	1	1	—	0	0	0	30	5	180
Carex rostrata	Stokes	1	1	—	0	0	0	5	5	248
Carex rostrata	Stokes	2	1	—	727	727	727	16	5	216
Carex rostrata	Stokes	4	2	—	15	31	23	12	5	174
Carex rostrata	Stokes	4	1	—	124	124	124	16	5	216
Carex rotundata	Wahlenb.	1	1	—	0	0	0	10	5	63
Carex rupestris	All.	1	3	—	0	0	0	6	5	39
Carex saxatilis	L.	1	1	—	0	0	0	10	5	63
Carex sempervirens	Vill.	1	2	—	0	0	0	5	4	91
Carex spicata	Hudson	1	1	—	0	0	0	5	5	69
Carex strigosa	Hudson	4	1	—	27	27	27	10	5	33
Carex sylvatica	Hudson	1	1	—	0	0	0	20	5	67
Carex sylvatica	Hudson	1	1	—	0	0	0	10	5	255
Carex sylvatica	Hudson	2	1	—	36	36	36	5	5	213
Carex sylvatica	Hudson	3	1	—	67	67	67	32	5	169
Carex sylvatica	Hudson	3	2	—	178	200	189	16	5	168
Carex sylvatica	Hudson	4	1	—	41	41	41	10	5	126
Carex sylvatica	Hudson	4	2	—	22	808	415	12.5	5	126
Carex tenuiflora	Wahlenb.	1	1	—	0	0	0	20	5	191

Species	Author									
Carex tenuiflora	Wahlenb.	4	3	—	31	51	41	20	5	191
Carex vesicaria	L.	1	9	—	0	0	0	3	5	208
Carex vesicaria	L.	1	2	—	0	0	0	6	5	208
Carex vesicaria	L.	1	1	—	0	0	0	10	5	255
Carex vesicaria	L.	2	1	—	3470	3470	3470	16	5	216
Carex vesicaria	L.	3	1	—	85	85	85	6	5	208
Carex vesicaria	L.	3	1	—	1767	1767	1767	16	5	216
Carex vesicaria	L.	4	1	—	38	38	38	10	5	255
Carex viridula subsp. *oedocarpa*	(Andersson) B. Schmid	4	1	—	333	333	333	7	5	232
Carex viridula subsp. *viridula*	Michaux	1	1	—	0	0	0	16	5	216
Carex viridula subsp. *viridula*	Michaux	3	1	>6	0	0	0	40	1	212
Carex viridula subsp. *viridula*	Michaux	3	1	—	1157	1157	1157	16	5	216
Carex × grahamii	Boott	1	1	—	0	0	0	10	5	149
Carex × grahamii	Boott	2	1	—	120	120	120	10	5	149
Carex × grahamii	Boott	4	3	—	200	460	313	10	5	149
Cladium mariscus	(L.) Pohl	1	1	—	0	0	0	15	4	148
Cyperus flavescens	L.	4	5	—	99	4253	1344	13.5	5	163
Cyperus fuscus	L.	2	1	—	60	60	60	6.5	5	128
Eleocharis acicularis	(L.) Roemer & Schultes	1	1	—	0	0	0	5	5	117
Eleocharis acicularis	(L.) Roemer & Schultes	2	1	—	97	97	97	0.538	5	269
Eleocharis multicaulis	(Smith) Desv.	2	1	—	337	337	337	5	5	222
Eleocharis palustris	(L.) Roemer & Schultes	1	1	—	0	0	0	10	5	27
Eleocharis palustris	(L.) Roemer & Schultes	1	1	—	0	0	0	3	5	96
Eleocharis palustris	(L.) Roemer & Schultes	1	1	—	0	0	0	20	4	225
Eleocharis palustris	(L.) Roemer & Schultes	1	2	—	0	0	0	3	7	235
Eleocharis palustris	(L.) Roemer & Schultes	2	2	—	12	24	18	4	5	217
Eleocharis palustris	(L.) Roemer & Schultes	2	1	—	1600	1600	1600	20	4	225
Eleocharis palustris	(L.) Roemer & Schultes	4	1	—	11	11	11	5	5	166
Eleocharis palustris	(L.) Roemer & Schultes	4	2	—	8	24	16	4	5	217
Eleocharis uniglumis	(Link) Schultes	1	2	—	0	0	0	7	5	112
Eriophorum angustifolium	Honck.	1	1	—	0	0	0	10	5	9
Eriophorum angustifolium	Honck.	1	3	—	0	0	0	3	5	96
Eriophorum angustifolium	Honck.	1	4	—	0	0	0	10	5	149
Eriophorum angustifolium	Honck.	1	2	—	0	0	0	3	7	235
Eriophorum angustifolium	Honck.	1	1	—	0	0	0	10	5	255
Eriophorum angustifolium	Honck.	2	2	2	0	0	0	10	3	216
Eriophorum angustifolium	Honck.	4	1	—	417	417	417	3	5	96
Eriophorum gracile	Koch ex Roth	1	1	—	0	0	0	5	5	248
Eriophorum latifolium	Hoppe	1	8	—	0	0	0	10	5	149
Eriophorum russeolum	Fries ex Hartman	1	1	—	0	0	0	16	5	130
Eriophorum vaginatum	L.	1	1	—	0	0	0	10	5	63
Eriophorum vaginatum	L.	1	1	—	0	0	0	3	5	96
Eriophorum vaginatum	L.	1	4	—	80	80	80	30	5	180
Eriophorum vaginatum	L.	3	1	—	345	345	345	25	5	144
Eriophorum vaginatum	L.	3	1	—	443	443	443	30	5	78
Isolepis setacea	(L.) R. Br.	1	2	—	26	26	26	2	7	99

Species	Authority	Seed bank type	Number of records	Longevity (y)	Minimum density (seeds m⁻²)	Maximum density (seeds m⁻²)	Mean density (seeds m⁻²)	Depth (cm)	Method	Source code
Isolepis setacea	(L.) R. Br.	2	1	—	28	28	28	15	5	32
Isolepis setacea	(L.) R. Br.	2	1	—	78	78	78	2	7	99
Isolepis setacea	(L.) R. Br.	2	1	—	126	126	126	20	4	111
Isolepis setacea	(L.) R. Br.	2	2	—	82	204	143	5	5	222
Isolepis setacea	(L.) R. Br.	3	1	—	228	228	228	20	5	67
Isolepis setacea	(L.) R. Br.	3	2	—	377	1636	1007	30	5	45
Isolepis setacea	(L.) R. Br.	4	1	—	101	101	101	17.8	5	155
Isolepis setacea	(L.) R. Br.	4	1	—	480	480	480	5	5	69
Kobresia myosuroides	(Vill.) Fiori	1	1	—	0	0	0	5	5	55
Rhynchospora alba	(L.) M. Vahl	1	1	—	0	0	0	30	5	180
Rhynchospora alba	(L.) M. Vahl	2	1	—	23000	23000	23000	30	5	180
Rhynchospora alba	(L.) M. Vahl	3	1	—	710	710	710	30	5	180
Rhynchospora fusca	(L.) W.T. Aiton	2	1	—	4676	4676	4676	21	5	160
Rhynchospora fusca	(L.) W.T. Aiton	4	1	—	22	22	22	5	5	117
Schoenoplectus lacustris	(L.) Palla	1	1	—	0	0	0	5	5	248
Schoenoplectus tabernaemontani	(C. Gmelin) Palla	4	1	—	660	660	660	5	5	262
Schoenoplectus tabernaemontani	(C. Gmelin) Palla	4	1	—	1040	1040	1040	50	5	216
Schoenus ferrugineus	L.	4	2	—	120	201	161	12	5	174
Scirpoides holoschoenus	(L.) Sojak	1	1	—	0	0	0	3	5	132
Scirpoides holoschoenus	(L.) Sojak	2	1	—	7521	7521	7521	50	5	216
Scirpus sylvaticus	L.	1	4	—	0	0	0	10	5	149
Scirpus sylvaticus	L.	1	4	—	0	0	0	3	5	208
Scirpus sylvaticus	L.	1	2	—	0	0	0	6	5	208
Scirpus sylvaticus	L.	1	1	—	0	0	0	10	5	255
Scirpus sylvaticus	L.	2	2	—	3200	5600	4400	20	4	225
Scirpus sylvaticus	L.	3	1	—	85	85	85	6	5	208
Scirpus sylvaticus	L.	4	3	—	75	2238	884	10	5	255
Scirpus sylvaticus	L.	4	2	—	538	2547	1543	3	5	208
Scirpus sylvaticus	L.	4	2	—	3200	4000	3600	20	4	225
Trichophorum cespitosum	(L.) Hartman	1	2	—	0	0	0	22.5	5	45
Trichophorum cespitosum	(L.) Hartman	1	1	—	0	0	0	2	5	153
Trichophorum cespitosum	(L.) Hartman	1	2	—	0	0	0	17.8	5	156
Trichophorum cespitosum	(L.) Hartman	1	1	—	0	0	0	7	5	232
Trichophorum cespitosum	(L.) Hartman	2	1	—	112	112	112	5	5	92
Trichophorum cespitosum	(L.) Hartman	4	1	—	151	151	151	17.8	5	155
Trichophorum cespitosum	(L.) Hartman	4	1	—	1070	1070	1070	5	5	92
Diapensiaceae	—				—				—	
Diapensia lapponica	L.	1	1	—	0	0	0	3	4	158
Dioscoraceae	—				—				—	
Tamus communis	L.	1	2	—	0	0	0	3	5	96

Species	Author									
Tamus communis	L.	1	1	—	0	0	0	15	5	257
Dipsacaceae										
Dipsacus fullonum	L.	1	1	0	0	0	0	20	1	219
Dipsacus fullonum	L.	3	1	>5	0	0	0	7.5	2	195
Dipsacus fullonum	L.	4	1	—	67	67	67	3	5	96
Knautia arvensis	(L.) Coulter	1	1	—	0	0	0	6.5	5	68
Knautia arvensis	(L.) Coulter	1	1	—	0	0	0	13	5	68
Knautia arvensis	(L.) Coulter	1	1	—	0	0	0	3	5	96
Knautia arvensis	(L.) Coulter	1	2	—	0	0	0	20	5	97
Knautia arvensis	(L.) Coulter	1	1	—	0	0	0	10	5	110
Knautia arvensis	(L.) Coulter	1	1	—	0	0	0	5	6	170
Knautia arvensis	(L.) Coulter	1	3	—	0	0	0	6	5	181
Knautia arvensis	(L.) Coulter	1	1	—	0	0	0	10	5	181
Knautia arvensis	(L.) Coulter	1	2	—	0	0	0	6.5	7	182
Knautia arvensis	(L.) Coulter	1	1	—	0	0	0	13	5	183
Knautia arvensis	(L.) Coulter	1	13	—	0	0	0	10	5	255
Knautia arvensis	(L.) Coulter	3	3	>35	78	78	78	24	5	168
Knautia arvensis	(L.) Coulter	4	1	—	50	50	50	10	5	255
Knautia dipsacifolia	Kreutzer	1	1	—	0	0	0	10	5	255
Scabiosa columbaria	L.	1	1	—	0	0	0	6	5	48
Scabiosa columbaria	L.	1	1	—	0	0	0	6	5	67
Scabiosa columbaria	L.	1	1	—	0	0	0	3	5	96
Scabiosa columbaria	L.	2	2	—	0	0	0	6	5	181
Scabiosa columbaria	L.	1	1	—	0	0	0	10	5	181
Scabiosa columbaria	L.	1	3	—	0	0	0	10	5	255
Scabiosa columbaria	L.	1	1	—	4	4	4	3	7	235
Scabiosa columbaria	L.	2	1	>3	0	0	0	6	3	179
Scabiosa columbaria	L.	2	1	—	9	9	9	3	7	235
Scabiosa columbaria	L.	2	2	—	64	90	37	6.5	7	182
Scabiosa columbaria	L.	2	1	—	75	75	75	6	5	211
Scabiosa columbaria	L.	4	1	—	40	40	40	6	5	181
Scabiosa columbaria	L.	4	2	—	108	133	83	3	5	96
Scabiosa ochroleuca	L.	4	1	—	714	714	714	8.5	4	254
Succisa pratensis	Moench	1	1	—	0	0	0	6	5	48
Succisa pratensis	Moench	1	1	—	0	0	0	7	5	58
Succisa pratensis	Moench	1	4	—	0	0	0	10	5	149
Succisa pratensis	Moench	1	2	—	0	0	0	10	5	152
Succisa pratensis	Moench	1	2	—	0	0	0	12	5	174
Succisa pratensis	Moench	1	1	—	0	0	0	16	5	216
Succisa pratensis	Moench	1	1	—	0	0	0	10	5	252
Succisa pratensis	Moench	1	7	—	0	0	0	10	5	255
Succisa pratensis	Moench	1	1	—	3	3	3	3	7	235
Succisa pratensis	Moench	1	1	—	65	65	65	30	5	45
Succisa pratensis	Moench	1	1	—	200	200	200	30	5	180
Succisa pratensis	Moench	2	4	>1	0	0	0	—	1	275
Succisa pratensis	Moench	2	1	1	0	0	0	—	1	275

Species	Authority	Seed bank type	Number of records	Longevity (y)	Minimum density (seeds m⁻²)	Maximum density (seeds m⁻²)	Mean density (seeds m⁻²)	Depth (cm)	Method	Source code
Succisa pratensis	Moench	3	2	—	210	2200	1205	30	5	180
Succisa pratensis	Moench	4	3	—	12	49	37	12	5	174
Droseraceae	—		—	—	—	—	—	—	—	
Drosera intermedia	Hayne	2	1	—	2297	2297	2297	21	5	160
Drosera intermedia	Hayne	4	1	—	140	140	140	5	5	117
Drosera rotundifolia	L.	2	1	—	52	52	52	16	5	216
Elatinaceae	—		—	—	—	—	—	—	—	
Elatine triandra	Schkuhr	3	1	>50	0	0	0	50	3	115
Empetraceae	—		—	—	—	—	—	—	—	
Empetrum nigrum	L.	1	4	—	0	0	0	5	5	92
Empetrum nigrum	L.	1	1	—	0	0	0	3	5	96
Empetrum nigrum	L.	1	1	—	0	0	0	2	5	153
Empetrum nigrum	L.	1	1	—	0	0	0	3	4	158
Empetrum nigrum	L.	1	2	—	0	0	0	3	7	235
Empetrum nigrum	L.	4	2	—	13	13	13	3	4	158
Empetrum nigrum	L.	4	1	—	49	49	49	25	5	144
Empetrum nigrum subsp. *hermaphroditum*	(Hagerup) Bocher	3	1	>5	0	0	0	2	3	83
Empetrum nigrum subsp. *hermaphroditum*	(Hagerup) Bocher	4	1	—	15	15	15	3	5	82
Empetrum nigrum subsp. *hermaphroditum*	(Hagerup) Bocher	4	1	—	31	31	31	5	5	82
Empetrum nigrum subsp. *hermaphroditum*	(Hagerup) Bocher	4	2	—	243	396	320	8	5	253
Ericaceae	—		—	—	—	—	—	—	—	
Andromeda polifolia	L.	1	1	—	—	—	—	16	5	216
Arctostaphylos alpinus	(L.) Sprengel	1	2	—	0	0	0	20	5	191
Arctostaphylos uva-ursi	(L.) Sprengel	1	1	—	0	0	0	10	5	7
Arctostaphylos uva-ursi	(L.) Sprengel	1	1	—	0	0	0	5	5	76
Arctostaphylos uva-ursi	(L.) Sprengel	4	1	—	158	158	158	5	5	92
Calluna vulgaris	(L.) Hull	1	1	—	0	0	0	20	5	67
Calluna vulgaris	(L.) Hull	1	1	—	0	0	0	12	5	174
Calluna vulgaris	(L.) Hull	1	3	—	0	0	0	10	5	255
Calluna vulgaris	(L.) Hull	1	1	—	26	26	26	2	7	99
Calluna vulgaris	(L.) Hull	1	1	—	6523	6523	6523	15	5	45
Calluna vulgaris	(L.) Hull	2	1	>3	0	0	0	10	3	178
Calluna vulgaris	(L.) Hull	2	1	—	12	12	12	12	5	174
Calluna vulgaris	(L.) Hull	2	1	—	50	50	50	10	5	255
Calluna vulgaris	(L.) Hull	2	1	—	143	143	143	16	5	216
Calluna vulgaris	(L.) Hull	2	3	—	117	1600	639	3	5	96
Calluna vulgaris	(L.) Hull	2	1	—	929	929	929	3	7	235
Calluna vulgaris	(L.) Hull	2	1	—	1360	1360	1360	5	5	69

Species	Authority									
Calluna vulgaris	(L.) Hull	2	1	—	2013	2013	2013	15	5	45
Calluna vulgaris	(L.) Hull	2	2	—	3563	5952	4758	22.5	5	45
Calluna vulgaris	(L.) Hull	2	3	—	5580	11310	7740	30	5	180
Calluna vulgaris	(L.) Hull	2	3	—	6370	17900	11230	5	5	92
Calluna vulgaris	(L.) Hull	2	4	—	600	31060	14285	15	5	257
Calluna vulgaris	(L.) Hull	3	1	>5	39830	39830	39830	9	3	84
Calluna vulgaris	(L.) Hull	3	1	>30	0	0	0	2	5	83
Calluna vulgaris	(L.) Hull	3	3	>20	260	260	260	6	5	28
Calluna vulgaris	(L.) Hull	3	1	>20	112	1592	619	5	5	222
Calluna vulgaris	(L.) Hull	3	1	—	3122	3122	3122	5	5	222
Calluna vulgaris	(L.) Hull	3	11	>20	7060	7060	7060	15	5	265
Calluna vulgaris	(L.) Hull	3	1	—	160	62980	12655	15	5	257
Calluna vulgaris	(L.) Hull	3	1	>59	16372	16372	16372	7.5	5	84
Calluna vulgaris	(L.) Hull	3	6	>25	16926	16926	16926	9.5	5	84
Calluna vulgaris	(L.) Hull	3	1	>68	19032	19032	19032	12	5	84
Calluna vulgaris	(L.) Hull	3	1	—	1210	43230	21728	30	5	180
Calluna vulgaris	(L.) Hull	4	1	>30	26702	26702	26702	10	5	84
Calluna vulgaris	(L.) Hull	4	3	—	321	321	321	12	5	174
Calluna vulgaris	(L.) Hull	4	5	—	1024	1024	1024	5	5	69
Calluna vulgaris	(L.) Hull	4	6	—	517	1700	1192	2	5	153
Calluna vulgaris	(L.) Hull	4	6	—	38	3625	1300	10	5	255
Calluna vulgaris	(L.) Hull	4	4	—	735	2490	1580	5	5	222
Calluna vulgaris	(L.) Hull	4	5	—	203	6274	2003	17.8	5	155
Calluna vulgaris	(L.) Hull	4	6	—	76	9472	3650	17.8	5	156
Calluna vulgaris	(L.) Hull	4	1	—	15100	52900	37460	5	5	92
Calluna vulgaris	(L.) Hull	4	1	—	28000	68000	46167	5	5	154
Cassiope tetragona	(L.) D. Don	1	1	—	0	0	0	10	5	9
Cassiope tetragona	(L.) D. Don	1	1	—	0	0	0	5	5	71
Daboecia cantabrica	(Hudson) K. Koch	1	2	—	0	0	0	30	5	180
Erica cinerea	L.	1	1	—	1765	1765	1765	15	5	45
Erica cinerea	L.	1	1	—	3251	3251	3251	22.5	5	45
Erica cinerea	L.	2	4	—	341	1440	717	5	5	92
Erica cinerea	L.	2	1	—	1098	1098	1098	22.5	5	45
Erica cinerea	L.	3	1	—	7	7	7	10	5	93
Erica cinerea	L.	4	2	—	110	143	127	2	5	153
Erica cinerea	L.	4	2	—	101	1662	882	17.8	5	155
Erica cinerea	L.	4	2	—	758	1212	985	17.8	5	156
Erica cinerea	L.	4	3	—	906	1130	1023	5	5	92
Erica herbacea	L.	1	2	—	0	0	0	5	4	91
Erica tetralix	L.	1	1	—	0	0	0	15	5	45
Erica tetralix	L.	1	1	—	0	0	0	22.5	5	45
Erica tetralix	L.	1	2	—	0	0	0	5	5	92
Erica tetralix	L.	1	3	—	0	0	0	3	5	96
Erica tetralix	L.	1	1	—	0	0	0	2	5	153
Erica tetralix	L.	1	1	—	0	0	0	3	7	235
Erica tetralix	L.	2	2	>3	0	0	0	10	3	178
Erica tetralix	L.	2	3	—	56	143	95	5	5	92

Species	Authority	Seed bank type	Number of records	Longevity (y)	Minimum density (seeds m⁻²)	Maximum density (seeds m⁻²)	Mean density (seeds m⁻²)	Depth (cm)	Method	Source code
Erica tetralix	L.	3	2	>30	78	78	78	6	5	28
Erica tetralix	L.	3	1	>20	255	255	255	5	5	222
Erica tetralix	L.	4	1	—	295	295	295	5	5	92
Erica tetralix	L.	4	3	—	202	758	438	17.8	5	156
Erica tetralix	L.	4	6	—	101	1788	630	17.8	5	155
Erica tetralix	L.	4	5	—	357	6010	2433	5	5	222
Ledum palustre	L.	1	2	—	0	0	0	10	5	63
Ledum palustre	L.	1	1	—	0	0	0	5	5	82
Ledum palustre	L.	1	2	—	0	0	0	8	5	253
Ledum palustre	L.	2	1	—	1295	1295	1295	25	5	144
Rhododendron lapponicum	(L.) Wahlenb.	1	2	—	0	0	0	20	5	191
Rhododendron ponticum	L.	1	2	—	0	0	0	15	5	257
Vaccinium myrtillus	L.	1	2	—	0	0	0	15	5	45
Vaccinium myrtillus	L.	1	1	—	0	0	0	12	5	84
Vaccinium myrtillus	L.	1	4	—	0	0	0	5	5	92
Vaccinium myrtillus	L.	1	2	—	0	0	0	3	5	96
Vaccinium myrtillus	L.	1	1	—	0	0	0	5	5	123
Vaccinium myrtillus	L.	1	3	—	0	0	0	2	5	153
Vaccinium myrtillus	L.	1	1	—	0	0	0	17.8	5	156
Vaccinium myrtillus	L.	1	4	—	0	0	0	5	6	170
Vaccinium myrtillus	L.	1	2	—	0	0	0	12	5	171
Vaccinium myrtillus	L.	1	1	—	0	0	0	30	5	180
Vaccinium myrtillus	L.	1	2	—	0	0	0	16	5	216
Vaccinium myrtillus	L.	1	3	—	0	0	0	7	5	232
Vaccinium myrtillus	L.	1	1	—	0	0	0	3	7	235
Vaccinium myrtillus	L.	1	7	—	0	0	0	10	5	255
Vaccinium myrtillus	L.	1	7	—	0	0	0	15	5	257
Vaccinium myrtillus	L.	1	2	—	0	0	0	5	5	264
Vaccinium myrtillus	L.	2	1	—	11	11	11	3	7	235
Vaccinium myrtillus	L.	2	4	—	80	2780	1510	15	5	257
Vaccinium myrtillus	L.	3	1	>5	0	0	0	2	3	83
Vaccinium myrtillus	L.	4	2	—	51	51	51	3	5	82
Vaccinium myrtillus	L.	4	1	—	41	87	64	5	5	92
Vaccinium myrtillus	L.	4	1	—	112	112	112	2	5	82
Vaccinium myrtillus	L.	4	3	—	126	455	244	17.8	5	156
Vaccinium myrtillus	L.	4	3	—	250	485	403	5	5	82
Vaccinium myrtillus	L.	4	3	—	301	605	428	17.8	5	155
Vaccinium oxycoccos	L.	1	1	—	0	0	0	3	5	96
Vaccinium oxycoccos	L.	1	2	—	0	0	0	30	5	180
Vaccinium oxycoccos	L.	1	1	—	0	0	0	16	5	216
Vaccinium oxycoccos	L.	4	2	—	1667	7400	4534	3	5	96
Vaccinium uliginosum	L.	1	1	—	0	0	0	25	5	144

Species	Authority									
Vaccinium uliginosum	L.	1	1	—	0	0	0	3	4	158
Vaccinium uliginosum	L.	1	1	—	0	0	0	30	5	180
Vaccinium uliginosum	L.	1	2	—	0	0	0	20	5	191
Vaccinium uliginosum	L.	4	1	—	16	16	16	8	5	253
Vaccinium vitis-idaea	L.	1	2	—	0	0	0	10	5	63
Vaccinium vitis-idaea	L.	1	1	—	0	0	0	5	5	76
Vaccinium vitis-idaea	L.	1	2	—	0	0	0	5	5	82
Vaccinium vitis-idaea	L.	1	7	—	0	0	0	5	5	92
Vaccinium vitis-idaea	L.	1	1	—	0	0	0	3	5	96
Vaccinium vitis-idaea	L.	1	1	—	0	0	0	25	5	144
Vaccinium vitis-idaea	L.	1	2	—	0	0	0	2	5	153
Vaccinium vitis-idaea	L.	1	2	—	0	0	0	3	4	158
Vaccinium vitis-idaea	L.	1	4	—	0	0	0	5	6	170
Vaccinium vitis-idaea	L.	1	2	—	0	0	0	12	5	171
Vaccinium vitis-idaea	L.	1	3	—	0	0	0	30	5	180
Vaccinium vitis-idaea	L.	1	2	—	0	0	0	16	5	216
Vaccinium vitis-idaea	L.	1	1	—	0	0	0	7	5	232
Vaccinium vitis-idaea	L.	1	2	—	0	0	0	10	5	255
Vaccinium vitis-idaea	L.	3	1	>5	0	0	0	2	3	83
Vaccinium vitis-idaea	L.	4	1	—	6	6	6	10	5	7
Vaccinium vitis-idaea	L.	4	1	—	13	13	13	3	4	158
Vaccinium vitis-idaea	L.	4	1	—	15	15	15	2	5	82
Vaccinium vitis-idaea	L.	4	2	—	10	29	20	8	5	253
Vaccinium vitis-idaea	L.	4	1	—	41	41	41	5	5	82
Eriocaulaceae	—			—			—		—	—
Eriocaulon aquaticum	(Hill) Druce	2	1	—	5351	5351	5351	21	5	160
Eriocaulon aquaticum	(Hill) Druce	4	1	—	702	702	702	5	5	117
Euphorbiaceae	—			—			—		—	—
Euphorbia amygdaloides	L.	3	2	—	60	80	70	15	5	257
Euphorbia brittingeri	Opiz ex Samp.	1	1	—	0	0	0	6	5	181
Euphorbia brittingeri	Opiz ex Samp.	1	2	—	0	0	0	6.5	7	182
Euphorbia brittingeri	Opiz ex Samp.	2	1	—	8	8	8	6.5	7	182
Euphorbia cyparissias	L.	1	1	—	0	0	0	6.5	7	182
Euphorbia cyparissias		2	1	—	13	13	13	6.5	5	182
Euphorbia cyparissias		2	1	>20	112	112	112	13	5	183
Euphorbia cyparissias		3	1	>30	96	96	96	13	5	183
Euphorbia cyparissias		3	1	—	136	136	136	10	5	181
Euphorbia cyparissias		4	3	—	48	304	133	6	5	181
Euphorbia cyparissias		4	1	>1	408	408	408	8.5	4	254
Euphorbia esula	L.	2	1	>3	4433	4433	4433	2.5	4	26
Euphorbia esula	L.	2	1	>4	11383	11383	11383	2.5	4	26
Euphorbia esula	L.	3	1	>4	0	0	0	15	1	34
Euphorbia esula	L.	3	1	>8	0	0	0	45	1	34
Euphorbia esula	L.	3	1	>4	15	15	15	2.5	4	26
Euphorbia esula	L.	3	1	—	3522	3522	3522	2.5	4	26

Species	Authority	Seed bank type	Number of records	Longevity (y)	Minimum density (seeds m⁻²)	Maximum density (seeds m⁻²)	Mean density (seeds m⁻²)	Depth (cm)	Method	Source code
Euphorbia esula	L.	4	3	—	4860	15150	8488	2.5	4	26
Euphorbia esula	L.	4	1	>3	11383	11383	11383	2.5	4	26
Euphorbia exigua	L.	1	1	—	0	0	0	10	4	22
Euphorbia exigua	L.	1	1	>3	0	0	0	25	4	36
Euphorbia exigua	L.	2	1	>3	0	0	0	10	3	135
Euphorbia exigua	L.	2	1	>3	0	0	0	25	3	135
Euphorbia exigua	L.	3	1	>5	0	0	0	30	3	14
Euphorbia exigua	L.	3	1	>30	17	17	17	6	5	211
Euphorbia exigua	L.	3	1	—	258	258	258	30	5	29
Euphorbia exigua	L.	3	1	—	303	303	303	30	4	12
Euphorbia exigua	L.	3	1	>10	495	495	495	30	5	29
Euphorbia exigua	L.	3	1	—	1820	1820	1820	12	5	81
Euphorbia exigua	L.	4	7	—	9	536	159	15	4	30
Euphorbia exigua	L.	4	7	—	33	880	215	10	5	15
Euphorbia exigua	L.	4	1	—	240	240	240	10	5	86
Euphorbia exigua	L.	4	1	—	256	256	256	25	4	136
Euphorbia exigua	L.	4	34	—	20	8997	475	30	4	13
Euphorbia exigua	L.	4	1	—	510	510	510	25	4	3
Euphorbia helioscopia	L.	1	2	—	0	0	0	10	4	22
Euphorbia helioscopia	L.	1	1	—	53	53	53	8	7	275
Euphorbia helioscopia	L.	2	1	—	25	25	25	10	4	22
Euphorbia helioscopia	L.	2	1	—	151	151	151	17.8	5	155
Euphorbia helioscopia	L.	2	1	—	350	350	350	10	5	255
Euphorbia helioscopia	L.	3	1	>5	0	0	0	2.5	1	203
Euphorbia helioscopia	L.	3	1	>5	0	0	0	2.5	2	203
Euphorbia helioscopia	L.	3	1	>5	0	0	0	7.5	1	203
Euphorbia helioscopia	L.	3	1	>5	0	0	0	7.5	2	203
Euphorbia helioscopia	L.	3	1	>5	0	0	0	15	1	203
Euphorbia helioscopia	L.	3	1	>5	0	0	0	15	2	203
Euphorbia helioscopia	L.	3	1	>19	11	11	11	6	5	211
Euphorbia helioscopia	L.	3	1	—	43	43	43	30	5	45
Euphorbia helioscopia	L.	3	1	>20	44	44	44	24	5	168
Euphorbia helioscopia	L.	3	1	—	96	96	96	70	5	219
Euphorbia helioscopia	L.	3	1	—	126	126	126	17.8	5	155
Euphorbia helioscopia	L.	4	1	—	7	7	7	20	4	111
Euphorbia helioscopia	L.	4	9	—	26	366	103	30	4	13
Euphorbia peplis	L.	1	2	—	0	0	0	10	4	22
Euphorbia peplus	L.	2	1	—	200	200	200	30	5	162
Euphorbia peplus	L.	2	1	—	226	226	226	65	5	162
Euphorbia peplus	L.	2	1	—	250	250	250	17	5	162
Euphorbia peplus	L.	2	1	—	375	375	375	32	5	162
Euphorbia peplus	L.	2	1	—	487	487	487	57	5	162

THE DATABASE / 113

Species	Authority									
Euphorbia peplus	L.	2	3	—	875	875	875	10	5	162
Euphorbia peplus	L.	3	1	>20	100	100	100	15	5	162
Euphorbia platyphyllos	L.	4	2	—	86	380	233	30	4	13
Euphorbia portlandica	L.	3	1	—	86	86	86	30	5	29
Mercurialis annua	L.	1	1	—	0	0	0	10	4	22
Mercurialis annua	L.	1	1	—	0	0	0	30	5	66
Mercurialis annua	L.	1	2	—	53	105	79	8	7	275
Mercurialis annua	L.	2	1	>2	0	0	0	20	1	219
Mercurialis annua	L.	2	1	—	15	15	15	10	4	22
Mercurialis annua	L.	2	2	—	42	42	42	15	5	272
Mercurialis annua	L.	2	1	—	135	135	135	8	7	275
Mercurialis annua	L.	3	1	>30	85	85	85	6	5	211
Mercurialis annua	L.	3	1	>12	180	180	180	50	5	219
Mercurialis annua	L.	3	1	—	320	320	320	70	5	219
Mercurialis annua	L.	3	1	—	324	324	324	30	5	219
Mercurialis annua	L.	3	1	—	800	800	800	50	5	66
Mercurialis annua	L.	3	1	—	1042	1042	1042	20	5	97
Mercurialis annua	L.	4	1	—	15	15	15	10	4	22
Mercurialis annua	L.	4	3	—	20	179	75	30	4	13
Mercurialis annua	L.	4	1	—	1013	1013	1013	15	5	207
Mercurialis perennis	L.	1	1	—	0	0	0	15	5	32
Mercurialis perennis	L.	1	2	—	0	0	0	6	5	49
Mercurialis perennis	L.	1	3	—	0	0	0	3	5	96
Mercurialis perennis	L.	1	1	—	0	0	0	10	5	126
Mercurialis perennis	L.	1	1	—	0	0	0	12.5	5	126
Mercurialis perennis	L.	1	1	—	0	0	0	30	5	173
Mercurialis perennis	L.	1	1	—	0	0	0	10	5	181
Mercurialis perennis	L.	1	2	—	0	0	0	5	5	220
Mercurialis perennis	L.	1	2	—	0	0	0	3	7	235
Mercurialis perennis	L.	1	5	—	0	0	0	15	5	257
Fagaceae										
Castanea sativa	Miller	1	1	—	0	0	0	15	5	257
Fagus sylvatica	L.	1	1	—	0	0	0	20	5	67
Fagus sylvatica	L.	1	4	—	0	0	0	3	5	96
Fagus sylvatica	L.	1	2	—	0	0	0	16	5	168
Fagus sylvatica	L.	1	1	—	0	0	0	24	5	168
Fagus sylvatica	L.	1	3	—	0	0	0	15	5	257
Quercus petraea	(Mattuschka) Liebl.	1	3	—	0	0	0	10	5	255
Quercus petraea	(Mattuschka) Liebl.	1	4	—	0	0	0	15	5	257
Quercus robur	L.	1	2	—	0	0	0	3	5	96
Quercus robur	L.	1	1	—	0	0	0	10	5	152
Quercus robur	L.	1	2	—	0	0	0	4	5	176
Quercus robur	L.	1	2	—	0	0	0	16	5	216
Quercus robur	L.	1	1	—	0	0	0	5	5	248
Quercus robur	L.	1	2	—	0	0	0	10	5	255

Fumariaceae

Species	Authority	Seed bank type	Number of records	Longevity (y)	Minimum density (seeds m⁻²)	Maximum density (seeds m⁻²)	Mean density (seeds m⁻²)	Depth (cm)	Method	Source code
Ceratocapnos claviculata	(L.) Liden	4	1	—	5	5	5	15	5	32
Corydalis cava	(L.) Schweigger & Koerte	1	1	—	0	0	0	12.5	5	126
Corydalis intermedia	(L.) Mérat	1	1	—	0	0	0	10	5	126
Corydalis solida	(L.) Clairv.	1	1	—	0	0	0	10	5	172
Fumaria capreolata	L.	1	1	—	0	0	0	10	4	22
Fumaria capreolata	L.	4	1	—	40	40	40	10	4	22
Fumaria officinalis	L.	1	1	—	0	0	0	20	4	111
Fumaria officinalis	L.	1	1	—	0	0	0	20	5	111
Fumaria officinalis	L.	2	2	—	38	75	57	10	5	255
Fumaria officinalis	L.	2	1	—	2217	2217	2217	30	5	45
Fumaria officinalis	L.	3	1	>5	0	0	0	7.5	2	196
Fumaria officinalis	L.	3	1	>5	0	0	0	2.5	1	203
Fumaria officinalis	L.	3	1	>5	0	0	0	2.5	2	203
Fumaria officinalis	L.	3	1	>5	0	0	0	7.5	1	203
Fumaria officinalis	L.	3	1	>5	0	0	0	7.5	2	203
Fumaria officinalis	L.	3	1	>5	0	0	0	15	1	203
Fumaria officinalis	L.	3	1	>5	0	0	0	15	2	203
Fumaria officinalis	L.	3	1	>11	0	0	0	25	3	209
Fumaria officinalis	L.	3	1	>660	17	17	17	55	5	161
Fumaria officinalis	L.	3	1	—	24	24	24	100	5	219
Fumaria officinalis	L.	3	1	>660	25	25	25	50	5	161
Fumaria officinalis	L.	3	1	>20	92	92	92	20	5	40
Fumaria officinalis	L.	3	1	>40	97	97	97	30	5	45
Fumaria officinalis	L.	3	1	—	104	104	104	70	5	219
Fumaria officinalis	L.	3	1	>60	140	140	140	30	5	45
Fumaria officinalis	L.	3	1	>24	194	194	194	30	5	45
Fumaria officinalis	L.	3	1	>8	258	258	258	30	5	45
Fumaria officinalis	L.	4	1	—	20	20	20	30	4	13
Fumaria officinalis	L.	4	1	—	50	50	50	10	5	255
Fumaria officinalis	L.	4	1	—	716	716	716	15	5	207

Gentianaceae

Species	Authority	Seed bank type	Number of records	Longevity (y)	Minimum density (seeds m⁻²)	Maximum density (seeds m⁻²)	Mean density (seeds m⁻²)	Depth (cm)	Method	Source code
Blackstonia perfoliata	(L.) Hudson	4	1	—	167	167	167	3	5	96
Centaurium erythraea	Rafn	2	1	—	5	5	5	15	5	32
Centaurium erythraea	Rafn	2	1	—	27	27	27	6	5	49
Centaurium erythraea	Rafn	2	1	—	72	72	72	20	5	67
Centaurium erythraea	Rafn	2	1	—	367	367	367	3	5	96
Centaurium erythraea	Rafn	4	2	—	72	82	77	10	5	110
Centaurium erythraea	Rafn	4	1	—	167	167	167	1	5	96
Centaurium littorale	Rafn	4	1	—	1888	1888	1888	5	5	69
Centaurium littorale	(Turner ex Smith) Gilmour	3	1	—	50000	50000	50000	20	5	210

Species	Authority									
Centaurium pulchellum	(Sw.) Druce	2	2	—	28	59	44	15	5	272
Centaurium pulchellum	(Sw.) Druce	2	1	—	120	120	120	6.5	5	128
Centaurium pulchellum	(Sw.) Druce	3	2	—	2500	5000	3750	20	5	210
Centaurium pulchellum	(Sw.) Druce	4	1	—	28	28	28	15	5	272
Centaurium pulchellum	(Sw.) Druce	4	1	—	720	720	720	6.5	5	128
Gentiana nivalis	L.	1	1	—	0	0	0	5	5	55
Gentiana nivalis	L.	1	2	—	0	0	0	5	4	91
Gentiana punctata	L.	1	2	—	0	0	0	5	4	91
Gentiana verna	L.	1	1	—	0	0	0	6.5	5	55
Gentiana verna	L.	1	2	—	0	0	0	5	7	182
Gentianella amarella	(L.) Boerner	4	3	—	22	157	72	10	5	122
Gentianella amarella	(L.) Boerner	4	2	—	88	2522	1305	1.5	5	110
Gentianella amarella	(L.) Boerner	1	1	—	0	0	0	5	5	121
Gentianella aspera	(Hegetschw. & Heer) Dost 1 ex Skalicky	1	2	—	0	0	0	6	5	55
Gentianella ciliata	(L.) Borkh.	1	2	—	0	0	0	10	5	181
Gentianella ciliata	(L.) Borkh.	1	1	—	0	0	0	5	5	255
Gentianella germanica	(Willd.) Boerner	1	1	—	0	0	0	6.5	5	55
Gentianella germanica	(Willd.) Boerner	1	1	—	0	0	0	10	7	182
Gentianella germanica	(Willd.) Boerner	1	1	—	51	51	51	6	5	255
Gentianella germanica	(Willd.) Boerner	1	1	—	0	0	0	6	5	211
Gentianella germanica	(Willd.) Boerner	2	2	>3	56	64	60	6	3	179
Gentianella germanica	(Willd.) Boerner	4	1	—	0	0	0	5	5	181
Gentianella tenella	(Rottb.) Boerner	1	1	—	0	0	0	5	5	55

Geraniaceae

Species	Authority									
Erodium cicutarium	(L.) L'Her.	1	1	—	0	0	0	1	5	96
Erodium cicutarium	(L.) L'Her.	1	1	—	0	0	0	20	4	111
Erodium cicutarium	(L.) L'Her.	1	1	—	0	0	0	20	5	111
Erodium cicutarium	(L.) L'Her.	1	3	—	0	0	0	3	5	132
Erodium cicutarium	(L.) L'Her.	3	1	>5	0	0	0	7.5	2	195
Erodium cicutarium	(L.) L'Her.	4	1	—	38	38	38	3	5	132
Erodium cicutarium	(L.) L'Her.	4	1	—	83	83	83	3	5	96
Erodium moschatum	(L.) L'Her.	1	1	—	0	0	0	10	4	22
Geranium dissectum	L.	1	1	—	0	0	0	25	5	2
Geranium dissectum	L.	1	3	—	0	0	0	10	4	22
Geranium dissectum	L.	2	1	0	0	0	0	2	5	95
Geranium dissectum	L.	2	2	1	0	0	0	25	1	125
Geranium dissectum	L.	2	1	1	0	0	0	26	1	133
Geranium dissectum	L.	2	1	4	0	0	0	39	1	133
Geranium dissectum	L.	3	2	>5	0	0	0	13	1	133
Geranium dissectum	L.	4	2	—	38	75	57	10	5	255
Geranium dissectum	L.	4	3	—	0	0	0	7.5	2	199
Geranium dissectum	L.	4	1	—	48	48	48	5	5	69
Geranium dissectum	L.	4	3	—	66	766	302	30	4	13
Geranium dissectum	L.	3	1	—	700	700	700	10	5	140
Geranium endressii	Gay	3	1	—	3100	3100	3100	20	5	69
Geranium molle	L.	1	4	—	0	0	0	3	5	96

Species	Authority	Seed bank type	Number of records	Longevity (y)	Minimum density (seeds m^{-2})	Maximum density (seeds m^{-2})	Mean density (seeds m^{-2})	Depth (cm)	Method	Source code
Geranium molle	L.	1	1	—	0	0	0	20	4	111
Geranium molle	L.	1	1	—	0	0	0	20	5	111
Geranium molle	L.	1	1	0	0	0	0	13	1	133
Geranium molle	L.	1	2	0	0	0	0	26	1	133
Geranium molle	L.	1	1	0	0	0	0	39	1	133
Geranium molle	L.	1	1	—	0	0	0	10	5	252
Geranium molle	L.	1	1	—	53	53	53	8	7	275
Geranium molle	L.	3	1	>5	0	0	0	7.5	2	199
Geranium palustre	L.	1	1	—	0	0	0	10	5	255
Geranium pratense	L.	1	1	—	0	0	0	3	5	96
Geranium pratense	L.	1	1	—	0	0	0	13	5	183
Geranium pratense	L.	2	2	2	0	0	0	7.5	2	199
Geranium pusillum	L.	1	1	—	0	0	0	25	4	36
Geranium pusillum	L.	1	1	—	0	0	0	3	5	96
Geranium pusillum	L.	3	1	21	0	0	0	25	1	125
Geranium pusillum	L.	3	3	>5	0	0	0	7.5	2	198
Geranium pusillum	L.	3	1	>30	75	75	75	40	5	162
Geranium pusillum	L.	4	2	—	80	768	424	5	5	69
Geranium pyrenaicum	Burman f.	2	1	4	0	0	0	7.5	2	199
Geranium robertianum	L.	1	2	—	0	0	0	6	5	49
Geranium robertianum	L.	1	1	—	0	0	0	3	5	96
Geranium robertianum	L.	1	1	—	0	0	0	10	5	126
Geranium robertianum	L.	1	1	—	0	0	0	12.5	5	126
Geranium robertianum	L.	1	1	—	0	0	0	4	5	176
Geranium robertianum	L.	1	1	—	0	0	0	20	4	225
Geranium robertianum	L.	1	1	—	0	0	0	15	5	257
Geranium robertianum	L.	2	1	3	0	0	0	7.5	2	199
Geranium robertianum	L.	2	4	—	2	5	4	5	5	19
Geranium robertianum	L.	2	1	—	49	49	49	11.5	5	134
Geranium robertianum	L.	3	1	>5	0	0	0	7.5	2	199
Geranium robertianum	L.	4	1	—	9	9	9	5	5	19
Geranium rotundifolium	L.	1	1	—	0	0	0	30	5	66
Geranium sanguineum	L.	1	12	—	0	0	0	10	5	152
Geranium sanguineum	L.	1	1	—	0	0	0	3	7	235
Geranium sylvaticum	L.	1	1	—	0	0	0	5	5	82
Geranium sylvaticum	L.	1	8	—	0	0	0	10	5	150
Geranium sylvaticum	L.	1	1	—	0	0	0	5	6	170
Geranium sylvaticum	L.	1	1	—	0	0	0	12	5	171
Geranium sylvaticum	L.	1	4	—	0	0	0	10	5	255
Geranium sylvaticum	L.	4	1	—	6	6	6	5	5	213

Gramineae (Poaceae)											
Agrostis canina	L	1	2	—	0	0	0	5	5	92
Agrostis canina	L	1	1	—	0	0	0	3	5	96
Agrostis canina	L	1	5	—	0	0	0	2	7	99
Agrostis canina	L	1	5	—	0	0	0	6	5	99
Agrostis canina	L	1	1	—	0	0	0	2	5	153
Agrostis canina	L	1	3	—	0	0	0	3	5	208
Agrostis canina	L	1	1	—	0	0	0	5	5	222
Agrostis canina	L	1	1	—	0	0	0	20	4	225
Agrostis canina	L	1	1	—	18	18	18	10	5	93
Agrostis canina	L	1	2	—	375	375	375	10	4	88
Agrostis canina	L	2	1	—	47	47	47	15	5	32
Agrostis canina	L	2	1	—	56	56	56	3	7	235
Agrostis canina	L	2	3	—	51	122	85	5	5	222
Agrostis canina	L	2	3	—	50	183	94	3	5	96
Agrostis canina	L	2	1	—	112	112	112	5	5	92
Agrostis canina	L	2	1	—	572	572	572	40	5	11
Agrostis canina	L	2	5	—	594	1557	1127	6	5	208
Agrostis canina	L	2	2	—	2300	2600	2450	20	5	4
Agrostis canina	L	2	1	—	5980	5980	5980	6	5	11
Agrostis canina	L	3	1	—	679	679	679	6	5	208
Agrostis canina	L	3	1	—	1377	1377	1377	6	5	48
Agrostis canina	L	3	1	—	1850	1850	1850	16	5	216
Agrostis canina	L	4	1	—	100	100	100	5	5	98
Agrostis canina	L	4	1	—	313	313	313	5	4	88
Agrostis canina	L	4	4	—	153	469	321	5	5	222
Agrostis canina	L	4	3	—	50	1067	439	3	5	96
Agrostis canina	L	4	4	—	300	700	500	15	5	249
Agrostis canina	L	4	4	—	158	1180	528	5	5	92
Agrostis canina	L	4	3	—	651	736	689	3	5	208
Agrostis canina	L	4	1	—	780	780	780	50	5	216
Agrostis capillaris	L	1	1	—	0	0	0	40	5	11
Agrostis capillaris	L	1	1	—	0	0	0	6.5	5	68
Agrostis capillaris	L	1	5	—	0	0	0	5	5	69
Agrostis capillaris	L	1	5	—	0	0	0	3	5	96
Agrostis capillaris	L	1	5	—	0	0	0	2	7	99
Agrostis capillaris	L	1	5	—	0	0	0	6	5	99
Agrostis capillaris	L	1	1	—	0	0	0	5	6	170
Agrostis capillaris	L	1	1	—	0	0	0	12	5	171
Agrostis capillaris	L	1	3	—	0	0	0	12	5	174
Agrostis capillaris	L	1	3	—	0	0	0	20	4	225
Agrostis capillaris	L	1	1	—	0	0	0	7	5	232
Agrostis capillaris	L	1	3	—	156	156	156	10	5	27
Agrostis capillaris	L	1	1	—	208	728	468	6	5	11
Agrostis capillaris	L	2	1	4	0	0	0	13	1	133
Agrostis capillaris	L	2	2	4	0	0	0	26	1	133

Species	Authority	Seed bank type	Number of records	Longevity (y)	Minimum density (seeds m^{-2})	Maximum density (seeds m^{-2})	Mean density (seeds m^{-2})	Depth (cm)	Method	Source code
Agrostis capillaris	L.	2	1	4	0	0	0	39	1	133
Agrostis capillaris	L.	2	1	—	13	13	13	5	5	234
Agrostis capillaris	L.	2	6	—	10	65	26	3	7	235
Agrostis capillaris	L.	2	1	—	51	51	51	6	5	49
Agrostis capillaris	L.	2	2	—	41	112	77	5	5	222
Agrostis capillaris	L.	2	1	—	102	102	102	12	5	174
Agrostis capillaris	L.	2	1	—	108	108	108	13	5	68
Agrostis capillaris	L.	2	1	—	128	128	128	9.5	5	84
Agrostis capillaris	L.	2	16	—	83	483	190	3	5	96
Agrostis capillaris	L.	2	1	—	304	304	304	20	5	67
Agrostis capillaris	L.	2	1	—	359	359	359	6	5	48
Agrostis capillaris	L.	2	12	—	38	2163	502	10	5	255
Agrostis capillaris	L.	2	3	—	156	1768	745	6	5	11
Agrostis capillaris	L.	2	1	—	1053	1053	1053	9	5	84
Agrostis capillaris	L.	2	2	—	1755	2123	1939	6	5	208
Agrostis capillaris	L.	2	2	—	180	5300	2740	15	5	257
Agrostis capillaris	L.	3	1	>7	0	0	0	0	3	188
Agrostis capillaris	L.	3	1	—	108	108	108	12.5	5	126
Agrostis capillaris	L.	3	1	>40	120	120	120	20	5	173
Agrostis capillaris	L.	3	2	—	117	183	150	3	5	96
Agrostis capillaris	L.	3	1	—	215	215	215	30	5	29
Agrostis capillaris	L.	3	4	—	80	760	425	15	5	257
Agrostis capillaris	L.	3	1	—	700	700	700	20	5	69
Agrostis capillaris	L.	3	1	—	1896	1896	1896	6	5	208
Agrostis capillaris	L.	4	2	—	17	23	20	7	5	58
Agrostis capillaris	L.	4	3	—	9	174	65	12	5	174
Agrostis capillaris	L.	4	1	—	179	179	179	5	4	53
Agrostis capillaris	L.	4	2	—	122	429	276	5	5	222
Agrostis capillaris	L.	4	1	—	287	287	287	2	5	153
Agrostis capillaris	L.	4	1	—	379	379	379	5	5	233
Agrostis capillaris	L.	4	1	—	676	676	676	6	5	99
Agrostis capillaris	L.	4	13	—	50	3217	717	3	5	96
Agrostis capillaris	L.	4	5	—	17	3698	965	7	5	232
Agrostis capillaris	L.	4	2	—	611	3373	1992	5	5	123
Agrostis capillaris	L.	4	8	—	300	4400	2013	15	5	249
Agrostis capillaris	L.	4	1	—	2337	2337	2337	15	5	267
Agrostis capillaris	L.	4	6	—	397	5310	2454	5	5	92
Agrostis capillaris	L.	4	34	—	88	17750	5237	10	5	255
Agrostis capillaris	L.	4	5	—	48	16544	6118	5	5	69
Agrostis castellana	Boiss. & Reuter	2	1	—	125	125	125	3	5	132
Agrostis gigantea	Roth	1	1	—	0	0	0	3	5	96
Agrostis gigantea	Roth	1	1	—	0	0	0	20	4	225

Species	Authority									
Agrostis gigantea	Roth	2	1	1	0	0	0	0	1	75
Agrostis gigantea	Roth	2	1	1	0	0	0	0	2	75
Agrostis gigantea	Roth	2	1	1	0	0	0	5	1	75
Agrostis gigantea	Roth	2	1	>2	0	0	0	5	2	194
Agrostis gigantea	Roth	2	1	—	100	100	100	7.5	5	96
Agrostis gigantea	Roth	3	1	>4	0	0	0	3	3	266
Agrostis gigantea	Roth	4	2	—	333	467	400	0	5	96
Agrostis gigantea	Roth	4	2	—	2833	4833	3833	3	5	74
Agrostis scabra	Willd.	1	2	—	1	1	1	20	4	127
Agrostis scabra	Willd.	2	2	—	88	236	162	5	5	76
Agrostis scabra	Willd.	4	1	—	444	444	444	5	5	117
Agrostis stolonifera	L.	1	2	—	0	0	0	10	4	8
Agrostis stolonifera	L.	1	3	—	0	0	0	25	4	36
Agrostis stolonifera	L.	1	5	—	0	0	0	5	5	53
Agrostis stolonifera	L.	1	7	—	0	0	0	1	5	57
Agrostis stolonifera	L.	1	5	—	0	0	0	5	5	69
Agrostis stolonifera	L.	1	7	—	0	0	0	3	5	96
Agrostis stolonifera	L.	1	12	—	0	0	0	20	5	97
Agrostis stolonifera	L.	1	2	—	0	0	0	6	5	99
Agrostis stolonifera	L.	1	4	—	0	0	0	10	5	110
Agrostis stolonifera	L.	1	3	—	0	0	0	7	5	112
Agrostis stolonifera	L.	1	1	—	0	0	0	11.5	5	134
Agrostis stolonifera	L.	1	3	—	0	0	0	3	5	208
Agrostis stolonifera	L.	1	1	—	0	0	0	20	4	225
Agrostis stolonifera	L.	1	1	—	0	0	0	3	7	235
Agrostis stolonifera	L.	1	3	—	0	0	0	5	5	244
Agrostis stolonifera	L.	1	1	—	0	0	0	10	5	252
Agrostis stolonifera	L.	1	5	—	0	0	0	10	5	255
Agrostis stolonifera	L.	1	1	—	0	0	0	25	5	260
Agrostis stolonifera	L.	1	3	—	0	120	71	2	7	99
Agrostis stolonifera	L.	2	5	—	150	150	150	20	5	4
Agrostis stolonifera	L.	2	1	—	156	156	156	10	5	27
Agrostis stolonifera	L.	2	4	—	368	396	382	6	5	208
Agrostis stolonifera	L.	2	6	—	364	572	468	6	5	11
Agrostis stolonifera	L.	2	2	—	13	13	13	5	5	234
Agrostis stolonifera	L.	2	1	—	12	29	20	3	7	235
Agrostis stolonifera	L.	2	2	—	33	59	46	5	5	123
Agrostis stolonifera	L.	2	2	—	135	135	135	15	5	32
Agrostis stolonifera	L.	2	1	—	160	160	160	6.5	5	128
Agrostis stolonifera	L.	2	9	—	50	300	185	3	5	96
Agrostis stolonifera	L.	2	1	—	235	235	235	7	5	112
Agrostis stolonifera	L.	2	1	—	416	416	416	10	5	27
Agrostis stolonifera	L.	2	12	—	38	4050	469	10	5	255
Agrostis stolonifera	L.	2	4	—	600	600	600	10	4	20
Agrostis stolonifera	L.	2	1	—	624	624	624	30	5	29
Agrostis stolonifera	L.	2	1	—	625	625	625	20	5	97

Species	Authority	Seed bank type	Number of records	Longevity (y)	Minimum density (seeds m⁻²)	Maximum density (seeds m⁻²)	Mean density (seeds m⁻²)	Depth (cm)	Method	Source code
Agrostis stolonifera	L.	2	4	—	260	1352	702	6	5	11
Agrostis stolonifera	L.	2	2	—	900	7300	4100	1	5	57
Agrostis stolonifera	L.	3	1	—	99	99	99	11.5	5	134
Agrostis stolonifera	L.	3	1	—	200	200	200	20	5	4
Agrostis stolonifera	L.	3	1	—	285	285	285	30	5	107
Agrostis stolonifera	L.	3	1	>14	416	416	416	6	5	11
Agrostis stolonifera	L.	3	1	—	416	416	416	10	5	27
Agrostis stolonifera	L.	3	1	—	474	474	474	30	5	29
Agrostis stolonifera	L.	3	1	—	570	570	570	50	5	107
Agrostis stolonifera	L.	3	1	—	936	936	936	40	5	11
Agrostis stolonifera	L.	4	1	—	8	8	8	5	5	213
Agrostis stolonifera	L.	4	1	—	14	14	14	20	4	111
Agrostis stolonifera	L.	4	1	—	62	62	62	11.5	5	134
Agrostis stolonifera	L.	4	4	—	67	83	79	3	5	96
Agrostis stolonifera	L.	4	2	—	80	100	90	6.5	5	128
Agrostis stolonifera	L.	4	1	—	144	144	144	5	5	69
Agrostis stolonifera	L.	4	5	—	57	445	161	15	5	62
Agrostis stolonifera	L.	4	2	—	106	224	165	15	4	31
Agrostis stolonifera	L.	4	4	—	75	301	186	3	5	59
Agrostis stolonifera	L.	4	1	—	207	207	207	5	5	244
Agrostis stolonifera	L.	4	1	—	252	252	252	5	4	53
Agrostis stolonifera	L.	4	1	—	400	400	400	1	5	96
Agrostis stolonifera	L.	4	2	—	226	679	453	3	5	208
Agrostis stolonifera	L.	4	3	—	300	700	467	1	5	57
Agrostis stolonifera	L.	4	7	—	260	1456	594	6	5	99
Agrostis stolonifera	L.	4	3	—	82	1714	629	10	5	110
Agrostis stolonifera	L.	4	6	—	50	2238	957	10	5	255
Agrostis stolonifera	L.	4	4	—	29	3701	1128	7	5	58
Agrostis stolonifera	L.	4	1	—	4900	4900	4900	8	5	137
Agrostis vinealis	Schreber	1	4	—	0	0	0	7	5	232
Agrostis vinealis	Schreber	1	1	—	0	0	0	8.5	4	254
Agrostis vinealis	Schreber	2	1	>2	0	0	0	10	3	178
Agrostis vinealis	Schreber	2	1	—	67	67	67	3	5	96
Agrostis vinealis	Schreber	2	1	—	69	69	69	3	7	235
Agrostis vinealis	Schreber	4	1	—	65	65	65	7	5	232
Aira praecox	L.	2	1	>2	0	0	0	7.5	2	194
Aira praecox	L.	2	1	—	83	83	83	3	5	96
Aira praecox	L.	2	2	—	546	4030	2288	5	5	123
Aira praecox	L.	4	2	—	300	400	350	15	5	249
Aira praecox	L.	4	4	—	300	1517	834	3	5	96
Alopecurus aequalis	Sobol.	2	2	1	0	0	0	—	1	5
Alopecurus aequalis	Sobol.	2	1	2	0	0	0	—	1	5

Species	Source	1	2	3	4	5	6	7	8	9
Alopecurus aequalis	Sobol.	3	1	>10	6	6	6	10	4	241
Alopecurus geniculatus	L.	1	1	—	0	0	0	5	5	69
Alopecurus geniculatus	L.	1	3	—	0	0	0	6	5	208
Alopecurus geniculatus	L.	1	1	—	3	3	3	3	7	235
Alopecurus geniculatus	L.	1	3	—	0	101	42	2	7	99
Alopecurus geniculatus	L.	1	3	—	78	78	78	6	5	28
Alopecurus geniculatus	L.	2	1	>2	156	156	156	6	5	11
Alopecurus geniculatus	L.	2	1	>3	0	0	0	7.5	2	194
Alopecurus geniculatus	L.	2	1	>3	0	0	0	5	3	227
Alopecurus geniculatus	L.	2	1	—	0	0	0	20	3	227
Alopecurus geniculatus	L.	2	1	—	19	19	19	3	7	235
Alopecurus geniculatus	L.	2	1	—	50	50	50	3	5	96
Alopecurus geniculatus	L.	2	1	—	156	156	156	6	5	11
Alopecurus geniculatus	L.	2	1	—	156	156	156	40	5	11
Alopecurus geniculatus	L.	2	1	—	360	360	360	5	5	244
Alopecurus geniculatus	L.	2	3	—	41	1041	391	5	5	222
Alopecurus geniculatus	L.	2	1	—	570	570	570	20	5	107
Alopecurus geniculatus	L.	2	2	—	1026	1026	1026	50	5	107
Alopecurus geniculatus	L.	2	2	—	1352	1716	1534	6	5	99
Alopecurus geniculatus	L.	3	2	—	1326	3094	2210	2	7	99
Alopecurus geniculatus	L.	3	2	>14	5	5	5	15	4	21
Alopecurus geniculatus	L.	3	3	—	312	416	364	6	5	11
Alopecurus geniculatus	L.	3	6	—	342	456	399	30	5	107
Alopecurus geniculatus	L.	3	1	—	260	780	589	6	5	11
Alopecurus geniculatus	L.	4	1	—	1508	16328	6023	10	5	27
Alopecurus geniculatus	L.	4	3	—	16328	16328	16328	6	5	99
Alopecurus geniculatus	L.	4	9	—	28	28	28	20	4	111
Alopecurus geniculatus	L.	4	8	—	100	617	333	3	5	96
Alopecurus geniculatus	L.	4		—	360	360	360	5	5	244
Alopecurus geniculatus	L.	4		—	92	1184	582	6	5	222
Alopecurus geniculatus	L.	4		—	728	34996	9848		5	99
Alopecurus myosuroides	Hudson	1	2	2	333	333	333	50	5	260
Alopecurus myosuroides	Hudson	2	1	—	0	0	0	30	3	14
Alopecurus myosuroides	Hudson	2	1	>2	0	0	0	0	1	75
Alopecurus myosuroides	Hudson	2	1	>2	0	0	0	0	2	75
Alopecurus myosuroides	Hudson	2	1	>2	0	0	0	5	1	75
Alopecurus myosuroides	Hudson	2	1	>2	0	0	0	5	2	75
Alopecurus myosuroides	Hudson	2	1	4	0	0	0	13	1	133
Alopecurus myosuroides	Hudson	2	2	4	0	0	0	26	1	133
Alopecurus myosuroides	Hudson	2	1	>3	0	0	0	39	1	133
Alopecurus myosuroides	Hudson	2	1	>3	0	0	0	10	3	135
Alopecurus myosuroides	Hudson	2	1	>2	0	0	0	25	3	135
Alopecurus myosuroides	Hudson	2	1	—	0	0	0	12.5	3	159
Alopecurus myosuroides	Hudson	2	6	—	6	724	164	25	4	36
Alopecurus myosuroides	Hudson	2	1	—	387	387	387	30	4	12
Alopecurus myosuroides	Hudson	2	6	—	292	8572	2309	8	4	159
Alopecurus myosuroides	Hudson	3	1	>11	0	0	0	25	3	209

Species	Authority	Seed bank type	Number of records	Longevity (y)	Minimum density (seeds m^{-2})	Maximum density (seeds m^{-2})	Mean density (seeds m^{-2})	Depth (cm)	Method	Source code
Alopecurus myosuroides	Hudson	3	2	>4	128	369	249	8	4	159
Alopecurus myosuroides	Hudson	3	1	—	7222	7222	7222	50	5	260
Alopecurus myosuroides	Hudson	4	2	—	82	174	128	25	4	36
Alopecurus myosuroides	Hudson	4	9	—	55	756	194	10	5	15
Alopecurus myosuroides	Hudson	4	2	—	185	454	320	25	4	185
Alopecurus myosuroides	Hudson	4	3	—	255	560	374	15	5	142
Alopecurus myosuroides	Hudson	4	39	—	20	3430	618	30	4	13
Alopecurus myosuroides	Hudson	4	5	—	500	5333	2167	20	5	74
Alopecurus myosuroides	Hudson	4	1	—	2571	2571	2571	25	5	260
Alopecurus myosuroides	Hudson	4	7	—	1046	4381	3147	15	4	30
Alopecurus myosuroides	Hudson	4	2	—	8668	56107	32388	8	4	159
Alopecurus pratensis	L.	1	1	—	0	0	0	5	5	19
Alopecurus pratensis	L.	1	2	—	0	0	0	25	4	36
Alopecurus pratensis	L.	1	2	—	0	0	0	30	5	45
Alopecurus pratensis	L.	1	1	—	0	0	0	15	5	62
Alopecurus pratensis	L.	1	2	—	0	0	0	13	5	68
Alopecurus pratensis	L.	.	2	—	0	0	0	5	5	69
Alopecurus pratensis	L.	1	1	—	0	0	0	10	4	88
Alopecurus pratensis	L.	1	1	—	0	0	0	3	5	96
Alopecurus pratensis	L.	1	3	—	0	0	0	6.5	5	128
Alopecurus pratensis	L.	1	2	—	0	0	0	12	5	174
Alopecurus pratensis	L.	1	3	—	0	0	0	3	5	208
Alopecurus pratensis	L.	1	12	—	0	0	0	20	4	225
Alopecurus pratensis	L.	1	1	—	0	0	0	3	7	235
Alopecurus pratensis	L.	1	17	—	0	0	0	10	5	255
Alopecurus pratensis	L.	1	1	—	54	54	54	30	5	29
Alopecurus pratensis	L.	1	9	—	142	142	142	6	5	208
Alopecurus pratensis	L.	1	2	—	364	364	364	10	5	27
Alopecurus pratensis	L.	1	1	—	400	400	400	6.5	5	68
Alopecurus pratensis	L.	1	14	—	400	7400	1840	5	5	251
Alopecurus pratensis	L.	2	1	1	0	0	0	7.5	2	194
Alopecurus pratensis	L.	2	1	—	49	49	49	12	5	174
Alopecurus pratensis	L.	2	2	—	156	156	156	6	5	11
Alopecurus pratensis	L.	2	1	—	225	225	225	6.5	5	68
Alopecurus pratensis	L.	2	1	—	250	250	250	10	5	255
Alopecurus pratensis	L.	2	1	—	400	400	400	5	5	251
Alopecurus pratensis	L.	2	1	—	513	513	513	30	5	107
Alopecurus pratensis	L.	2	2	—	500	2000	1250	10	4	88
Alopecurus pratensis	L.	3	1	—	32	32	32	30	5	45
Alopecurus pratensis	L.	3	1	—	171	171	171	50	5	107
Alopecurus pratensis	L.	3	1	—	375	375	375	10	4	88
Alopecurus pratensis	L.	3	1	—	400	400	400	5	5	251

Species	Author								
Alopecurus pratensis	L.	1	—	19	19	19	12	5	174
Alopecurus pratensis	L.	1	—	62	62	62	15	5	267
Alopecurus pratensis	L.	4	—	75	275	166	10	5	255
Alopecurus pratensis	L.	1	—	313	313	313	5	4	88
Ammophila arenaria	(L.) Link	6	—	0	0	0	10	4	20
Ammophila arenaria	(L.) Link	4	—	0	0	0	1	5	57
Ammophila arenaria	(L.) Link	1	—	800	800	800	10	4	20
Anisantha diandra	(Roth) Tutin ex Tzvelev	2	0	0	0	0	5	3	43
Anisantha diandra	(Roth) Tutin ex Tzvelev	1	0	0	0	0	10	3	43
Anisantha diandra	(Roth) Tutin ex Tzvelev	1	0	0	0	0	15	3	43
Anisantha diandra	(Roth) Tutin ex Tzvelev	1	0	0	0	0	5	1	89
Anisantha diandra	(Roth) Tutin ex Tzvelev	1	1	0	0	0	15	1	89
Anisantha diandra	(Roth) Tutin ex Tzvelev	3	1	0	0	0	0	3	43
Anisantha diandra	(Roth) Tutin ex Tzvelev	3	1	0	0	0	1	3	43
Anisantha diandra	(Roth) Tutin ex Tzvelev	1	2	0	0	0	5	3	43
Anisantha diandra	(Roth) Tutin ex Tzvelev	1	2	0	0	0	15	3	43
Anisantha diandra	(Roth) Tutin ex Tzvelev	1	2	0	0	0	5	3	43
Anisantha diandra	(Roth) Tutin ex Tzvelev	1	1	0	0	0	10	3	43
Anisantha diandra	(Roth) Tutin ex Tzvelev	1	—	0	0	0	15	3	43
Anisantha diandra	(Roth) Tutin ex Tzvelev	2	2	0	0	0	0	1	89
Anisantha diandra	(Roth) Tutin ex Tzvelev	2	—	1480	2000	1740	10	5	140
Anisantha rigida	(Roth) N. Hylander	2	1	0	0	0	10	4	22
Anisantha rigida	(Roth) N. Hylander	5	1	0	0	0	2	5	95
Anisantha rigida	(Roth) N. Hylander	2	1	0	0	0	0	1	79
Anisantha rigida	(Roth) N. Hylander	1	1	0	0	0	10	1	79
Anisantha rigida	(Roth) N. Hylander	2	1	0	0	0	20	1	79
Anisantha rigida	(Roth) N. Hylander	1	—	0	0	0	30	1	79
Anisantha sterilis	(L.) Nevski	2	0	0	0	0	30	5	66
Anisantha sterilis	(L.) Nevski	1	0	0	0	0	6	5	67
Anisantha sterilis	(L.) Nevski	1	—	0	0	0	0	1	75
Anisantha sterilis	(L.) Nevski	2	0	0	0	0	7.5	1	75
Anisantha sterilis	(L.) Nevski	1	0	0	0	0	3	5	96
Anisantha sterilis	(L.) Nevski	1	0	0	0	0	7.5	2	194
Anisantha sterilis	(L.) Nevski	1	—	458	458	458	25	3	209
Anisantha sterilis	(L.) Nevski	2	—	100	917	509	15	5	142
Anisantha tectorum	(L.) Nevski	1	0	0	0	0	3	5	96
Anisantha tectorum	(L.) Nevski	1	0	0	0	0	10	4	22
Anisantha tectorum	(L.) Nevski	1	0	0	0	0	10	1	102
Anisantha tectorum	(L.) Nevski	1	0	0	0	0	25	1	102
Anisantha tectorum	(L.) Nevski	1	—	0	0	0	50	1	102
Anisantha tectorum	(L.) Nevski	1	1	0	0	0	100	1	102
Anisantha tectorum	(L.) Nevski	3	1	0	0	0	3	5	132
Anisantha tectorum	(L.) Nevski	5	3	0	0	0	7	2	44
Anisantha tectorum	(L.) Nevski	3	2	0	0	0	7	2	44
Anisantha tectorum	(L.) Nevski	2	3	0	0	0	7	2	44
Anisantha tectorum	(L.) Nevski	1	2	0	0	0	25	3	209

Species	Authority	Seed bank type	Number of records	Longevity (y)	Minimum density (seeds m⁻²)	Maximum density (seeds m⁻²)	Mean density (seeds m⁻²)	Depth (cm)	Method	Source code
Anthoxanthum aristatum	Boiss.	4	1	—	38	38	38	10	5	255
Anthoxanthum odoratum	L.	1	3	—	0	0	0	10	5	27
Anthoxanthum odoratum	L.	1	1	—	0	0	0	25	4	36
Anthoxanthum odoratum	L.	1	1	—	0	0	0	15	5	62
Anthoxanthum odoratum	L.	1	1	—	0	0	0	5	5	69
Anthoxanthum odoratum	L.	1	1	—	0	0	0	9	5	84
Anthoxanthum odoratum	L.	1	1	—	0	0	0	3	5	96
Anthoxanthum odoratum	L.	1	4	—	0	0	0	6.5	5	128
Anthoxanthum odoratum	L.	1	2	—	0	0	0	10	5	149
Anthoxanthum odoratum	L.	1	1	—	0	0	0	10	5	152
Anthoxanthum odoratum	L.	1	2	—	0	0	0	17.8	5	156
Anthoxanthum odoratum	L.	1	1	—	0	0	0	6.5	7	182
Anthoxanthum odoratum	L.	1	3	—	0	0	0	6	5	208
Anthoxanthum odoratum	L.	1	1	—	0	0	0	5	5	222
Anthoxanthum odoratum	L.	1	5	—	0	0	0	20	4	225
Anthoxanthum odoratum	L.	1	1	—	0	0	0	7	5	232
Anthoxanthum odoratum	L.	1	2	—	0	0	0	5	5	250
Anthoxanthum odoratum	L.	1	11	—	0	0	0	10	5	255
Anthoxanthum odoratum	L.	1	1	—	3	3	3	3	7	235
Anthoxanthum odoratum	L.	1	2	—	0	26	13	2	7	99
Anthoxanthum odoratum	L.	1	1	—	14	14	14	10	5	252
Anthoxanthum odoratum	L.	1	1	—	60	60	60	13	5	68
Anthoxanthum odoratum	L.	1	7	—	113	113	113	3	5	208
Anthoxanthum odoratum	L.	1	4	—	32	506	178	30	5	45
Anthoxanthum odoratum	L.	1	1	—	180	180	180	6	5	48
Anthoxanthum odoratum	L.	1	8	—	78	1092	484	6	5	28
Anthoxanthum odoratum	L.	1	4	—	500	500	500	10	4	88
Anthoxanthum odoratum	L.	1	1	—	572	572	572	6	5	11
Anthoxanthum odoratum	L.	1	1	—	900	900	900	20	5	69
Anthoxanthum odoratum	L.	1	13	—	300	1700	1300	5	5	251
Anthoxanthum odoratum	L.	2	1	>2	0	0	0	7.5	2	194
Anthoxanthum odoratum	L.	2	5	>1	0	0	0	—	1	275
Anthoxanthum odoratum	L.	2	3	—	8	34	20	3	7	235
Anthoxanthum odoratum	L.	2	1	—	32	32	32	30	5	45
Anthoxanthum odoratum	L.	2	1	—	41	41	41	5	5	92
Anthoxanthum odoratum	L.	2	1	—	43	43	43	15	5	32
Anthoxanthum odoratum	L.	2	1	—	75	75	75	30	5	29
Anthoxanthum odoratum	L.	2	6	—	38	213	96	10	5	255
Anthoxanthum odoratum	L.	2	1	—	275	275	275	6.5	5	68
Anthoxanthum odoratum	L.	2	2	—	500	1000	750	5	5	251
Anthoxanthum odoratum	L.	2	3	—	130	2954	1459	2	7	99
Anthoxanthum odoratum	L.	2	3	—	1248	2340	1612	10	5	27

Species	Author									
Anthoxanthum odoratum	L.	2	3	—	780	7020	3137	6	5	11
Anthoxanthum odoratum	L.	3	1	—	75	75	75	30	5	45
Anthoxanthum odoratum	L.	3	1	—	100	100	100	8	5	108
Anthoxanthum odoratum	L.	3	1	—	600	600	600	5	5	251
Anthoxanthum odoratum	L.	4	1	—	1196	1196	1196	6	5	99
Anthoxanthum odoratum	L.	4	1	—	6	6	6	5	5	233
Anthoxanthum odoratum	L.	4	1	—	27	27	27	2	5	153
Anthoxanthum odoratum	L.	4	1	—	37	37	37	15	5	267
Anthoxanthum odoratum	L.	4	1	—	50	50	50	3	5	96
Anthoxanthum odoratum	L.	4	1	—	60	60	60	6.5	5	128
Anthoxanthum odoratum	L.	4	8	—	15	278	100	12	5	174
Anthoxanthum odoratum	L.	4	9	—	38	213	106	10	5	255
Anthoxanthum odoratum	L.	4	2	—	51	163	107	5	5	222
Anthoxanthum odoratum	L.	4	2	—	101	152	127	17.8	5	156
Anthoxanthum odoratum	L.	4	2	—	112	214	163	5	5	92
Anthoxanthum odoratum	L.	4	2	—	39	296	168	5	5	123
Anthoxanthum odoratum	L.	4	1	—	192	192	192	5	5	69
Anthoxanthum odoratum	L.	4	1	—	198	198	198	3	5	208
Anthoxanthum odoratum	L.	4	1	—	390	390	390	10	5	152
Anthoxanthum odoratum	L.	4	1	—	500	500	500	5	4	88
Anthoxanthum odoratum	L.	4	1	—	500	500	500	15	5	249
Anthoxanthum odoratum	L.	4	4	—	101	1109	649	17.8	5	155
Anthoxanthum odoratum	L.	4	2	—	1200	2000	1600	20	4	225
Anthoxanthum odoratum	L.	4	10	—	208	18928	5060	6	5	99
Anthoxanthum odoratum	L.	4	1	—	13367	13367	13367	8.5	4	254
Apera spica-venti	(L.) P. Beauv.	1	2	—	0	0	0	25	4	36
Apera spica-venti	(L.) P. Beauv.	1	1	—	0	0	0	10	5	255
Apera spica-venti	(L.) P. Beauv.	2	1	—	3	3	3	5	5	244
Apera spica-venti	(L.) P. Beauv.	2	4	—	6	126	61	25	4	36
Apera spica-venti	(L.) P. Beauv.	2	1	—	1020	1020	1020	25	4	3
Apera spica-venti	(L.) P. Beauv.	3	1	>11	0	0	0	25	3	209
Apera spica-venti	(L.) P. Beauv.	4	4	—	24	190	116	25	4	36
Apera spica-venti	(L.) P. Beauv.	4	1	—	176	176	176	5	5	69
Apera spica-venti	(L.) P. Beauv.	4	2	—	163	488	326	10	5	255
Apera spica-venti	(L.) P. Beauv.	4	7	—	55	2898	747	10	5	15
Arrhenatherum elatius	(L.) P. Beauv. ex J.S. Presl & C. Presl	1	3	—	0	0	0	25	4	36
Arrhenatherum elatius	(L.) P. Beauv. ex J.S. Presl & C. Presl	1	1	—	0	0	0	7	5	58
Arrhenatherum elatius	(L.) P. Beauv. ex J.S. Presl & C. Presl	1	1	—	0	0	0	15	5	62
Arrhenatherum elatius	(L.) P. Beauv. ex J.S. Presl & C. Presl	1	1	—	0	0	0	30	5	66
Arrhenatherum elatius	(L.) P. Beauv. ex J.S. Presl & C. Presl	1	2	—	0	0	0	6	5	67
Arrhenatherum elatius	(L.) P. Beauv. ex J.S. Presl & C. Presl	1	1	—	0	0	0	6.5	5	68
Arrhenatherum elatius	(L.) P. Beauv. ex J.S. Presl & C. Presl	1	1	—	0	0	0	13	5	68
Arrhenatherum elatius	(L.) P. Beauv. ex J.S. Presl & C. Presl	1	1	0	0	0	0	0	1	75
Arrhenatherum elatius	(L.) P. Beauv. ex J.S. Presl & C. Presl	1	1	0	0	0	0	7.5	1	75
Arrhenatherum elatius	(L.) P. Beauv. ex J.S. Presl & C. Presl	1	4	—	0	0	0	10	4	88
Arrhenatherum elatius	(L.) P. Beauv. ex J.S. Presl & C. Presl	1	9	—	0	0	0	3	5	96
Arrhenatherum elatius	(L.) P. Beauv. ex J.S. Presl & C. Presl	1	5	—	0	0	0	2	7	99

Species	Authority	Seed bank type	Number of records	Longevity (y)	Minimum density (seeds m⁻²)	Maximum density (seeds m⁻²)	Mean density (seeds m⁻²)	Depth (cm)	Method	Source code
Arrhenatherum elatius	(L.) P. Beauv. ex J.S. Presl & C. Presl	1	1	—	0	0	0	10	5	110
Arrhenatherum elatius	(L.) P. Beauv. ex J.S. Presl & C. Presl	1	5	—	0	0	0	6.5	5	128
Arrhenatherum elatius	(L.) P. Beauv. ex J.S. Presl & C. Presl	1	1	—	0	0	0	6.5	7	182
Arrhenatherum elatius	(L.) P. Beauv. ex J.S. Presl & C. Presl	1	3	—	0	0	0	3	5	208
Arrhenatherum elatius	(L.) P. Beauv. ex J.S. Presl & C. Presl	1	6	—	0	0	0	20	4	225
Arrhenatherum elatius	(L.) P. Beauv. ex J.S. Presl & C. Presl	1	11	—	0	0	0	5	5	251
Arrhenatherum elatius	(L.) P. Beauv. ex J.S. Presl & C. Presl	1	12	—	0	0	0	10	5	255
Arrhenatherum elatius	(L.) P. Beauv. ex J.S. Presl & C. Presl	1	1	—	0	0	0	15	5	257
Arrhenatherum elatius	(L.) P. Beauv. ex J.S. Presl & C. Presl	1	4	—	3	4	3	3	7	235
Arrhenatherum elatius	(L.) P. Beauv. ex J.S. Presl & C. Presl	1	1	—	22	22	22	10	5	252
Arrhenatherum elatius	(L.) P. Beauv. ex J.S. Presl & C. Presl	1	5	—	53	333	177	8	7	275
Arrhenatherum elatius	(L.) P. Beauv. ex J.S. Presl & C. Presl	2	1	>2	0	0	0	7.5	2	194
Arrhenatherum elatius	(L.) P. Beauv. ex J.S. Presl & C. Presl	2	5	>1	0	0	0	—	1	275
Arrhenatherum elatius	(L.) P. Beauv. ex J.S. Presl & C. Presl	2	1	—	16	16	16	3	7	235
Arrhenatherum elatius	(L.) P. Beauv. ex J.S. Presl & C. Presl	2	1	—	50	50	50	3	5	96
Arrhenatherum elatius	(L.) P. Beauv. ex J.S. Presl & C. Presl	2	4	—	128	391	286	8	7	275
Arrhenatherum elatius	(L.) P. Beauv. ex J.S. Presl & C. Presl	2	1	—	4792	4792	4792	20	5	97
Arrhenatherum elatius	(L.) P. Beauv. ex J.S. Presl & C. Presl	4	1	—	5	5	5	3	7	235
Arrhenatherum elatius	(L.) P. Beauv. ex J.S. Presl & C. Presl	4	2	—	16	104	60	10	5	110
Arrhenatherum elatius	(L.) P. Beauv. ex J.S. Presl & C. Presl	4	4	—	95	116	103	8	7	275
Arrhenatherum elatius	(L.) P. Beauv. ex J.S. Presl & C. Presl	4	1	—	127	127	127	15	5	142
Arrhenatherum elatius	(L.) P. Beauv. ex J.S. Presl & C. Presl	4	1	—	300	300	300	3	5	96
Avena fatua	L.	1	1	—	0	0	0	25	5	2
Avena fatua	L.	1	3	—	0	0	0	2	5	95
Avena fatua	L.	1	1	—	0	0	0	3	5	96
Avena fatua	L.	1	1	0	0	0	0	26	1	133
Avena fatua	L.	1	1	—	0	0	0	15	5	142
Avena fatua	L.	2	1	2	0	0	0	30	3	14
Avena fatua	L.	2	4	3	0	0	0	7	2	44
Avena fatua	L.	2	2	4	0	0	0	7	2	44
Avena fatua	L.	2	1	>2	0	0	0	2	1	47
Avena fatua	L.	2	1	>2	0	0	0	15	1	47
Avena fatua	L.	2	1	4	0	0	0	13	1	133
Avena fatua	L.	2	1	4	0	0	0	26	1	133
Avena fatua	L.	2	1	4	0	0	0	39	1	133
Avena fatua	L.	2	1	>3	0	0	0	0	1	167
Avena fatua	L.	2	1	>3	0	0	0	2.5	1	167
Avena fatua	L.	2	1	>3	0	0	0	7.5	1	167
Avena fatua	L.	2	1	>3	0	0	0	12.5	1	167
Avena fatua	L.	2	1	>3	0	0	0	20	1	167
Avena fatua	L.	2	1	>2	0	0	0	7.5	2	194
Avena fatua	L.	2	1	3	0	0	0	2.5	3	239

Species	Authority									
Avena fatua	L.	2	1	3	0	0	0	7.5	3	239
Avena fatua	L.	2	1	3	0	0	0	15	3	239
Avena fatua	L.	2	1	1	0	0	0	15	1	240
Avena fatua	L.	2	1	1	0	0	0	50	1	240
Avena fatua	L.	2	1	2	0	0	0	100	1	256
Avena fatua	L.	2	1	2	0	0	0	2.5	1	256
Avena fatua	L.	2	1	2	0	0	0	5	1	256
Avena fatua	L.	2	1	2	0	0	0	7.5	1	256
Avena fatua	L.	2	1	2	0	0	0	12.5	1	256
Avena fatua	L.	2	1	2	0	0	0	17.5	1	256
Avena fatua	L.	2	1	>3	2	2	2	25	5	70
Avena fatua	L.	2	1	>2	3	3	3	15	5	70
Avena fatua	L.	3	1	>1	10	10	10	15	5	70
Avena fatua	L.	3	1	>4	0	0	0	15	5	70
Avena fatua	L.	3	1	>5	0	0	0	15	5	70
Avena fatua	L.	3	1	>6	0	0	0	0	1	129
Avena fatua	L.	3	1	>4	0	0	0	5	1	129
Avena fatua	L.	3	1	>4	0	0	0	10	1	129
Avena fatua	L.	3	1	>4	0	0	0	20	1	129
Avena fatua	L.	3	1	>8	0	0	0	30	1	129
Avena fatua	L.	3	1	>8	0	0	0	50	3	236
Avena fatua	L.	3	1	>7	0	0	0	5	3	236
Avena fatua	L.	3	1	>7	1	1	1	15	6	237
Avena fatua	L.	4	1	>5	46	46	46	—	4	13
Avena fatua	L.	4	2	—	230	230	230	30	5	95
Avena fatua	L.	4	1	—	30	1590	810	2	5	140
Avena fatua	L.	1	1	0	0	0	0	10	1	240
Avena sativa	L.	1	1	0	0	0	0	15	1	240
Avena sativa	L.	1	1	0	0	0	0	50	1	240
Avena sativa	L.	1	1	—	0	0	0	100	5	255
Avena sativa	L.	4	1	—	4	4	4	10	4	111
Avena sativa	L.	1	1	>2	0	0	0	20	4	22
Avena sterilis	L.	2	1	4	0	0	0	10	2	194
Avena sterilis	L.	2	1	4	0	0	0	7.5	3	236
Avena sterilis	L.	2	1	—	0	0	0	5	3	236
Avena sterilis	L.	2	1	—	18	18	18	15	4	22
Avena sterilis	L.	4	3	—	957	2242	1422	10	4	65
Avena strigosa	Schreber	1	1	—	0	0	0	20	4	111
Avena strigosa	Schreber	1	1	—	0	0	0	20	5	111
Brachypodium pinnatum	(L.) P. Beauv.	1	4	—	0	0	0	3	5	96
Brachypodium pinnatum	(L.) P. Beauv.	1	3	—	0	0	0	20	5	97
Brachypodium pinnatum	(L.) P. Beauv.	1	2	—	0	0	0	6	5	181
Brachypodium pinnatum	(L.) P. Beauv.	1	1	—	0	0	0	10	5	181
Brachypodium pinnatum	(L.) P. Beauv.	1	1	—	0	0	0	13	5	183

Species	Authority	Seed bank type	Number of records	Longevity (y)	Minimum density (seeds m⁻²)	Maximum density (seeds m⁻²)	Mean density (seeds m⁻²)	Depth (cm)	Method	Source code
Brachypodium pinnatum	(L.) P. Beauv.	1	5	—	0	0	0	10	5	255
Brachypodium pinnatum	(L.) P. Beauv.	1	1	—	4	4	4	3	7	235
Brachypodium pinnatum	(L.) P. Beauv.	1	2	—	11	28	20	6.5	7	182
Brachypodium pinnatum	(L.) P. Beauv.	2	1	>2	0	0	0	7.5	2	194
Brachypodium pinnatum	(L.) P. Beauv.	4	1	—	67	67	67	3	5	96
Brachypodium pinnatum	(L.) P. Beauv.	4	1	—	136	136	136	6	5	67
Brachypodium pinnatum	(L.) P. Beauv.	4	2	—	198	355	277	10	5	110
Brachypodium sylvaticum	(Hudson) P. Beauv.	1	1	—	0	0	0	6	5	48
Brachypodium sylvaticum	(Hudson) P. Beauv.	1	1	—	0	0	0	6	5	49
Brachypodium sylvaticum	(Hudson) P. Beauv.	1	2	—	0	0	0	7	5	58
Brachypodium sylvaticum	(Hudson) P. Beauv.	1	4	—	0	0	0	3	5	96
Brachypodium sylvaticum	(Hudson) P. Beauv.	1	1	—	0	0	0	12.5	5	126
Brachypodium sylvaticum	(Hudson) P. Beauv.	1	1	—	0	0	0	3	7	235
Brachypodium sylvaticum	(Hudson) P. Beauv.	1	2	—	0	0	0	15	5	257
Brachypodium sylvaticum	(Hudson) P. Beauv.	2	1	—	17	17	17	7	5	58
Brachypodium sylvaticum	(Hudson) P. Beauv.	2	1	—	50	50	50	3	5	96
Briza media	L.	1	1	—	0	0	0	6	5	48
Briza media	L.	1	2	—	0	0	0	7	5	58
Briza media	L.	1	1	—	0	0	0	20	5	97
Briza media	L.	1	2	—	0	0	0	6.5	5	128
Briza media	L.	1	7	—	0	0	0	10	5	149
Briza media		1	6	—	0	0	0	12	5	174
Briza media		1	2	—	0	0	0	6	5	181
Briza media		1	1	—	0	0	0	8.5	4	254
Briza media		1	8	—	0	0	0	10	5	255
Briza media		1	3	—	0	3	2	3	7	235
Briza media		4	2	—	8	32	20	6.5	7	182
Briza media		4	1	—	25	25	25	10	5	110
Briza media		4	1	—	1575	1575	1575	10	5	255
Briza minor	L.	4	2	—	420	620	520	2	5	95
Bromopsis benekenii	(Lange) Holub	1	1	—	0	0	0	12.5	5	126
Bromopsis erecta	(Hudson) Fourr.	1	1	—	0	0	0	7	5	58
Bromopsis erecta	(Hudson) Fourr.	1	2	—	0	0	0	6	5	181
Bromopsis erecta	(Hudson) Fourr.	1	1	—	29	29	29	3	7	235
Bromopsis erecta	(Hudson) Fourr.	1	1	—	37	37	37	6.5	7	182
Bromopsis erecta	(Hudson) Fourr.	2	1	1	0	0	0	25	3	209
Bromopsis erecta	(Hudson) Fourr.	2	1	—	47	47	47	6.5	7	182
Bromopsis erecta	(Hudson) Fourr.	4	1	—	133	133	133	3	5	96
Bromopsis inermis	(Leysser) Holub	1	1	—	0	0	0	5	5	69
Bromopsis ramosa	(Hudson) Holub	1	2	—	0	0	0	3	5	96
Bromopsis ramosa	(Hudson) Holub	1	1	—	0	0	0	15	5	257
Bromus arvensis	L.	1	1	0	0	0	0	25	3	209

Species	Authority									
Bromus arvensis	L.	1	1	—	0	0	0	10	5	255
Bromus hordeaceus	L.	1	1	—	0	0	0	30	5	66
Bromus hordeaceus	L.	1	2	—	0	0	0	6	5	67
Bromus hordeaceus	L.	1	1	—	0	0	0	6.5	5	68
Bromus hordeaceus	L.	1	1	—	0	0	0	13	5	68
Bromus hordeaceus	L.	1	6	—	0	0	0	5	5	69
Bromus hordeaceus	L.	1	1	—	0	0	0	2	5	95
Bromus hordeaceus	L.	1	2	—	0	0	0	3	5	96
Bromus hordeaceus	L.	1	2	—	0	0	0	3	5	132
Bromus hordeaceus	L.	1	1	0	0	0	0	26	1	133
Bromus hordeaceus	L.	1	1	0	0	0	0	25	3	209
Bromus hordeaceus	L.	2	1	1	0	0	0	13	1	133
Bromus hordeaceus	L.	2	1	1	0	0	0	26	1	133
Bromus hordeaceus	L.	2	1	1	0	0	0	39	1	133
Bromus hordeaceus	L.	2	1	>2	0	0	0	7.5	2	194
Bromus hordeaceus	L.	2	1	—	210	210	210	2	5	95
Bromus hordeaceus	L.	2	1	—	300	300	300	1	5	57
Bromus hordeaceus	L.	2	2	—	1488	1674	1581	5	5	139
Bromus hordeaceus	L.	3	1	>18	144	144	144	32	5	169
Bromus hordeaceus	L.	4	2	—	133	283	208	10	5	184
Bromus hordeaceus	L.	4	3	—	600	1870	1163	2	5	95
Bromus hordeaceus	L.	4	1	—	18110	18110	18110	10	5	140
Bromus secalinus	L.	1	1	0	0	0	0	15	1	240
Bromus secalinus	L.	1	1	0	0	0	0	50	1	240
Bromus secalinus	L.	1	1	0	0	0	0	100	1	240
Bromus secalinus	L.	1	1	—	0	0	0	10	5	255
Bromus secalinus	L.	1	1	—	0	0	0	25	5	260
Bromus secalinus	L.	2	1	1	0	0	0	25	3	209
Calamagrostis arundinacea	(L.) Roth	1	1	—	0	0	0	6	5	49
Calamagrostis arundinacea	(L.) Roth	1	1	—	0	0	0	5	6	170
Calamagrostis arundinacea	(L.) Roth	1	2	—	0	0	0	12	5	171
Calamagrostis arundinacea	(L.) Roth	1	2	—	0	0	0	16	5	216
Calamagrostis arundinacea	(L.) Roth	2	2	—	5	7	5	5	6	170
Calamagrostis canescens	(Wigg.) Roth	1	3	—	0	0	0	3	5	208
Calamagrostis canescens	(Wigg.) Roth	1	3	—	0	0	0	6	5	208
Calamagrostis canescens	(Wigg.) Roth	1	1	—	0	0	0	10	5	255
Calamagrostis canescens	(Wigg.) Roth	3	3	>4	0	0	0	10	3	216
Calamagrostis canescens	(Wigg.) Roth	4	2	—	52	72	62	16	5	216
Calamagrostis canescens	(Wigg.) Roth	4	3	—	108	1358	878	5	5	98
Calamagrostis epigejos	(L.) Roth	1	1	—	0	0	0	20	5	67
Calamagrostis epigejos	(L.) Roth	1	1	—	0	0	0	20	5	97
Calamagrostis epigejos	(L.) Roth	1	2	—	0	0	0	10	5	255
Calamagrostis epigejos	(L.) Roth	2	2	—	31	38	35	6	5	49
Calamagrostis epigejos	(L.) Roth	2	1	—	61	61	61	5	5	222
Catapodium marinum	(L.) C.E. Hubb.	1	1	—	0	0	0	10	5	110
Corynephorus canescens	(L.) P. Beauv.	2	1	>2	0	0	0	10	3	178
Corynephorus canescens	(L.) P. Beauv.	4	2	—	396	855	626	20	4	226

Species	Authority	Seed bank type	Number of records	Longevity (y)	Minimum density (seeds m⁻²)	Maximum density (seeds m⁻²)	Mean density (seeds m⁻²)	Depth (cm)	Method	Source code
Corynephorus canescens	(L.) P. Beauv.	4	1	—	700	700	700	15	5	249
Cynosurus cristatus	L.	1	2	—	0	0	0	6	5	28
Cynosurus cristatus	L.	1	5	—	0	0	0	5	4	53
Cynosurus cristatus	L.	1	1	—	0	0	0	15	5	62
Cynosurus cristatus	L.	1	1	—	0	0	0	6.5	5	68
Cynosurus cristatus	L.	1	1	—	0	0	0	13	5	68
Cynosurus cristatus	L.	1	4	—	0	0	0	5	5	69
Cynosurus cristatus	L.	1	2	—	0	0	0	10	4	88
Cynosurus cristatus	L.	1	3	—	0	0	0	20	5	97
Cynosurus cristatus	L.	1	18	—	0	0	0	6	5	99
Cynosurus cristatus	L.	1	1	—	0	0	0	6.5	7	182
Cynosurus cristatus	L.	1	1	—	0	0	0	3	7	235
Cynosurus cristatus	L.	1	1	—	0	0	0	5	5	244
Cynosurus cristatus	L.	1	6	—	0	0	0	5	5	251
Cynosurus cristatus	L.	1	5	—	0	52	10	2	7	99
Cynosurus cristatus	L.	1	1	—	12	12	12	10	5	252
Cynosurus cristatus	L.	1	7	—	0	366	74	30	5	45
Cynosurus cristatus	L.	1	6	—	156	156	156	10	5	27
Cynosurus cristatus	L.	3	1	—	320	320	320	8	5	108
Cynosurus cristatus	L.	4	1	—	50	50	50	15	5	267
Cynosurus cristatus	L.	4	1	—	95	95	95	5	5	244
Cynosurus cristatus	L.	4	2	—	126	151	139	17.8	5	155
Cynosurus cristatus	L.	4	1	—	313	313	313	5	4	88
Cynosurus cristatus	L.	4	1	—	1104	1104	1104	5	5	69
Cynosurus echinatus	L.	2	1	2	0	0	0	25	3	209
Dactylis glomerata	L.	1	1	—	0	0	0	6	5	49
Dactylis glomerata	L.	1	1	—	0	0	0	5	4	53
Dactylis glomerata	L.	1	1	—	0	0	0	15	5	62
Dactylis glomerata	L.	1	1	—	0	0	0	6	5	67
Dactylis glomerata	L.	1	1	—	0	0	0	20	5	67
Dactylis glomerata	L.	1	2	—	0	0	0	6.5	5	68
Dactylis glomerata	L.	1	2	—	0	0	0	13	5	68
Dactylis glomerata	L.	1	9	—	0	0	0	5	5	69
Dactylis glomerata	L.	1	1	—	0	0	0	5	4	88
Dactylis glomerata	L.	1	13	—	0	0	0	3	5	96
Dactylis glomerata	L.	1	7	—	0	0	0	20	5	97
Dactylis glomerata	L.	1	1	—	0	0	0	12.5	5	126
Dactylis glomerata	L.	1	4	—	0	0	0	6.5	5	128
Dactylis glomerata	L.	1	1	—	0	0	0	15	5	142
Dactylis glomerata	L.	1	8	—	0	0	0	10	5	152
Dactylis glomerata	L.	1	3	—	0	0	0	12	5	174
Dactylis glomerata	L.	1	2	—	0	0	0	6	5	181

Species										
Dactylis glomerata	L.	1	2	—	0	0	0	10	5	184
Dactylis glomerata	L.	1	6	—	0	0	0	20	4	225
Dactylis glomerata	L.	1	1	—	0	0	0	5	5	244
Dactylis glomerata	L.	1	24	—	0	0	0	10	5	255
Dactylis glomerata	L.	1	5	—	0	9	5	3	7	235
Dactylis glomerata	L.	1	1	—	54	54	54	30	5	29
Dactylis glomerata	L.	1	7	—	53	281	123	8	7	275
Dactylis glomerata	L.	1	1	—	300	300	300	20	5	69
Dactylis glomerata	L.	1	13	—	300	1300	783	5	5	251
Dactylis glomerata	L.	1	4	—	875	875	875	10	4	88
Dactylis glomerata	L.	2	1	2	0	0	0	8	1	61
Dactylis glomerata	L.	2	1	2	0	0	0	20	1	61
Dactylis glomerata	L.	2	1	2	0	0	0	30	1	61
Dactylis glomerata	L.	2	1	2	0	0	0	25	1	125
Dactylis glomerata	L.	2	1	1	0	0	0	26	1	133
Dactylis glomerata	L.	2	1	4	0	0	0	13	1	133
Dactylis glomerata	L.	2	1	4	0	0	0	26	1	133
Dactylis glomerata	L.	2	1	4	0	0	0	39	1	133
Dactylis glomerata	L.	2	1	2	0	0	0	0	3	188
Dactylis glomerata	L.	2	1	>2	0	0	0	7.5	2	194
Dactylis glomerata	L.	2	5	>1	6	6	6	—	1	275
Dactylis glomerata	L.	2	1	—	6	6	6	3	7	235
Dactylis glomerata	L.	2	1	—	13	13	13	8	5	108
Dactylis glomerata	L.	2	1	—	55	55	55	6	5	49
Dactylis glomerata	L.	2	4	—	54	237	121	30	5	45
Dactylis glomerata	L.	2	3	—	500	900	767	5	5	251
Dactylis glomerata	L.	2	2	—	234	1465	850	10	5	252
Dactylis glomerata	L.	2	1	—	1200	1200	1200	20	4	225
Dactylis glomerata	L.	3	1	—	32	32	32	30	5	45
Dactylis glomerata	L.	3	2	—	62	99	81	11.5	5	134
Dactylis glomerata	L.	4	1	—	7	7	7	20	4	111
Dactylis glomerata	L.	4	2	—	9	86	48	12	5	174
Dactylis glomerata	L.	4	2	—	17	110	64	7	5	58
Dactylis glomerata	L.	4	4	—	75	675	256	10	5	255
Dactylis glomerata	L.	4	3	—	13	781	286	10	5	110
Dactylis glomerata	L.	4	1	—	458	458	458	5	5	98
Danthonia decumbens	(L.) DC.	1	1	—	0	0	0	5	5	69
Danthonia decumbens	(L.) DC.	1	1	—	0	0	0	5	5	123
Danthonia decumbens	(L.) DC.	1	1	—	0	0	0	2	5	153
Danthonia decumbens	(L.) DC.	1	6	—	0	0	0	10	5	255
Danthonia decumbens	(L.) DC.	1	2	—	0	8	4	3	7	235
Danthonia decumbens	(L.) DC.	1	1	—	97	97	97	22.5	5	45
Danthonia decumbens	(L.) DC.	2	1	>2	0	0	0	7.5	2	194
Danthonia decumbens	(L.) DC.	2	1	—	12	12	12	3	7	235
Danthonia decumbens	(L.) DC.	2	1	—	50	50	50	10	5	255
Danthonia decumbens	(L.) DC.	2	1	—	706	706	706	5	5	233

Species	Authority	Seed bank type	Number of records	Longevity (y)	Minimum density (seeds m⁻²)	Maximum density (seeds m⁻²)	Mean density (seeds m⁻²)	Depth (cm)	Method	Source code
Danthonia decumbens	(L.) DC.	2	1	—	958	958	958	6	5	48
Danthonia decumbens	(L.) DC.	4	1	—	65	65	65	7	5	232
Danthonia decumbens	(L.) DC.	4	1	—	118	118	118	5	5	123
Danthonia decumbens	(L.) DC.	4	4	—	63	363	166	10	5	255
Danthonia decumbens	(L.) DC.	4	2	—	110	370	240	7	5	58
Danthonia decumbens	(L.) DC.	4	7	—	77	706	241	17.8	5	155
Danthonia decumbens	(L.) DC.	4	1	—	816	816	816	8.5	4	254
Deschampsia cespitosa	(L.) P. Beauv.	1	1	—	0	0	0	10	5	8
Deschampsia cespitosa	(L.) P. Beauv.	1	1	—	0	0	0	22.5	5	45
Deschampsia cespitosa	(L.) P. Beauv.	1	1	—	0	0	0	6	5	49
Deschampsia cespitosa	(L.) P. Beauv.	1	3	—	0	0	0	5	5	69
Deschampsia cespitosa	(L.) P. Beauv.	1	1	—	0	0	0	10	4	88
Deschampsia cespitosa	(L.) P. Beauv.	1	1	—	0	0	0	3	5	96
Deschampsia cespitosa	(L.) P. Beauv.	1	1	—	0	0	0	10	5	110
Deschampsia cespitosa	(L.) P. Beauv.	1	1	—	0	0	0	10	5	126
Deschampsia cespitosa	(L.) P. Beauv.	1	1	—	0	0	0	6.5	5	128
Deschampsia cespitosa	(L.) P. Beauv.	1	2	—	0	0	0	10	5	149
Deschampsia cespitosa	(L.) P. Beauv.	1	1	—	0	0	0	10	5	150
Deschampsia cespitosa	(L.) P. Beauv.	1	1	—	0	0	0	10	5	152
Deschampsia cespitosa	(L.) P. Beauv.	1	1	—	0	0	0	5	6	170
Deschampsia cespitosa	(L.) P. Beauv.	1	1	—	0	0	0	12	5	171
Deschampsia cespitosa	(L.) P. Beauv.	1	3	—	0	0	0	3	5	208
Deschampsia cespitosa	(L.) P. Beauv.	1	2	—	0	0	0	6	5	208
Deschampsia cespitosa	(L.) P. Beauv.	1	1	—	0	0	0	20	4	225
Deschampsia cespitosa	(L.) P. Beauv.	1	14	—	0	0	0	10	5	252
Deschampsia cespitosa	(L.) P. Beauv.	1	1	—	0	0	0	10	5	255
Deschampsia cespitosa	(L.) P. Beauv.	1	2	—	0	0	0	15	5	257
Deschampsia cespitosa	(L.) P. Beauv.	1	2	—	3	7	5	3	7	235
Deschampsia cespitosa	(L.) P. Beauv.	1	1	—	26	26	26	2	7	99
Deschampsia cespitosa	(L.) P. Beauv.	1	1	—	56	56	56	16	5	168
Deschampsia cespitosa	(L.) P. Beauv.	1	4	—	300	300	300	5	5	251
Deschampsia cespitosa	(L.) P. Beauv.	2	1	>2	0	0	0	7.5	2	194
Deschampsia cespitosa	(L.) P. Beauv.	2	2	—	1	1	1	5	6	170
Deschampsia cespitosa	(L.) P. Beauv.	2	1	—	36	36	36	5	5	220
Deschampsia cespitosa	(L.) P. Beauv.	2	2	—	33	39	36	3	7	235
Deschampsia cespitosa	(L.) P. Beauv.	2	2	—	50	133	92	3	5	96
Deschampsia cespitosa	(L.) P. Beauv.	2	1	—	175	175	175	10	5	150
Deschampsia cespitosa	(L.) P. Beauv.	2	1	—	375	375	375	10	5	255
Deschampsia cespitosa	(L.) P. Beauv.	2	1	—	781	781	781	10	5	110
Deschampsia cespitosa	(L.) P. Beauv.	2	1	—	1641	1641	1641	6	5	208
Deschampsia cespitosa	(L.) P. Beauv.	3	1	—	400	400	400	5	5	251
Deschampsia cespitosa	(L.) P. Beauv.	3	1	—	416	416	416	6	5	11

Species	Author									
Deschampsia cespitosa	(L.) P. Beauv.	3	3	—	113	1387	925	6	5	208
Deschampsia cespitosa	(L.) P. Beauv.	4	1	—	8	8	8	5	5	213
Deschampsia cespitosa	(L.) P. Beauv.	4	2	—	43	79	61	12.5	5	126
Deschampsia cespitosa	(L.) P. Beauv.	4	1	—	67	67	67	3	5	96
Deschampsia cespitosa	(L.) P. Beauv.	4	2	—	60	160	110	10	5	149
Deschampsia cespitosa	(L.) P. Beauv.	4	1	—	261	261	261	15	5	267
Deschampsia cespitosa	(L.) P. Beauv.	4	5	—	40	1353	480	6	5	39
Deschampsia cespitosa	(L.) P. Beauv.	4	5	—	146	1081	491	10	5	150
Deschampsia cespitosa	(L.) P. Beauv.	4	2	—	646	1064	855	6	5	151
Deschampsia cespitosa	(L.) P. Beauv.	4	1	—	1204	1204	1204	10	5	126
Deschampsia cespitosa	(L.) P. Beauv.	4	7	—	88	5200	1584	10	5	255
Deschampsia flexuosa	(L.) Trin.	1	1	—	0	0	0	1	5	56
Deschampsia flexuosa	(L.) Trin.	1	2	—	0	0	0	3	5	82
Deschampsia flexuosa	(L.) Trin.	1	1	—	0	0	0	5	5	82
Deschampsia flexuosa	(L.) Trin.	1	1	—	0	0	0	7.5	5	84
Deschampsia flexuosa	(L.) Trin.	1	1	—	0	0	0	9	5	84
Deschampsia flexuosa	(L.) Trin.	1	5	—	0	0	0	10	5	84
Deschampsia flexuosa	(L.) Trin.	1	5	—	0	0	0	5	5	92
Deschampsia flexuosa	(L.) Trin.	1	1	—	0	0	0	3	5	96
Deschampsia flexuosa	(L.) Trin.	1	3	—	0	0	0	17.5	5	126
Deschampsia flexuosa	(L.) Trin.	1	2	—	0	0	0	2	5	153
Deschampsia flexuosa	(L.) Trin.	1	3	0	0	0	0	10	3	178
Deschampsia flexuosa	(L.) Trin.	1	4	—	0	0	0	5	5	220
Deschampsia flexuosa	(L.) Trin.	1	4	—	0	0	0	7	5	232
Deschampsia flexuosa	(L.) Trin.	1	2	—	0	0	0	10	5	255
Deschampsia flexuosa	(L.) Trin.	1	1	—	0	0	0	15	5	257
Deschampsia flexuosa	(L.) Trin.	2	3	—	0	3	2	3	7	235
Deschampsia flexuosa	(L.) Trin.	2	1	—	7	7	7	3	7	235
Deschampsia flexuosa	(L.) Trin.	3	1	—	50	50	50	10	5	255
Deschampsia flexuosa	(L.) Trin.	4	1	>5	0	0	0	2	3	83
Deschampsia flexuosa	(L.) Trin.	4	4	—	26	26	26	2	5	82
Deschampsia flexuosa	(L.) Trin.	4	4	—	76	202	126	17.8	5	156
Deschampsia flexuosa	(L.) Trin.	4	6	—	101	301	172	17.8	5	155
Deschampsia flexuosa	(L.) Trin.	4	5	—	56	367	189	5	5	92
Deschampsia flexuosa	(L.) Trin.	4	9	—	38	1625	459	10	5	255
Digitaria ischaemum	(Schreber ex Schweigger) Muhlenb.	1	1	—	0	0	0	25	5	2
Digitaria ischaemum	(Schreber ex Schweigger) Muhlenb.	1	1	—	0	0	0	3	5	59
Digitaria ischaemum	(Schreber ex Schweigger) Muhlenb.	2	1	—	64	64	64	5	5	69
Digitaria ischaemum	(Schreber ex Schweigger) Muhlenb.	2	1	—	86	86	86	3	5	59
Digitaria ischaemum	(Schreber ex Schweigger) Muhlenb.	3	1	—	14	14	14	16	5	23
Digitaria ischaemum	(Schreber ex Schweigger) Muhlenb.	4	1	—	334	334	334	3	5	59
Digitaria sanguinalis	(L.) Scop.	1	1	—	0	0	0	25	4	36
Digitaria sanguinalis	(L.) Scop.	2	1	3	0	0	0	25	3	209
Digitaria sanguinalis	(L.) Scop.	2	1	—	14	14	14	25	4	36
Digitaria sanguinalis	(L.) Scop.	2	1	—	20	20	20	10	5	143
Digitaria sanguinalis	(L.) Scop.	2	3	—	829	1243	1103	15	5	175
Digitaria sanguinalis	(L.) Scop.	2	4	—	1542	5800	4189	15	5	272

Species	Authority	Seed bank type	Number of records	Longevity (y)	Minimum density (seeds m^{-2})	Maximum density (seeds m^{-2})	Mean density (seeds m^{-2})	Depth (cm)	Method	Source code
Digitaria sanguinalis	(L.) Scop.	2	2	—	4400	10400	7400	7	4	50
Digitaria sanguinalis	(L.) Scop.	3	1	>10	0	0	0	23	1	38
Digitaria sanguinalis	(L.) Scop.	3	1	>5	0	0	0	23	1	64
Digitaria sanguinalis	(L.) Scop.	3	1	—	646	646	646	15	5	175
Digitaria sanguinalis	(L.) Scop.	4	1	—	84	84	84	25	4	36
Digitaria sanguinalis	(L.) Scop.	4	2	—	53	213	133	30	4	13
Digitaria sanguinalis	(L.) Scop.	4	4	—	120	584	281	15	5	272
Digitaria sanguinalis	(L.) Scop.	4	1	—	350	350	350	8	5	137
Digitaria sanguinalis	(L.) Scop.	4	5	—	27	2983	1152	13.5	5	163
Digitaria sanguinalis	(L.) Scop.	4	2	—	33300	37700	35500	7	4	50
Echinochloa crusgalli	(L.) P. Beauv.	2	1	4	0	0	0	23	1	64
Echinochloa crusgalli	(L.) P. Beauv.	2	1	3	0	0	0	20	1	228
Echinochloa crusgalli	(L.) P. Beauv.	2	2	—	32	86	59	3	5	59
Echinochloa crusgalli	(L.) P. Beauv.	2	4	—	64	163	102	15	5	272
Echinochloa crusgalli	(L.) P. Beauv.	2	1	—	816	816	816	8.5	4	254
Echinochloa crusgalli	(L.) P. Beauv.	3	1	11	0	0	0	10	1	52
Echinochloa crusgalli	(L.) P. Beauv.	3	1	13	0	0	0	20	1	52
Echinochloa crusgalli	(L.) P. Beauv.	3	1	9	0	0	0	2.5	1	52
Echinochloa crusgalli	(L.) P. Beauv.	3	1	>11	0	0	0	25	3	209
Echinochloa crusgalli	(L.) P. Beauv.	3	1	>4	0	0	0	15	1	258
Echinochloa crusgalli	(L.) P. Beauv.	3	1	—	135	135	135	15	5	175
Echinochloa crusgalli	(L.) P. Beauv.	4	6	—	20	220	56	30	4	13
Echinochloa crusgalli	(L.) P. Beauv.	4	1	—	407	407	407	25	4	136
Echinochloa crusgalli	(L.) P. Beauv.	4	1	—	1900	1900	1900	10	5	147
Echinochloa crusgalli	(L.) P. Beauv.	4	3	—	132	6818	4050	15	5	272
Elymus caninus	(L.) L.	1	1	—	0	0	0	12.5	5	126
Elymus caninus	(L.) L.	2	1	>2	0	0	0	7.5	2	194
Elytrigia atherica	(Link) Kerguelen ex Carreras Martinez	1	5	—	0	0	0	1	5	57
Elytrigia atherica	(Link) Kerguelen ex Carreras Martinez	4	3	—	500	1200	733	1	5	57
Elytrigia repens	(L.) Desv. ex Nevski	1	1	—	0	0	0	10	5	8
Elytrigia repens	(L.) Desv. ex Nevski	1	1	—	0	0	0	6	5	11
Elytrigia repens	(L.) Desv. ex Nevski	1	2	—	0	0	0	25	4	36
Elytrigia repens	(L.) Desv. ex Nevski	1	1	—	0	0	0	5	4	53
Elytrigia repens	(L.) Desv. ex Nevski	1	8	—	0	0	0	1	5	57
Elytrigia repens	(L.) Desv. ex Nevski	1	2	—	0	0	0	3	5	59
Elytrigia repens	(L.) Desv. ex Nevski	1	1	—	0	0	0	15	5	62
Elytrigia repens	(L.) Desv. ex Nevski	1	1	—	0	0	0	50	5	66
Elytrigia repens	(L.) Desv. ex Nevski	1	5	—	0	0	0	5	5	69
Elytrigia repens	(L.) Desv. ex Nevski	1	1	—	0	0	0	10	4	88
Elytrigia repens	(L.) Desv. ex Nevski	1	4	—	0	0	0	15	5	94
Elytrigia repens	(L.) Desv. ex Nevski	1	25	—	0	0	0	3	5	96
Elytrigia repens	(L.) Desv. ex Nevski	1	2	—	0	0	0	20	5	97

Species	Author									
Elytrigia repens	(L.) Desv. ex Nevski	1	1	—	0	0	0	20	4	111
Elytrigia repens	(L.) Desv. ex Nevski	1	1	—	0	0	0	20	5	111
Elytrigia repens	(L.) Desv. ex Nevski	1	1	—	0	0	0	38	5	186
Elytrigia repens	(L.) Desv. ex Nevski	1	1	—	0	0	0	3	5	208
Elytrigia repens	(L.) Desv. ex Nevski	1	3	—	0	0	0	20	4	225
Elytrigia repens	(L.) Desv. ex Nevski	1	4	—	0	0	0	5	5	251
Elytrigia repens	(L.) Desv. ex Nevski	1	1	—	0	0	0	10	5	252
Elytrigia repens	(L.) Desv. ex Nevski	1	19	—	0	0	0	10	5	255
Elytrigia repens	(L.) Desv. ex Nevski	1	2	—	0	5	2	3	7	235
Elytrigia repens	(L.) Desv. ex Nevski	2	8	>3	53	158	81	8	7	275
Elytrigia repens	(L.) Desv. ex Nevski	2	3	1	0	0	0	7	2	44
Elytrigia repens	(L.) Desv. ex Nevski	2	2	3	0	0	0	7	2	44
Elytrigia repens	(L.) Desv. ex Nevski	2	2	4	0	0	0	7	2	44
Elytrigia repens	(L.) Desv. ex Nevski	2	1	>2	0	0	0	7	2	47
Elytrigia repens	(L.) Desv. ex Nevski	2	1	>2	0	0	0	2	1	47
Elytrigia repens	(L.) Desv. ex Nevski	2	1	1	0	0	0	15	1	240
Elytrigia repens	(L.) Desv. ex Nevski	2	1	3	0	0	0	15	1	240
Elytrigia repens	(L.) Desv. ex Nevski	2	1	—	0	0	0	50	1	240
Elytrigia repens	(L.) Desv. ex Nevski	2	1	5	510	510	510	8.5	4	254
Elytrigia repens	(L.) Desv. ex Nevski	2	1	10	1196	1196	1196	6	5	11
Elytrigia repens	(L.) Desv. ex Nevski	3	1	>4	0	0	0	7	2	44
Elytrigia repens	(L.) Desv. ex Nevski	3	1	—	0	0	0	100	1	240
Elytrigia repens	(L.) Desv. ex Nevski	3	1	—	0	0	0	0	3	266
Elytrigia repens	(L.) Desv. ex Nevski	4	2	—	102	102	102	15	5	142
Elytrigia repens	(L.) Desv. ex Nevski	4	1	—	100	125	113	10	5	255
Elytrigia repens	(L.) Desv. ex Nevski	4	1	—	220	220	220	30	4	13
Elytrigia repens	(L.) Desv. ex Nevski	4	1	—	4000	4000	4000	20	4	225
Festuca arundinacea		1	3	—	0	0	0	13	5	68
Festuca arundinacea	Schreber	1	1	—	0	0	0	5	5	69
Festuca arundinacea	Schreber	1	5	—	0	0	0	3	5	96
Festuca arundinacea	Schreber	1	2	—	0	0	0	20	5	97
Festuca arundinacea	Schreber	1	1	0	0	0	0	6.5	5	128
Festuca arundinacea	Schreber	1	1	—	0	0	0	26	1	133
Festuca arundinacea	Schreber	1	2	—	0	0	0	20	4	225
Festuca arundinacea	Schreber	1	1	—	0	0	0	3	7	235
Festuca arundinacea	Schreber	1	1	1	100	100	100	6.5	5	68
Festuca arundinacea	Schreber	2	1	1	299	299	299	6	5	48
Festuca arundinacea	Schreber	2	2	1	0	0	0	13	1	133
Festuca arundinacea	Schreber	2	1	2	0	0	0	26	1	133
Festuca arundinacea	Schreber	2	1	—	0	0	0	39	1	133
Festuca arundinacea	Schreber	1	2	1	0	0	0	0	3	188
Festuca filiformis	Pourret	1	1	—	0	0	0	3	5	208
Festuca filiformis	Pourret	4	1	—	31	31	31	5	5	222
Festuca gigantea	(L.) Villars	1	2	—	0	0	0	6	5	49
Festuca gigantea	(L.) Villars	1	4	—	0	0	0	3	5	96
Festuca gigantea	(L.) Villars	1	1	—	0	0	0	20	4	225
Festuca gigantea	(L.) Villars	1	2	—	0	0	0	15	5	257

Species	Authority	Seed bank type	Number of records	Longevity (y)	Minimum density (seeds m⁻²)	Maximum density (seeds m⁻²)	Mean density (seeds m⁻²)	Depth (cm)	Method	Source code
Festuca gigantea	(L.) Villars	2	1	—	50	50	50	3	5	96
Festuca heterophylla	Lam.	1	1	—	0	0	0	6	5	49
Festuca ovina	L.	1	4	—	0	0	0	6	5	28
Festuca ovina	L.	1	1	—	0	0	0	15	5	45
Festuca ovina	L.	1	2	—	0	0	0	7	5	58
Festuca ovina	L.	1	1	—	0	0	0	6	5	67
Festuca ovina	L.	1	1	—	0	0	0	5	5	69
Festuca ovina	L.	1	1	—	0	0	0	3	5	96
Festuca ovina	L.	1	1	—	0	0	0	10	5	149
Festuca ovina	L.	1	1	—	0	0	0	17.8	5	156
Festuca ovina	L.	1	2	—	0	0	0	6	5	181
Festuca ovina	L.	1	1	—	0	0	0	7	5	232
Festuca ovina	L.	1	12	—	0	0	0	5	5	233
Festuca ovina	L.	1	4	—	0	21	8	10	5	255
Festuca ovina	L.	1	2	—	8	14	11	3	7	235
Festuca ovina	L.	1	1	—	24	24	24	6.5	7	182
Festuca ovina	L.	1	1	—	32	32	32	13	5	183
Festuca ovina	L.	1	1	—	43	43	43	22.5	5	45
Festuca ovina	L.	2	1	—	13	13	13	9	5	84
Festuca ovina	L.	2	1	—	54	54	54	3	7	235
Festuca ovina	L.	2	1	—	778	778	778	15	5	45
Festuca ovina	L.	3	1	—	32	32	32	6	5	48
Festuca ovina	L.	4	1	—	22	22	22	30	5	29
Festuca ovina	L.	4	1	—	23	23	23	10	5	110
Festuca ovina	L.	4	4	—	26	39	34	7	5	58
Festuca ovina	L.	4	1	—	81	81	81	7	5	232
Festuca ovina	L.	4	5	—	33	257	121	13.5	5	163
Festuca ovina	L.	4	1	—	179	179	179	6	5	39
Festuca ovina	L.	4	1	—	183	183	183	5	4	53
Festuca ovina	L.	4	5	—	76	859	243	3	5	96
Festuca ovina	L.	4	1	—	300	300	300	17.8	5	156
Festuca ovina	L.	4	2	—	191	1157	674	15	5	249
Festuca ovina	L.	4	7	—	427	5236	1273	5	5	123
Festuca pratensis	Hudson	1	4	—	0	0	0	17.8	5	155
Festuca pratensis	Hudson	1	1	—	0	0	0	10	5	27
Festuca pratensis	Hudson	1	1	—	0	0	0	30	5	45
Festuca pratensis	Hudson	1	1	—	0	0	0	5	4	53
Festuca pratensis	Hudson	1	4	—	0	0	0	3	5	59
Festuca pratensis	Hudson	1	2	—	0	0	0	13	5	68
Festuca pratensis	Hudson	1	10	—	0	0	0	5	5	69
Festuca pratensis	Hudson	1	1	—	0	0	0	5	4	88
Festuca pratensis	Hudson	1	3	—	0	0	0	20	5	97

Species	Source									
Festuca pratensis	Hudson	1	13	—	0	0	0	6	5	99
Festuca pratensis	Hudson	1	1	—	0	0	0	6.5	5	128
Festuca pratensis	Hudson	1	1	0	0	0	0	26	1	133
Festuca pratensis	Hudson	1	2	—	0	0	0	10	5	149
Festuca pratensis	Hudson	1	6	—	0	0	0	12	5	174
Festuca pratensis	Hudson	1	3	—	0	0	0	3	5	208
Festuca pratensis	Hudson	1	6	—	0	0	0	6	5	208
Festuca pratensis	Hudson	1	3	—	0	0	0	20	4	225
Festuca pratensis	Hudson	1	1	0	0	0	0	3	7	235
Festuca pratensis	Hudson	1	1	0	0	0	0	50	1	240
Festuca pratensis	Hudson	1	6	—	0	0	0	100	1	240
Festuca pratensis	Hudson	1	2	—	100	100	100	10	5	255
Festuca pratensis	Hudson	1	4	—	0	351	101	6.5	5	68
Festuca pratensis	Hudson	1	6	—	312	312	312	2	7	99
Festuca pratensis	Hudson	2	1	1	0	0	0	6	5	11
Festuca pratensis	Hudson	2	1	1	0	0	0	13	1	133
Festuca pratensis	Hudson	2	1	1	0	0	0	26	1	133
Festuca pratensis	Hudson	2	1	1	0	0	0	39	1	133
Festuca pratensis	Hudson	2	2	—	75	75	75	15	7	240
Festuca pratensis	Hudson	2	2	—	35	122	79	2	5	99
Festuca pratensis	Hudson	4	1	—	11	11	11	10	4	252
Festuca pratensis	Hudson	4	1	—	306	306	306	20	5	111
Festuca pratensis	Hudson	4	5	—	156	2704	718	7	5	58
Festuca pratensis	Hudson	4	1	—	0	0	0	6	5	99
Festuca rubra	L.	1	8	—	0	0	0	40	5	11
Festuca rubra	L.	1	1	—	0	0	0	10	5	11
Festuca rubra	L.	1	6	—	0	0	0	25	4	27
Festuca rubra	L.	1	1	—	0	0	0	6	5	36
Festuca rubra	L.	1	2	—	0	0	0	1	5	48
Festuca rubra	L.	1	5	—	0	0	0	3	5	57
Festuca rubra	L.	1	3	—	0	0	0	15	5	59
Festuca rubra	L.	1	1	—	0	0	0	5	5	62
Festuca rubra	L.	1	10	—	0	0	0	5	4	69
Festuca rubra	L.	1	1	—	0	0	0	10	4	88
Festuca rubra	L.	1	4	—	0	0	0	3	5	88
Festuca rubra	L.	1	23	—	0	0	0	20	5	96
Festuca rubra	L.	1	7	—	0	0	0	6	5	97
Festuca rubra	L.	1	11	—	0	0	0	10	4	99
Festuca rubra	L.	1	1	—	0	0	0	10	5	103
Festuca rubra	L.	1	1	—	0	0	0	7	5	110
Festuca rubra	L.	1	1	—	0	0	0	6.5	5	112
Festuca rubra	L.	1	1	—	0	0	0	12	5	128
Festuca rubra	L.	1	5	—	0	0	0	3	5	174
Festuca rubra	L.	1	5	—	0	0	0	5	5	208
Festuca rubra	L.	1	2	—	0	0	0	20	4	222
Festuca rubra	L.	1	3	—	0	0	0	5	5	225
Festuca rubra	L.	1	1	—	0	0	0	5	5	234

Species	Authority	Seed bank type	Number of records	Longevity (y)	Minimum density (seeds m⁻²)	Maximum density (seeds m⁻²)	Mean density (seeds m⁻²)	Depth (cm)	Method	Source code
Festuca rubra	L.	1	22	—	0	0	0	10	5	255
Festuca rubra	L.	1	7	—	0	12	5	3	7	235
Festuca rubra	L.	1	3	—	0	32	11	30	5	45
Festuca rubra	L.	1	5	—		52	16	2	7	99
Festuca rubra	L.	1	2	—	31	33	32	10	5	252
Festuca rubra	L.	1	2	—	50	50	50	6.5	5	68
Festuca rubra	L.	1	2	—	72	72	72	13	5	68
Festuca rubra	L.	1	4	—	53	210	118	8	7	275
Festuca rubra	L.	1	9	—	300	500	375	5	5	251
Festuca rubra	L.	2	1	1	0	0	0	13	1	133
Festuca rubra	L.	2	2	1	0	0	0	26	1	133
Festuca rubra	L.	2	1	1	0	0	0	39	1	133
Festuca rubra	L.	2	1	2	0	0	0	0	3	188
Festuca rubra	L.	2	1	1	0	0	0	7.5	2	194
Festuca rubra	L.	2	1	—	15	15	15	12	5	174
Festuca rubra	L.	2	2	—	50	88	69	10	5	255
Festuca rubra	L.	2	1	—	120	120	120	6.5	5	128
Festuca rubra	L.	2	1	—	320	320	320	10	5	149
Festuca rubra	L.	2	2	—	161	592	377	30	5	45
Festuca rubra	L.	2	3	—	400	500	433	1	5	57
Festuca rubra	L.	2	5	—	300	4800	2020	5	5	251
Festuca rubra	L.	3	2	—	75	97	86	30	5	45
Festuca rubra	L.	3	2	—	300	400	350	5	5	251
Festuca rubra	L.	4	1	—	16	16	16	10	5	110
Festuca rubra	L.	4	1	—	37	37	37	2	5	153
Festuca rubra	L.	4	1	—	85	85	85	3	5	208
Festuca rubra	L.	4	2	—	12	159	86	12	5	174
Festuca rubra	L.	4	1	—	92	92	92	7	5	58
Festuca rubra	L.	4	1	—	126	126	126	17.8	5	156
Festuca rubra	L.	4	12	—	38	438	142	10	5	255
Festuca rubra	L.	4	5	—	120	200	172	6.5	5	128
Festuca rubra	L.	4	1	—	298	298	298	15	5	267
Festuca rubra	L.	4	3	—	176	471	333	7	5	112
Festuca rubra	L.	4	2	—	375	378	377	17.8	5	155
Festuca rubra	L.	4	2	—	360	480	420	10	5	149
Festuca rubra	L.	4	6	—	400	1100	633	1	5	57
Festuca rubra	L.	4	2	—	400	1400	900	15	5	249
Festuca vivipara	(L.) Smith	1	1	—	0	0	0	5	5	55
Glyceria fluitans	(L.) R. Br.	1	1	—	0	0	0	6	5	11
Glyceria fluitans	(L.) R. Br.	1	5	—	0	0	0	6	5	99
Glyceria fluitans	(L.) R. Br.	1	5	—	0	0	0	3	5	208
Glyceria fluitans	(L.) R. Br.	1	1	—	0	0	0	6	5	208

Species	Author									
Glyceria fluitans	(L.) R. Br.	1	2	—	0	0	0	20	4	225
Glyceria fluitans	(L.) R. Br.	1	5	—	0	153	70	2	7	99
Glyceria fluitans	(L.) R. Br.	1	7	—	53	105	70	8	7	275
Glyceria fluitans	(L.) R. Br.	2	2	—	43	86	65	30	5	45
Glyceria fluitans	(L.) R. Br.	2	1	—	85	85	85	6	5	208
Glyceria fluitans	(L.) R. Br.	2	2	—	113	283	198	3	5	208
Glyceria fluitans	(L.) R. Br.	2	1	—	208	208	208	40	5	11
Glyceria fluitans	(L.) R. Br.	2	1	—	520	520	520	6	5	11
Glyceria fluitans	(L.) R. Br.	2	1	—	627	627	627	50	5	107
Glyceria fluitans	(L.) R. Br.	2	1	—	3536	3536	3536	10	5	27
Glyceria fluitans	(L.) R. Br.	3	3	—	156	260	191	6	5	11
Glyceria fluitans	(L.) R. Br.	3	4	—	170	311	226	6	5	208
Glyceria fluitans	(L.) R. Br.	3	1	—	312	312	312	40	5	11
Glyceria fluitans	(L.) R. Br.	3	6	—	208	1664	615	6	5	99
Glyceria fluitans	(L.) R. Br.	3	1	—	798	798	798	20	5	107
Glyceria fluitans	(L.) R. Br.	3	1	—	969	969	969	30	5	107
Glyceria fluitans	(L.) R. Br.	3	6	—	1248	38376	17957	10	5	27
Glyceria fluitans	(L.) R. Br.	4	2	—	83	83	83	3	5	96
Glyceria fluitans	(L.) R. Br.	4	1	—	142	142	142	3	5	208
Glyceria fluitans	(L.) R. Br.	4	1	—	171	171	171	8	7	275
Glyceria fluitans	(L.) R. Br.	4	1	—	208	208	208	6	5	99
Glyceria fluitans	(L.) R. Br.	4	3	—	25	583	217	5	5	98
Glyceria maxima	(Hartman) O. Holmb.	1	1	—	0	0	0	3	5	96
Glyceria maxima	(Hartman) O. Holmb.	1	6	—	0	0	0	3	5	208
Glyceria maxima	(Hartman) O. Holmb.	1	3	—	0	0	0	6	5	208
Glyceria maxima	(Hartman) O. Holmb.	1	1	—	0	0	0	5	5	248
Glyceria maxima	(Hartman) O. Holmb.	2	1	>2	0	0	0	7.5	2	194
Glyceria maxima	(Hartman) O. Holmb.	2	1	—	1635	1635	1635	50	5	216
Glyceria notata	Chevall.	2	1	>2	0	0	0	7.5	2	194
Helictotrichon pratense	(L.) Besser	1	1	—	0	0	0	6	5	67
Helictotrichon pratense	(L.) Besser	1	1	—	0	0	0	5	5	122
Helictotrichon pratense	(L.) Besser	1	1	—	0	0	0	13	5	183
Helictotrichon pratense	(L.) Besser	1	7	—	0	0	0	10	5	255
Helictotrichon pratense	(L.) Besser	2	2	—	0	25	12	3	7	235
Helictotrichon pratense	(L.) Besser	4	1	—	539	539	539	6	5	48
Helictotrichon pratense	(L.) Besser	4	2	—	9	9	9	5	5	122
Helictotrichon pubescens	(Hudson) Pilger	1	1	—	63	63	63	10	5	255
Helictotrichon pubescens	(Hudson) Pilger	1	2	—	0	0	0	5	5	69
Helictotrichon pubescens	(Hudson) Pilger	1	4	—	0	0	0	3	5	96
Helictotrichon pubescens	(Hudson) Pilger	1	2	—	0	0	0	20	5	97
Helictotrichon pubescens	(Hudson) Pilger	1	3	—	0	0	0	5	5	122
Helictotrichon pubescens	(Hudson) Pilger	1	4	—	0	0	0	6.5	5	128
Helictotrichon pubescens	(Hudson) Pilger	1	1	—	0	0	0	10	5	149
Helictotrichon pubescens	(Hudson) Pilger	1	1	—	0	0	0	6.5	7	182
Helictotrichon pubescens	(Hudson) Pilger	1	14	—	0	0	0	10	5	255
Helictotrichon pubescens	(Hudson) Pilger	1	2	—	0	6	3	3	7	235

Species	Authority	Seed bank type	Number of records	Longevity (y)	Minimum density (seeds m^{-2})	Maximum density (seeds m^{-2})	Mean density (seeds m^{-2})	Depth (cm)	Method	Source code
Helictotrichon pubescens	(Hudson) Pilger	2	1	—	60	60	60	6.5	5	128
Helictotrichon pubescens	(Hudson) Pilger	2	1	—	125	125	125	10	5	255
Helictotrichon pubescens	(Hudson) Pilger	4	1	—	80	80	80	10	5	149
Hierochloe alpina	(Willd.) Roemer & Schultes	1	1	—	0	0	0	3	4	158
Holcus lanatus	L.	1	1	—	0	0	0	10	5	27
Holcus lanatus	L.	1	1	—	0	0	0	22.5	5	45
Holcus lanatus	L.	1	4	—	0	0	0	5	4	53
Holcus lanatus	L.	1	1	—	0	0	0	5	5	69
Holcus lanatus	L.	1	1	—	0	0	0	5	5	92
Holcus lanatus	L.	1	1	—	0	0	0	1	5	96
Holcus lanatus	L.	1	9	—	0	0	0	3	5	96
Holcus lanatus	L.	1	5	—	0	0	0	20	5	97
Holcus lanatus	L.	1	3	—	0	0	0	6	5	99
Holcus lanatus	L.	1	4	—	0	0	0	6.5	5	128
Holcus lanatus	L.	1	2	—	0	0	0	12	5	174
Holcus lanatus	L.	1	1	—	0	0	0	3	5	208
Holcus lanatus	L.	1	3	—	0	0	0	5	5	222
Holcus lanatus	L.	1	7	—	0	0	0	20	4	225
Holcus lanatus	L.	1	1	—	0	0	0	5	5	244
Holcus lanatus	L.	1	14	—	0	0	0	10	5	255
Holcus lanatus	L.	1	1	—	7	7	7	12	4	51
Holcus lanatus	L.	1	2	—	11	11	11	6.5	7	182
Holcus lanatus	L.	1	1	—	33	33	33	2	7	99
Holcus lanatus	L.	1	4	—	53	79	66	8	7	275
Holcus lanatus	L.	1	1	—	124	124	124	20	5	67
Holcus lanatus	L.	1	1	—	150	150	150	20	5	4
Holcus lanatus	L.	1	1	—	269	269	269	30	5	45
Holcus lanatus	L.	1	2	—	375	375	375	10	4	88
Holcus lanatus	L.	1	9	—	300	1100	725	5	5	251
Holcus lanatus	L.	2	1	4	0	0	0	13	1	133
Holcus lanatus	L.	2	2	4	0	0	0	26	1	133
Holcus lanatus	L.	2	2	4	0	0	0	39	1	133
Holcus lanatus	L.	2	1	>2	0	0	0	7.5	2	194
Holcus lanatus	L.	2	5	>1	0	0	0	—	1	275
Holcus lanatus	L.	2	1	—	10	10	10	25	4	36
Holcus lanatus	L.	2	1	—	11	11	11	15	5	32
Holcus lanatus	L.	2	7	—	5	189	47	3	7	235
Holcus lanatus	L.	2	2	—	38	63	51	10	5	255
Holcus lanatus	L.	2	2	—	113	311	212	3	5	208
Holcus lanatus	L.	2	1	—	248	248	248	30	5	29
Holcus lanatus	L.	2	7	—	50	600	290	3	5	96
Holcus lanatus	L.	2	2	—	108	680	394	13	5	68

Species										
Holcus lanatus	L.	2	1	—	400	400	400	1	5	57
Holcus lanatus	L.	2	4	—	105	999	499	8	7	275
Holcus lanatus	L.	2	7	—	140	1959	715	30	5	45
Holcus lanatus	L.	2	1	—	760	760	760	15	5	257
Holcus lanatus	L.	2	2	—	798	798	798	20	5	107
Holcus lanatus	L.	2	1	—	344	1812	1078	10	5	252
Holcus lanatus	L.	2	2	—	1138	1138	1138	6	5	48
Holcus lanatus	L.	2	2	—	300	2075	1188	6.5	4	68
Holcus lanatus	L.	2	1	—	500	1875	1188	10	4	88
Holcus lanatus	L.	2	8	—	1200	1200	1200	20	5	225
Holcus lanatus	L.	2	4	—	1300	1300	1300	10	7	27
Holcus lanatus	L.	2	1	—	624	2158	1407	6	5	28
Holcus lanatus	L.	2	9	—	432	2943	1414	2	5	99
Holcus lanatus	L.	2	1	—	1612	1612	1612	40	5	11
Holcus lanatus	L.	2	7	—	832	5772	2380	6	5	11
Holcus lanatus	L.	2	1	—	3500	3500	3500	20	5	69
Holcus lanatus	L.	2	1	—	600	16900	5957	5	5	251
Holcus lanatus	L.	3	2	—	108	108	108	8	5	108
Holcus lanatus	L.	3	1	—	113	113	113	3	5	208
Holcus lanatus	L.	3	1	—	129	151	140	30	5	29
Holcus lanatus	L.	3	6	—	228	228	228	50	5	107
Holcus lanatus	L.	3	1	—	388	388	388	30	5	45
Holcus lanatus	L.	3	1	>12	636	636	636	50	5	219
Holcus lanatus	L.	3	1	—	156	6812	2383	10	5	27
Holcus lanatus	L.	3	1	—	3120	3120	3120	6	5	99
Holcus lanatus	L.	4		—	6	6	6	10	5	120
Holcus lanatus	L.	4		—	60	60	60	10	5	110
Holcus lanatus	L.	4		—	60	60	60	6.5	5	128
Holcus lanatus	L.	4		—	18	185	71	12	5	174
Holcus lanatus	L.	4	5	—	74	148	111	15	5	62
Holcus lanatus	L.	4	2	—	71	199	119	5	5	92
Holcus lanatus	L.	4	3	—	145	145	145	5	5	244
Holcus lanatus	L.	4	1	—	26	275	151	5	5	122
Holcus lanatus	L.	4	2	—	158	158	158	8	7	275
Holcus lanatus	L.	4	1	—	162	162	162	5	5	234
Holcus lanatus	L.	4	1	—	31	449	206	5	5	222
Holcus lanatus	L.	4	7	—	38	1300	229	10	5	255
Holcus lanatus	L.	4	16	—	142	396	236	3	5	208
Holcus lanatus	L.	4	3	—	261	261	261	15	5	267
Holcus lanatus	L.	4	1	—	200	420	310	5	5	250
Holcus lanatus	L.	4	2	—	83	2850	587	3	5	96
Holcus lanatus	L.	4	10	—	48	976	608	5	5	69
Holcus lanatus	L.	4	3	—	17	1726	618	7	5	58
Holcus lanatus	L.	4	3	—	553	2164	1208	17.8	5	155
Holcus lanatus	L.	4	1	—	3125	3125	3125	5	4	88
Holcus lanatus	L.	4	5	—	1200	10000	3760	20	4	225
Holcus lanatus	L.	4	15	—	312	16536	3966	6	5	99

142 / THE DATABASE

Species	Authority	Seed bank type	Number of records	Longevity (y)	Minimum density (seeds m^{-2})	Maximum density (seeds m^{-2})	Mean density (seeds m^{-2})	Depth (cm)	Method	Source code
Holcus mollis	L.	1	1	—	0	0	0	25	4	36
Holcus mollis	L.	1	13	—	0	0	0	3	5	96
Holcus mollis	L.	1	3	—	0	0	0	5	5	222
Holcus mollis	L.	1	4	—	0	0	0	20	4	225
Holcus mollis	L.	1	7	—	0	0	0	10	5	255
Holcus mollis	L.	1	2	—	0	0	0	15	5	257
Holcus mollis	L.	1	4	—	0	3	1	3	7	235
Holcus mollis	L.	4	1	—	101	101	101	17.8	5	155
Hordelymus europaeus	(L.) Jessen	1	2	—	0	0	0	20	5	67
Hordeum distichon	L.	1	1	0	0	0	0	25	1	125
Hordeum jubatum	L.	1	1	—	0	0	0	5	5	242
Hordeum jubatum	L.	2	1	2	0	0	0	7	2	44
Hordeum jubatum	L.	2	2	3	0	0	0	7	2	44
Hordeum jubatum	L.	2	1	>2	0	0	0	2	1	47
Hordeum jubatum	L.	2	1	>2	0	0	0	15	1	47
Hordeum jubatum	L.	4	1	—	1	1	1	5	5	166
Hordeum jubatum	L.	4	3	—	70	714	454	15	4	46
Hordeum murinum	(Link) Arcang.	2	1	1	0	0	0	7.5	2	194
Hordeum murinum subsp. leporinum		1	1	—	0	0	0	3	5	132
Hordeum secalinum	Schreber	1	1	—	0	0	0	5	5	244
Hordeum secalinum	Schreber	1	1	—	0	0	0	10	5	252
Hordeum secalinum	Schreber	1	2	—	300	500	400	5	5	251
Hordeum secalinum	Schreber	3	1	—	5	5	5	15	4	21
Hordeum vulgare	L.	1	1	0	0	0	0	15	1	240
Hordeum vulgare	L.	1	1	0	0	0	0	50	1	240
Hordeum vulgare	L.	1	1	0	0	0	0	100	1	240
Hordeum vulgare	L.	4	1	—	17	17	17	7	5	58
Hordeum vulgare	L.	4	1	—	91	91	91	20	5	111
Koeleria macrantha	(Ledeb.) Schultes	1	1	—	0	0	0	6	5	67
Koeleria macrantha	(Ledeb.) Schultes	1	1	—	0	0	0	1	5	96
Koeleria macrantha	(Ledeb.) Schultes	1	2	—	0	0	0	10	5	110
Koeleria macrantha	(Ledeb.) Schultes	1	1	—	0	0	0	2.5	5	114
Koeleria macrantha	(Ledeb.) Schultes	1	1	—	0	0	0	5	5	122
Koeleria macrantha	(Ledeb.) Schultes	1	1	—	0	0	0	13	5	183
Koeleria macrantha	(Ledeb.) Schultes	1	2	—	3	4	3	3	7	235
Koeleria macrantha	(Ledeb.) Schultes	2	1	—	719	719	719	6	5	48
Koeleria macrantha	(Ledeb.) Schultes	4	2	—	7	12	10	20	4	226
Koeleria macrantha	(Ledeb.) Schultes	4	3	—	17	217	86	2.5	5	114
Koeleria macrantha	(Ledeb.) Schultes	4	1	—	1633	1633	1633	8.5	4	254
Koeleria pyramidata	(Lam.) Beauv.	1	1	—	0	0	0	6	5	67
Koeleria pyramidata	(Lam.) Beauv.	1	1	—	0	0	0	6	5	181
Koeleria pyramidata	(Lam.) Beauv.	1	4	—	0	0	0	10	5	255

Species	Author									
Koeleria pyramidata	(Lam.) Beauv.	4	1	—	32	32	32	6	5	181
Koeleria pyramidata	(Lam.) Beauv.	4	2	—	38	75	57	10	5	255
Leersia oryzoides	(L.) Sw.	2	1	—	7866	7866	7866	35	5	247
Leersia oryzoides	(L.) Sw.	3	1	—	250	250	250	35	5	217
Leersia oryzoides	(L.) Sw.	4	7	—	8	48	24	7	4	50
Lolium multiflorum	Lam.	1	4	1	0	0	0	26	1	133
Lolium multiflorum	Lam.	2	1	4	0	0	0	13	1	133
Lolium multiflorum	Lam.	2	1	4	0	0	0	26	1	133
Lolium multiflorum	Lam.	2	1	4	0	0	0	39	1	133
Lolium multiflorum	Lam.	2	1	>2	0	0	0	7.5	2	194
Lolium multiflorum	Lam.	3	1	>7	0	0	0	0	3	188
Lolium multiflorum	Lam.	3	1	6	0	0	0	0	3	188
Lolium multiflorum	Lam.	4	5	—	530	1940	974	2	5	95
Lolium multiflorum	Lam.	4	1	—	11200	11200	11200	10	5	140
Lolium perenne	L.	1	2	—	0	0	0	6	5	11
Lolium perenne	L.	1	4	—	0	0	0	6	5	28
Lolium perenne	L.	1	1	—	0	0	0	25	4	36
Lolium perenne	L.	1	6	—	0	0	0	5	4	53
Lolium perenne	L.	1	2	—	0	0	0	15	5	62
Lolium perenne	L.	1	7	—	0	0	0	5	5	69
Lolium perenne	L.	1	1	—	0	0	0	5	4	88
Lolium perenne	L.	1	4	—	0	0	0	10	4	88
Lolium perenne	L.	1	4	—	0	0	0	3	5	96
Lolium perenne	L.	1	2	—	0	0	0	20	5	97
Lolium perenne	L.	1	10	—	0	0	0	6	5	99
Lolium perenne	L.	1	9	—	0	0	0	5	5	222
Lolium perenne	L.	1	1	—	0	0	0	5	5	244
Lolium perenne	L.	1	1	—	0	0	0	5	5	248
Lolium perenne	L.	1	5	—	0	32	6	30	5	45
Lolium perenne	L.	1	5	—	0	52	17	2	7	99
Lolium perenne	L.	1	1	—	18	18	18	3	7	235
Lolium perenne	L.	1	3	1	53	105	70	8	7	275
Lolium perenne	L.	1	4	1	156	156	156	10	5	27
Lolium perenne	L.	1	14	1	300	1000	480	5	5	251
Lolium perenne	L.	2	1	—	0	0	0	8	1	60
Lolium perenne	L.	2	1	1	0	0	0	20	1	60
Lolium perenne	L.	2	1	1	0	0	0	30	1	60
Lolium perenne	L.	2	1	4	0	0	0	25	1	125
Lolium perenne	L.	2	1	4	0	0	0	26	1	133
Lolium perenne	L.	2	1	4	0	0	0	13	1	133
Lolium perenne	L.	2	1	2	0	0	0	26	1	133
Lolium perenne	L.	2	1	4	0	0	0	39	3	188
Lolium perenne	L.	2	1	>2	0	0	0	0	3	188
Lolium perenne	L.	2	1	—	0	0	0	7.5	2	194
Lolium perenne	L.	2	2	—	214	520	367	10	5	252

Species	Authority	Seed bank type	Number of records	Longevity (y)	Minimum density (seeds m^{-2})	Maximum density (seeds m^{-2})	Mean density (seeds m^{-2})	Depth (cm)	Method	Source code
Lolium perenne	L.	2	1	—	441	441	441	30	5	29
Lolium perenne	L.	2	2	—	700	800	750	5	5	251
Lolium perenne	L.	3	1	—	108	108	108	8	5	108
Lolium perenne	L.	3	1	>14	156	156	156	6	5	11
Lolium perenne	L.	3	4	—	57	1249	413	30	4	29
Lolium perenne	L.	4	1	—	11	11	11	20	4	111
Lolium perenne	L.	4	1	—	49	49	49	15	5	62
Lolium perenne	L.	4	1	—	64	64	64	5	5	244
Lolium perenne	L.	4	1	—	69	69	69	7	5	58
Lolium perenne	L.	4	1	—	75	75	75	15	5	267
Lolium perenne	L.	4	1	—	167	167	167	3	5	96
Lolium perenne	L.	4	1	—	184	184	184	8	7	275
Lolium perenne	L.	4	1	—	1691	1691	1691	30	4	13
Lolium perenne	L.	4	3	—	128	4400	1701	5	5	69
Lolium rigidum	Gaudin	1	2	—	0	0	0	10	4	22
Lolium rigidum	Gaudin	2	2	2	0	0	0	7	2	44
Lolium rigidum	Gaudin	2	1	3	0	0	0	7	2	44
Lolium rigidum	Gaudin	2	1	—	23	23	23	10	4	22
Lolium rigidum	Gaudin	4	1	—	49	49	49	10	4	22
Melica ciliata	L.	1	2	—	0	0	0	3	5	132
Melica nutans	L.	1	2	—	0	0	0	5	6	170
Melica nutans	L.	1	2	—	0	0	0	12	5	171
Melica nutans	L.	1	2	—	0	0	0	16	5	216
Melica nutans	L.	2	2	—	5	5	5	5	6	170
Melica uniflora	Retz.	1	4	—	0	0	0	6	5	49
Melica uniflora	Retz.	1	1	—	0	0	0	20	5	67
Melica uniflora	Retz.	1	1	—	0	0	0	3	5	96
Melica uniflora	Retz.	1	1	—	0	0	0	10	5	126
Melica uniflora	Retz.	1	2	—	0	0	0	12.5	5	126
Mibora minima	(L.) Desv.	2	2	—	175	175	175	3	5	132
Mibora minima	(L.) Desv.	4	2	—	38	300	169	3	5	132
Milium effusum	L.	1	2	—	0	0	0	20	5	67
Milium effusum	L.	1	1	—	0	0	0	3	5	96
Milium effusum	L.	1	1	—	0	0	0	17.5	5	126
Milium effusum	L.	2	1	—	18	18	18	3	7	235
Milium effusum	L.	4	2	—	1	1	1	4	5	176
Milium effusum	L.	4	2	—	22	72	47	12.5	5	126
Milium effusum	L.	4	2	—	43	58	51	10	5	126
Milium effusum	L.	4	3	—	26	166	105	5	5	220
Molinia caerulea	(L.) Moench	1	1	—	0	0	0	6	5	28
Molinia caerulea	(L.) Moench	1	2	—	0	0	0	15	5	45
Molinia caerulea	(L.) Moench	1	1	—	0	0	0	1	5	56

Species	Author									
Molinia caerulea	(L.) Moench	1	1	—	0	0	0	12	5	174
Molinia caerulea	(L.) Moench	1	1	—	0	0	0	16	5	216
Molinia caerulea	(L.) Moench	1	5	—	0	0	0	5	5	222
Molinia caerulea	(L.) Moench	1	1	—	0	0	0	10	5	255
Molinia caerulea	(L.) Moench	1	2	—	0	8	4	3	7	235
Molinia caerulea	(L.) Moench	1	2	—	0	32	16	22.5	5	45
Molinia caerulea	(L.) Moench	2	2	>3	0	0	0	10	3	178
Molinia caerulea	(L.) Moench	2	1	2	0	0	0	10	3	216
Molinia caerulea	(L.) Moench	2	1	—	83	83	83	3	5	96
Molinia caerulea	(L.) Moench	2	2	—	97	97	97	30	5	45
Molinia caerulea	(L.) Moench	2	2	>4	540	860	700	30	5	180
Molinia caerulea	(L.) Moench	3	3	—	0	0	0	10	3	216
Molinia caerulea	(L.) Moench	3	3	—	36	36	36	12.5	5	126
Molinia caerulea	(L.) Moench	3	2	—	290	8850	4570	30	5	180
Molinia caerulea	(L.) Moench	4	1	—	22	22	22	17.5	5	126
Molinia caerulea	(L.) Moench	4	5	—	40	123	66	12	5	174
Molinia caerulea	(L.) Moench	4	1	—	191	191	191	16	5	216
Molinia caerulea	(L.) Moench	4	6	—	80	480	200	10	5	149
Molinia caerulea	(L.) Moench	4	4	—	151	301	245	17.8	5	155
Molinia caerulea	(L.) Moench	4	1	—	300	300	300	3	5	96
Nardus stricta	L.	1	1	—	0	0	0	15	4	45
Nardus stricta	L.	1	2	—	0	0	0	5	5	91
Nardus stricta	L.	1	1	—	0	0	0	2	5	153
Nardus stricta	L.	1	4	—	0	0	0	17.8	5	156
Nardus stricta	L.	1	5	—	0	10	5	10	7	255
Nardus stricta	L.	1	2	—	65	65	65	3	5	235
Nardus stricta	L.	2	1	—	118	118	118	22.5	5	45
Nardus stricta	L.	2	2	—	80	240	160	30	5	45
Nardus stricta	L.	2	2	—	175	175	175	10	5	149
Nardus stricta	L.	4	1	—	60	60	60	10	5	255
Nardus stricta	L.	4	3	—	69	113	95	7	5	232
Nardus stricta	L.	4	2	—	126	151	139	17.8	5	155
Nardus stricta	L.	4	1	—	150	150	150	3	5	96
Nardus stricta	L.	4	5	—	50	525	173	10	5	255
Phalaris arundinacea	L.	1	2	—	0	0	0	3	5	96
Phalaris arundinacea	L.	1	1	—	0	0	0	6.5	5	128
Phalaris arundinacea	L.	1	3	—	0	0	0	3	5	208
Phalaris arundinacea	L.	1	2	—	0	0	0	20	4	225
Phalaris arundinacea	L.	1	2	—	0	0	0	10	5	255
Phalaris arundinacea	L.	1	3	21	53	105	70	8	7	275
Phalaris arundinacea	L.	1	3	21	142	368	255	6	5	208
Phalaris arundinacea	L.	2	1	30	9	9	9	3	7	235
Phalaris arundinacea	L.	3	1	—	0	0	0	15	1	240
Phalaris arundinacea	L.	3	1	—	0	0	0	100	1	240
Phalaris arundinacea	L.	4	1	—	25	25	25	5	5	98

Species	Authority	Seed bank type	Number of records	Longevity (y)	Minimum density (seeds m⁻²)	Maximum density (seeds m⁻²)	Mean density (seeds m⁻²)	Depth (cm)	Method	Source code
Phalaris arundinacea	L.	4	3	—	142	396	236	3	5	208
Phalaris arundinacea	L.	4	1	—	333	333	333	3	5	96
Phalaris arundinacea	L.	4	2	—	1819	5717	3768	50	5	216
Phalaris canariensis	L.	1	1	—	0	0	0	10	4	22
Phleum alpinum	L.	1	1	—	0	0	0	5	5	55
Phleum alpinum	L.	2	1	—	19	19	19	5	4	91
Phleum phleoides	(L.) Karsten	1	2	—	0	0	0	6	5	67
Phleum pratense	L.	1	1	—	0	0	0	10	5	8
Phleum pratense	L.	1	1	—	0	0	0	5	5	19
Phleum pratense	L.	1	6	—	0	0	0	6	5	28
Phleum pratense	L.	1	1	—	0	0	0	25	4	36
Phleum pratense	L.	1	1	—	0	0	0	5	4	53
Phleum pratense	L.	1	2	—	0	0	0	3	5	59
Phleum pratense	L.	1	1	—	0	0	0	15	5	62
Phleum pratense	L.	1	7	—	0	0	0	5	5	69
Phleum pratense	L.	1	1	—	0	0	0	5	4	88
Phleum pratense	L.	1	3	—	0	0	0	10	4	88
Phleum pratense	L.	1	1	—	0	0	0	3	5	96
Phleum pratense	L.	1	5	—	0	0	0	20	5	97
Phleum pratense	L.	1	3	—	0	0	0	2	7	99
Phleum pratense	L.	1	3	—	0	0	0	6	5	99
Phleum pratense	L.	1	8	—	0	0	0	20	4	225
Phleum pratense	L.	1	6	—	0	0	0	5	5	251
Phleum pratense	L.	2	1	1	0	0	0	10	5	255
Phleum pratense	L.	2	1	4	0	0	0	26	1	133
Phleum pratense	L.	2	1	4	0	0	0	13	1	133
Phleum pratense	L.	2	1	4	0	0	0	26	1	133
Phleum pratense	L.	2	1	>1	0	0	0	39	1	275
Phleum pratense	L.	2	5	—	129	224	179	2.5	5	114
Phleum pratense	L.	2	5	—	214	214	214	5	4	53
Phleum pratense	L.	3	1	21	0	0	0	25	1	125
Phleum pratense	L.	3	1	21	0	0	0	15	1	240
Phleum pratense	L.	3	1	21	0	0	0	50	1	240
Phleum pratense	L.	3	1	—	521	521	521	100	1	240
Phleum pratense	L.	4	1	—	4	4	0	20	5	97
Phleum pratense	L.	4	1	—	43	43	43	10	5	33
Phleum pratense	L.	4	1	—	83	83	83	3	5	59
Phleum pratense	L.	4	1	—	151	151	151	5	5	234
Phleum pratense	L.	4	1	—				17.8	5	155
Phragmites australis	(Cav.) Trin. ex Steudel	4	2	—	1110	4120	2615	15	5	94
Phragmites australis		1	2	—	0	0	0	3	5	96

Species	Authority									
Phragmites australis	(Cav.) Trin. ex Steudel	1	8	—	0	0	0	7	5	112
Phragmites australis	(Cav.) Trin. ex Steudel	1	1	—	0	0	0	15	4	148
Phragmites australis	(Cav.) Trin. ex Steudel	1	1	—	0	0	0	50	5	216
Phragmites australis	(Cav.) Trin. ex Steudel	1	4	—	0	0	0	5	5	222
Phragmites australis	(Cav.) Trin. ex Steudel	1	2	—	0	0	0	20	4	225
Phragmites australis	(Cav.) Trin. ex Steudel	1	1	—	0	0	0	35	5	247
Phragmites australis	(Cav.) Trin. ex Steudel	1	1	—	0	0	0	5	5	248
Phragmites australis	(Cav.) Trin. ex Steudel	1	1	—	0	0	0	10	5	255
Phragmites australis	(Cav.) Trin. ex Steudel	4	1	—	10	10	10	5	5	262
Phragmites australis	(Cav.) Trin. ex Steudel	4	1	—	18	18	18	5	5	166
Phragmites australis	(Cav.) Trin. ex Steudel	4	1	—	31	31	31	16	5	216
Phragmites australis	(Cav.) Trin. ex Steudel	4	1	—	64	64	64	4	5	217
Phragmites australis	(Cav.) Trin. ex Steudel	4	2	—	93	284	189	12	4	230
Poa alpina	L.	1	2	—	0	0	0	5	4	91
Poa alpina	L.	4	2	—	35	49	42	5	5	55
Poa angustifolia	L.	1	1	—	0	0	0	3	5	96
Poa angustifolia	L.	4	1	—	83	83	83	3	5	96
Poa annua	L.	1	2	—	0	0	0	10	4	22
Poa annua	L.	1	2	—	0	0	0	5	5	69
Poa annua	L.	1	2	—	0	0	0	3	5	96
Poa annua	L.	1	3	—	0	26	9	2	7	99
Poa annua	L.	1	1	—	32	32	32	30	5	29
Poa annua	L.	1	2	—	78	78	78	6	5	28
Poa annua	L.	2	1	>2	1	0	0	7.5	2	194
Poa annua	L.	2	1	—	1	1	1	5	6	170
Poa annua	L.	2	1	—	11	11	11	6	5	211
Poa annua	L.	2	8	—	6	44	22	25	4	36
Poa annua	L.	2	3	—	26	28	27	15	5	272
Poa annua	L.	2	1	—	50	50	50	4	5	162
Poa annua	L.	2	3	—	12	117	66	12	5	174
Poa annua	L.	2	1	—	87	87	87	10	5	126
Poa annua	L.	2	3	—	51	122	92	5	5	222
Poa annua	L.	2	6	—	60	180	117	10	5	149
Poa annua	L.	2	5	—	8	609	136	3	7	235
Poa annua	L.	2	2	—	130	189	160	2	7	99
Poa annua	L.	2	1	—	234	234	234	6	5	28
Poa annua	L.	2	7	—	38	1688	316	10	5	255
Poa annua	L.	2	6	—	50	1250	433	3	5	96
Poa annua	L.	2	1	—	500	500	500	10	4	88
Poa annua	L.	2	12	—	338	874	541	8	7	275
Poa annua	L.	2	1	—	595	595	595	50	5	260
Poa annua	L.	2	1	—	600	600	600	50	5	66
Poa annua	L.	2	1	—	627	627	627	30	5	107
Poa annua	L.	2	4	—	312	1092	650	6	5	99
Poa annua	L.	2	1	—	795	795	795	8	5	108
Poa annua	L.	2	1	—	875	875	875	10	5	162

Species	Authority	Seed bank type	Number of records	Longevity (y)	Minimum density (seeds m⁻²)	Maximum density (seeds m⁻²)	Mean density (seeds m⁻²)	Depth (cm)	Method	Source code
Poa annua	L.	2	1	—	950	950	950	40	5	162
Poa annua	L.	2	3	—	226	2239	1094	30	5	45
Poa annua	L.	2	1	—	1175	1175	1175	90	5	162
Poa annua	L.	2	1	—	3730	3730	3730	5	5	244
Poa annua	L.	2	2	—	150	14200	7175	20	5	4
Poa annua	L.	2	1	—	20600	20600	20600	20	5	69
Poa annua	L.	3	1	>5	0	0	0	7.5	2	196
Poa annua	L.	3	1	>5	0	0	0	2.5	1	203
Poa annua	L.	3	1	>5	0	0	0	2.5	2	203
Poa annua	L.	3	1	>5	0	0	0	7.5	1	203
Poa annua	L.	3	1	>5	0	0	0	7.5	2	203
Poa annua	L.	3	1	>5	0	0	0	15	1	203
Poa annua	L.	3	1	>5	0	0	0	15	2	203
Poa annua	L.	3	1	—	18	18	18	15	4	21
Poa annua	L.	3	1	—	22	22	22	12.5	5	126
Poa annua	L.	3	1	—	25	25	25	6	5	211
Poa annua	L.	3	1	>10	38	38	38	10	4	241
Poa annua	L.	3	1	—	80	80	80	15	5	257
Poa annua	L.	3	1	—	162	162	162	5	5	234
Poa annua	L.	3	1	—	172	172	172	30	5	29
Poa annua	L.	3	1	—	300	300	300	20	5	4
Poa annua	L.	3	1	—	500	500	500	10	4	88
Poa annua	L.	3	2	—	208	832	520	10	5	27
Poa annua	L.	3	2	—	208	884	546	6	5	99
Poa annua	L.	3	6	—	86	6297	2101	30	5	45
Poa annua	L.	4	3	—	20	72	50	15	5	272
Poa annua	L.	4	2	—	12	110	61	25	4	36
Poa annua	L.	4	5	—	23	196	73	7	5	58
Poa annua	L.	4	1	—	80	80	80	25	5	2
Poa annua	L.	4	1	—	150	150	150	10	5	255
Poa annua	L.	4	1	—	152	152	152	17.8	5	156
Poa annua	L.	4	1	—	158	158	158	8	7	275
Poa annua	L.	4	2	—	150	167	159	3	5	96
Poa annua	L.	4	2	—	90	405	248	25	4	185
Poa annua	L.	4	2	—	167	350	259	1	5	96
Poa annua	L.	4	2	—	416	728	572	6	5	99
Poa annua	L.	4	6	—	158	3286	952	25	5	260
Poa annua	L.	4	7	—	77	6802	1227	17.8	5	155
Poa annua	L.	4	8	—	71	8224	1862	5	5	222
Poa annua	L.	4	2	—	950	3600	2275	15	5	205
Poa annua	L.	4	2	—	1626	3484	2555	15	4	31
Poa annua	L.	4	9	—	667	7500	2648	20	5	74

Species	Authority									
Poa annua	L.	4	1	—	3100	3100	3100	20	5	118
Poa annua	L.	4	6	—	313	10722	3113	30	4	13
Poa annua	L.	4	1	—	3730	3730	3730	5	5	244
Poa annua	L.	4	1	—	4361	4361	4361	20	5	111
Poa annua	L.	4	1	—	4763	4763	4763	23	5	202
Poa annua	L.	4	1	—	4947	4947	4947	15	5	62
Poa annua	L.	4	10	—	48	18176	5582	5	5	69
Poa annua	L.	4	1	—	41620	41620	41620	15	5	207
Poa arctica	R. Br.	1	1	—	168000	168000	168000	8	5	137
Poa bulbosa	L.	1	1	—	0	0	0	10	5	63
Poa bulbosa	L.	4	3	—	38	313	192	3	5	132
Poa chaixii	Villars	1	6	—	0	0	0	3	5	132
Poa chaixii	Villars	2	2	—	38	113	76	10	5	255
Poa chaixii	Villars	4	1	—	438	438	438	10	5	255
Poa compressa	L.	1	1	—	0	0	0	3	5	59
Poa compressa	L.	1	1	—	0	0	0	6	5	67
Poa compressa	L.	4	3	—	53	53	53	10	5	33
Poa compressa	L.	4	1	—	43	140	86	3	5	59
Poa flexuosa	Smith	3	1	—	33	33	33	10	5	252
Poa humilis	Ehrh. ex Hoffm.	1	1	—	0	0	0	7	5	112
Poa laxa	Haenke	1	1	—	0	0	0	5	5	55
Poa nemoralis	L.	1	1	—	0	0	0	20	5	67
Poa nemoralis	L.	1	2	—	0	0	0	5	6	170
Poa nemoralis	L.	1	1	—	0	0	0	12	5	171
Poa nemoralis	L.	2	1	—	0	0	0	10	5	255
Poa nemoralis	L.	4	3	—	2	2	2	15	5	32
Poa palustris	L.	1	1	—	120	634	387	6	5	49
Poa palustris	L.	1	2	—	0	0	0	3	5	208
Poa palustris	L.	2	1	—	0	0	0	6	5	208
Poa palustris	L.	4	2	—	594	594	594	6	5	208
Poa palustris	L.	4	2	—	396	566	481	3	5	208
Poa pratensis	L.	1	2	—	5198	6497	5848	50	5	216
Poa pratensis	L.	1	2	—	0	0	0	10	5	8
Poa pratensis	L.	1	2	—	0	0	0	25	4	36
Poa pratensis	L.	1	2	—	0	0	0	5	4	53
Poa pratensis	L.	1	1	—	0	0	0	1	5	57
Poa pratensis	L.	1	2	—	0	0	0	3	5	59
Poa pratensis	L.	1	4	—	0	0	0	5	5	69
Poa pratensis	L.	1	1	—	0	0	0	9	5	84
Poa pratensis	L.	1	1	—	0	0	0	10	4	88
Poa pratensis	L.	1	24	—	0	0	0	3	5	96
Poa pratensis	L.	1	6	—	0	0	0	20	5	97
Poa pratensis	L.	1	12	—	0	0	0	6	5	99
Poa pratensis	L.	1	2	—	0	0	0	10	5	110
Poa pratensis	L.	1	2	—	0	0	0	6.5	5	128
Poa pratensis	L.	1	3	—	0	0	0	10	5	149

Species	Authority	Seed bank type	Number of records	Longevity (y)	Minimum density (seeds m^{-2})	Maximum density (seeds m^{-2})	Mean density (seeds m^{-2})	Depth (cm)	Method	Source code
Poa pratensis	L.	1	1	—	0	0	0	6	5	208
Poa pratensis	L.	1	2	—	0	0	0	5	5	222
Poa pratensis	L.	1	1	—	0	0	0	7	5	232
Poa pratensis	L.	1	6	—	0	0	0	3	7	235
Poa pratensis	L.	1	1	—	0	0	0	5	5	244
Poa pratensis	L.	1	4	—	0	0	0	5	5	251
Poa pratensis	L.	1	8	—	0	0	0	10	5	255
Poa pratensis	L.	1	1	—	54	54	54	15	5	175
Poa pratensis	L.	1	5	—	26	231	95	2	7	99
Poa pratensis	L.	1	10	—	53	175	110	8	7	275
Poa pratensis	L.	1	6	—	156	468	260	6	5	11
Poa pratensis	L.	1	4	—	78	1014	546	6	5	28
Poa pratensis	L.	1	1	—	600	600	600	20	5	69
Poa pratensis	L.	2	1	>2	0	0	0	20	1	219
Poa pratensis	L.	2	1	—	12	12	12	12	5	174
Poa pratensis	L.	2	2	—	34	36	35	6.5	7	182
Poa pratensis	L.	2	2	—	37	111	74	11.5	5	134
Poa pratensis	L.	2	1	—	75	75	75	15	5	162
Poa pratensis	L.	2	2	—	48	104	76	7.5	5	10
Poa pratensis	L.	2	1	—	86	86	86	30	5	29
Poa pratensis	L.	2	8	—	38	313	114	10	5	255
Poa pratensis	L.	2	2	—	171	228	200	30	5	107
Poa pratensis	L.	2	2	—	183	326	255	5	5	92
Poa pratensis	L.	2	2	—	163	357	260	5	4	53
Poa pratensis	L.	2	2	—	500	500	500	1	5	57
Poa pratensis	L.	2	4	—	156	988	546	6	5	11
Poa pratensis	L.	2	4	—	78	1950	611	6	5	28
Poa pratensis	L.	2	2	—	500	750	625	10	4	88
Poa pratensis	L.	2	3	—	400	1000	633	6	5	181
Poa pratensis	L.	2	1	—	688	688	688	5	5	251
Poa pratensis	L.	2	1	—	741	741	741	17	5	162
Poa pratensis	L.	2	1	—	988	988	988	20	5	107
Poa pratensis	L.	2	1	—	1256	1256	1256	40	5	11
Poa pratensis	L.	2	1	—	3213	3213	3213	13	5	183
Poa pratensis	L.	2	1	—				10	5	184
Poa pratensis	L.	3	1	>7	0	0	0	0	3	188
Poa pratensis	L.	3	1	>30	0	0	0	15	1	240
Poa pratensis	L.	3	1	>39	0	0	0	50	1	240
Poa pratensis	L.	3	1	>39	0	0	0	100	1	240
Poa pratensis	L.	3	1	>10	6	6	6	10	4	241
Poa pratensis	L.	3	1	—	50	50	50	3	5	96
Poa pratensis	L.	3	1	—	60	60	60	13	5	68

Species										
Poa pratensis	L	3	3	—	37	235	107	11.5	5	134
Poa pratensis	L	3	2	—	113	113	113	6	5	208
Poa pratensis	L	3	1	—	184	184	184	70	5	219
Poa pratensis	L	3	7	—	300	1700	571	5	5	251
Poa pratensis	L	3	2	—	521	1042	782	20	5	97
Poa pratensis	L	3	1	—	797	797	797	30	5	29
Poa pratensis	L	3	1	>20	848	848	848	13	5	183
Poa pratensis	L	3	2	—	228	1596	912	50	5	107
Poa pratensis	L	3	1	—	2080	2080	2080	6	5	99
Poa pratensis	L	4	1	—	32	32	32	3	5	59
Poa pratensis	L	4	1	—	36	36	36	5	5	244
Poa pratensis	L	4	1	—	48	48	48	5	5	1
Poa pratensis	L	4	1	—	50	50	50	15	5	267
Poa pratensis	L	4	1	—	64	64	64	6	5	181
Poa pratensis	L	4	1	—	101	101	101	17.8	5	155
Poa pratensis	L	4	1	—	105	105	105	8	7	275
Poa pratensis	L	4	1	—	111	111	111	11.5	5	134
Poa pratensis	L	4	1	—	150	150	150	3	5	96
Poa pratensis	L	4	12	—	75	325	198	10	5	255
Poa pratensis	L	4	1	—	240	240	240	6	5	67
Poa pratensis	L	4	1	—	250	250	250	5	4	88
Poa pratensis	L	4	7	—	80	480	261	5	5	69
Poa pratensis	L	4	1	—	400	400	400	1	5	57
Poa pratensis	L	4	2	—	400	800	600	15	5	249
Poa pratensis	L	4	4	—	290	2460	1142	12	5	174
Poa pratensis	L	4	5	—	312	2288	1279	6	5	99
Poa pratensis	L	4	1	—	2717	2717	2717	10	5	184
Poa pratensis	L	4	4	—	330	6380	3198	15	5	94
Poa trivialis	L	1	1	—	0	0	0	25	4	36
Poa trivialis	L	1	4	—	0	0	0	5	5	69
Poa trivialis	L	1	7	—	0	0	0	3	5	96
Poa trivialis	L	1	2	—	0	0	0	6	5	99
Poa trivialis	L	1	1	—	0	0	0	12	5	174
Poa trivialis	L	1	10	—	0	0	0	20	4	225
Poa trivialis	L	1	4	—	0	0	0	10	5	255
Poa trivialis	L	1	2	—	156	260	208	6	5	11
Poa trivialis	L	1	6	—	884	884	884	6	5	28
Poa trivialis	L	2	1	>2	0	0	0	7.5	2	194
Poa trivialis	L	2	1	—	11	11	11	15	5	32
Poa trivialis	L	2	1	—	36	36	36	13	5	68
Poa trivialis	L	2	1	—	53	53	53	20	4	111
Poa trivialis	L	2	5	—	10	189	68	3	7	235
Poa trivialis	L	2	4	—	63	250	132	10	5	255
Poa trivialis	L	2	3	—	180	260	227	30	5	149
Poa trivialis	L	2	1	—	228	228	228	30	5	107
Poa trivialis	L	2	2	—	175	349	262	10	5	252
Poa trivialis	L	2	5	—	142	792	317	3	5	208

Species	Authority	Seed bank type	Number of records	Longevity (y)	Minimum density (seeds m⁻²)	Maximum density (seeds m⁻²)	Mean density (seeds m⁻²)	Depth (cm)	Method	Source code
Poa trivialis	L.	2	9	—	50	3017	478	3	5	96
Poa trivialis	L.	2	1	—	494	494	494	6	5	28
Poa trivialis	L.	2	1	—	630	630	630	5	5	213
Poa trivialis	L.	2	1	—	750	750	750	10	4	88
Poa trivialis	L.	2	1	—	776	776	776	5	5	222
Poa trivialis	L.	2	1	—	1040	1040	1040	40	5	11
Poa trivialis	L.	2	5	—	455	2158	1068	2	7	99
Poa trivialis	L.	2	3	—	729	1354	1076	7	5	97
Poa trivialis	L.	2	8	—	398	2659	1313	20	5	45
Poa trivialis	L.	2	1	—	1350	1350	1350	30	5	131
Poa trivialis	L.	2	1	>1	1404	1404	1404	32	5	28
Poa trivialis	L.	2	1	—	1720	1720	1720	15	5	257
Poa trivialis	L.	2	1	—	2225	2225	2225	6.5	5	68
Poa trivialis	L.	2	7	—	468	3380	2273	6	5	11
Poa trivialis	L.	2	1	—	2400	2400	2400	5	5	251
Poa trivialis	L.	2	3	—	2184	4732	3397	10	5	27
Poa trivialis	L.	2	1	—	3560	3560	3560	10	5	131
Poa trivialis	L.	2	1	—	3600	3600	3600	20	5	69
Poa trivialis	L.	2	5	—	1840	10471	5785	6	5	208
Poa trivialis	L.	3	1	—	6	6	6	15	4	21
Poa trivialis	L.	3	4	—	100	340	225	15	5	257
Poa trivialis	L.	3	1	—	262	262	262	13	5	68
Poa trivialis	L.	3	1	—	340	340	340	3	5	208
Poa trivialis	L.	3	1	—	448	448	448	8	5	108
Poa trivialis	L.	3	6	—	300	700	500	5	5	251
Poa trivialis	L.	3	4	—	85	2462	708	6	5	208
Poa trivialis	L.	3	1	—	750	750	750	10	4	88
Poa trivialis	L.	3	4	—	521	1667	1172	20	5	97
Poa trivialis	L.	3	2	—	807	1593	1200	30	5	45
Poa trivialis	L.	3	3	—	1248	1560	1352	6	5	99
Poa trivialis	L.	3	5	—	2650	13000	5630	20	5	4
Poa trivialis	L.	4	9	—	2704	11180	6479	10	5	27
Poa trivialis	L.	4	1	—	8	104	53	25	4	36
Poa trivialis	L.	4	2	—	80	80	80	6.5	5	128
Poa trivialis	L.	4	1	—	58	108	83	5	5	98
Poa trivialis	L.	4	1	—	92	92	92	25	4	136
Poa trivialis	L.	4	1	—	148	148	148	15	5	62
Poa trivialis	L.	4	2	—	29	855	442	7	5	58
Poa trivialis	L.	4	1	—	472	472	472	15	5	267
Poa trivialis	L.	4	7	—	83	1617	491	3	5	96
Poa trivialis	L.	4	2	—	200	899	550	30	4	13
Poa trivialis	L.	4	1	—	563	563	563	5	4	88

Species	Author									
Poa trivialis	L.	4	10	—	41	2745	722	5	5	222
Poa trivialis	L.	4	25	—	38	5975	910	10	5	255
Poa trivialis	L.	4	3	—	277	2092	932	17.8	5	155
Poa trivialis	L.	4	4	—	15	3933	1017	12	5	174
Poa trivialis	L.	4	1	—	1075	1075	1075	5	5	244
Poa trivialis	L.	4	7	—	128	3600	1095	5	5	69
Poa trivialis	L.	4	6	—	643	2269	1327	5	4	53
Poa trivialis	L.	4	6	—	113	7584	3707	3	5	208
Poa trivialis	L.	4	14	—	260	20280	4052	6	5	99
Polypogon monspeliensis	(L.) Desf.	4	1	—	50	50	50	3	5	132
Polypogon monspeliensis	(L.) Desf.	4	7	—	6	3216	816	4	5	217
Puccinellia distans	(Jacq.) Parl.	2	7	—	600	1800	1171	10	4	20
Puccinellia distans	(Jacq.) Parl.	4	1	—	16	16	16	15	4	21
Puccinellia maritima	(Hudson) Parl.	1	6	—	0	0	0	10	4	20
Puccinellia maritima	(Hudson) Parl.	1	4	—	0	0	0	1	5	57
Puccinellia maritima	(Hudson) Parl.	2	9	—	0	0	0	10	4	103
Puccinellia maritima	(Hudson) Parl.	4	1	—	600	1600	1178	10	4	20
Puccinellia maritima	(Hudson) Parl.	4	3	—	3	3	3	15	4	21
Puccinellia maritima	(Hudson) Parl.	4	3	—	631	758	699	17.8	5	156
Scolochloa festucacea	(Willd.) Link	2	6	—	37	111	65	5	5	177
Scolochloa festucacea	(Willd.) Link	4	1	—	25	25	25	5	5	262
Scolochloa festucacea	(Willd.) Link	4	2	—	55	111	83	5	5	177
Secale cereale	L.	1	1	—	0	0	0	3	4	96
Secale cereale	L.	1	1	0	0	0	0	20	4	111
Secale cereale	L.	1	1	0	0	0	0	20	5	111
Secale cereale	L.	1	1	0	0	0	0	2.5	3	239
Secale cereale	L.	1	1	0	0	0	0	7.5	3	239
Secale cereale	L.	1	1	0	0	0	0	15	3	239
Secale cereale	L.	1	1	0	0	0	0	15	1	240
Secale cereale	L.	1	1	0	0	0	0	50	1	240
Secale cereale	L.	1	1	—	0	0	0	100	1	255
Sesleria caerulea	(L.) Ard.	1	1	—	0	0	0	10	5	58
Sesleria caerulea	(L.) Ard.	4	2	—	51	57	54	7	4	91
Setaria pumila	(Poiret) Schultes	1	2	—	0	0	0	5	4	36
Setaria pumila	(Poiret) Schultes	2	1	>3	0	0	0	25	1	224
Setaria pumila	(Poiret) Schultes	2	1	>3	0	0	0	2.5	1	224
Setaria pumila	(Poiret) Schultes	2	1	—	28	28	28	10.2	4	36
Setaria pumila	(Poiret) Schultes	3	1	13	0	0	0	25	1	52
Setaria pumila	(Poiret) Schultes	3	1	7	0	0	0	20	1	52
Setaria pumila	(Poiret) Schultes	3	1	9	0	0	0	10	3	124
Setaria pumila	(Poiret) Schultes	3	1	30	0	0	0	2.5	3	209
Setaria pumila	(Poiret) Schultes	3	1	>11	0	0	0	—	1	240
Setaria pumila	(Poiret) Schultes	3	1	>30	0	0	0	25	1	240
Setaria pumila	(Poiret) Schultes	3	1	16	0	0	0	50	1	240
Setaria pumila	(Poiret) Schultes	4	1	21	2	2	2	10	5	33

Species	Authority	Seed bank type	Number of records	Longevity (y)	Minimum density (seeds m⁻²)	Maximum density (seeds m⁻²)	Mean density (seeds m⁻²)	Depth (cm)	Method	Source code
Setaria verticillata	(L.) Beauv.	3	1	>30	0	0	0	15	1	240
Setaria verticillata	(L.) Beauv.	3	1	>39	0	0	0	50	1	240
Setaria verticillata	(L.) Beauv.	3	1	>39	0	0	0	100	1	240
Setaria viridis	(L.) P. Beauv.	1	2	—	0	0	0	11.5	5	134
Setaria viridis	(L.) P. Beauv.	2	6	1	0	0	0	7	2	44
Setaria viridis	(L.) P. Beauv.	2	2	2	0	0	0	7	2	44
Setaria viridis	(L.) P. Beauv.	2	1	3	0	0	0	7	2	44
Setaria viridis	(L.) P. Beauv.	2	1	2	0	0	0	2.5	1	256
Setaria viridis	(L.) P. Beauv.	2	1	2	0	0	0	5	1	256
Setaria viridis	(L.) P. Beauv.	2	1	2	0	0	0	7.5	1	256
Setaria viridis	(L.) P. Beauv.	2	1	2	0	0	0	17.5	1	256
Setaria viridis	(L.) P. Beauv.	3	1	>10	0	0	0	23	1	38
Setaria viridis	(L.) P. Beauv.	3	1	13	0	0	0	10	1	52
Setaria viridis	(L.) P. Beauv.	3	1	13	0	0	0	20	1	52
Setaria viridis	(L.) P. Beauv.	3	1	9	0	0	0	2.5	1	52
Setaria viridis	(L.) P. Beauv.	3	1	>39	0	0	0	50	1	240
Setaria viridis	(L.) P. Beauv.	3	1	21	0	0	0	15	1	240
Setaria viridis	(L.) P. Beauv.	3	1	21	0	0	0	100	1	240
Setaria viridis	(L.) P. Beauv.	3	1	>5	0	0	0	12.5	1	256
Setaria viridis	(L.) P. Beauv.	3	1	>5	0	0	0	25	1	256
Setaria viridis	(L.) P. Beauv.	3	1	—	384	384	384	11.5	5	134
Setaria viridis	(L.) P. Beauv.	4	1	—	3	3	3	10	5	33
Setaria viridis	(L.) P. Beauv.	4	2	—	17	36	27	10	5	8
Setaria viridis	(L.) P. Beauv.	4	4	—	20	220	87	30	4	13
Setaria viridis	(L.) P. Beauv.	4	3	—	372	5391	2494	7.5	5	10
Setaria viridis	(L.) P. Beauv.	4	1	—	2540	2540	2540	25	4	3
Sorghum halepense	(L.) Pers.	2	3	—	20	78	48	15	5	272
Sorghum halepense	(L.) Pers.	3	1	6	0	0	0	30	1	35
Sorghum halepense	(L.) Pers.	3	1	>5	0	0	0	23	1	64
Sorghum halepense	(L.) Pers.	4	4	—	78	1257	646	15	5	272
Spartina alterniflora	Lois.	1	1	—	2	2	2	0	6	90
Spartina anglica	C.E. Hubb.	2	1	—	2	2	2	2	5	113
Trisetum flavescens	(L.) P. Beauv.	1	2	—	0	0	0	6	5	67
Trisetum flavescens	(L.) P. Beauv.	1	1	—	0	0	0	6.5	5	68
Trisetum flavescens	(L.) P. Beauv.	1	1	—	0	0	0	13	5	68
Trisetum flavescens	(L.) P. Beauv.	1	4	—	0	0	0	5	5	69
Trisetum flavescens	(L.) P. Beauv.	1	1	—	0	0	0	5	4	88
Trisetum flavescens	(L.) P. Beauv.	1	3	—	0	0	0	10	4	88
Trisetum flavescens	(L.) P. Beauv.	1	1	—	0	0	0	3	5	96
Trisetum flavescens	(L.) P. Beauv.	1	5	—	0	0	0	20	5	97
Trisetum flavescens	(L.) P. Beauv.	1	3	—	0	0	0	10	5	110
Trisetum flavescens	(L.) P. Beauv.	1	1	—	0	0	0	5	5	122

Species	Author									
Trisetum flavescens	(L.) P. Beauv.	1	3	—	0	0	0	6.5	5	128
Trisetum flavescens	(L.) P. Beauv.	1	3	—	0	0	0	12	5	174
Trisetum flavescens	(L.) P. Beauv.	1	1	—	0	0	0	6	5	181
Trisetum flavescens	(L.) P. Beauv.	1	1	—	0	0	0	6.5	7	182
Trisetum flavescens	(L.) P. Beauv.	1	1	—	0	0	0	13	5	183
Trisetum flavescens	(L.) P. Beauv.	1	10	—	0	0	0	5	5	251
Trisetum flavescens	(L.) P. Beauv.	1	18	—	3	3	3	10	5	255
Trisetum flavescens	(L.) P. Beauv.	1	1	—	3	3	3	3	7	235
Trisetum flavescens	(L.) P. Beauv.	1	1	—	40	40	40	10	5	252
Trisetum flavescens	(L.) P. Beauv.	1	1	—	180	180	180	6	5	48
Trisetum flavescens	(L.) P. Beauv.	4	2	—	3	5	4	3	7	235
Trisetum flavescens	(L.) P. Beauv.	4	1	—	5	5	5	5	5	122
Trisetum flavescens	(L.) P. Beauv.	4	1	—	9	9	9	12	5	174
Trisetum flavescens	(L.) P. Beauv.	4	1	—	19	19	19	10	5	110
Trisetum spicatum	(L.) K. Richter	4	6	—	80	290	148	6	5	39
Triticum aestivum	L.	1	1	0	0	0	0	25	1	125
Triticum aestivum	L.	1	1	0	0	0	0	3	7	235
Triticum aestivum	L.	1	1	0	0	0	0	2.5	3	239
Triticum aestivum	L.	1	1	0	0	0	0	7.5	3	239
Triticum aestivum	L.	1	1	0	0	0	0	15	3	239
Triticum aestivum	L.	4	1	—	7	7	7	20	4	111
Vulpia bromoides	(L.) Gray	1	2	.	0	0	0	3	5	96
Vulpia bromoides	(L.) Gray	2	1	1	0	0	0	7.5	2	194
Vulpia ciliata	Dumort.	1	1	—	0	0	0	3	5	132
Vulpia ciliata	Dumort.	4	1	—	150	150	150	3	5	132
Vulpia fasciculata	(Forsskal) Fritsch	1	1	0	0	0	0	0	1	259
Vulpia fasciculata	(Forsskal) Fritsch	1	1	0	0	0	0	0.5	1	259
Vulpia fasciculata	(Forsskal) Fritsch	1	1	0	0	0	0	1	1	259
Vulpia fasciculata	(Forsskal) Fritsch	1	1	0	0	0	0	3	1	259
Vulpia fasciculata	(Forsskal) Fritsch	1	1	0	0	0	0	10	1	259
Vulpia myuros	(L.) C. Gmelin	1	3	—	0	0	0	2	5	95
Vulpia myuros	(L.) C. Gmelin	2	1	—	213	213	213	3	5	132
Vulpia myuros	(L.) C. Gmelin	4	1	—	75	75	75	3	5	132
Vulpia myuros	(L.) C. Gmelin	4	4	—	210	1590	1053	2	5	95
Vulpia unilateralis	(L.) Stace	4	2	0	38	38	38	3	5	132
Zea mays	L.	1	1	0	0	0	0	15	1	240
Zea mays	L.	1	1	0	0	0	0	50	1	240
Zea mays	L.	1	1	0	0	0	0	100	1	240
Grossulariaceae	—	—	—	—	—	—	—	—	—	—
Ribes uva-crispa	L.	1	3	—	0	0	0	6	5	49
Guttiferae (Clusiaceae)	—	—	—	—	—	—	—	—	—	—
Hypericum canadense	L.	2	1	—	946	946	946	21	5	160
Hypericum canadense	L.	4	1	—	162	162	162	5	5	117
Hypericum hircinum	L.	2	1	—	2189	2189	2189	21	5	160

Species	Authority	Seed bank type	Number of records	Longevity (y)	Minimum density (seeds m⁻²)	Maximum density (seeds m⁻²)	Mean density (seeds m⁻²)	Depth (cm)	Method	Source code
Hypericum hirsutum	L	1	1	—	0	0	0	3	5	96
Hypericum hirsutum	L	1	1	—	0	0	0	15	5	257
Hypericum hirsutum	L	1	1	—	4	4	4	3	7	235
Hypericum hirsutum	L	2	1	—	1	1	1	15	5	32
Hypericum hirsutum	L	2	1	—	14	14	14	3	7	235
Hypericum hirsutum	L	2	1	—	60	60	60	15	5	257
Hypericum hirsutum	L	3	1	>5	0	0	0	7.5	2	195
Hypericum hirsutum	L	3	1	—	2108	2108	2108	20	5	67
Hypericum hirsutum	L	4	1	—	217	217	217	3	5	96
Hypericum humifusum	L	1	1	—	9	9	9	10	5	93
Hypericum humifusum	L	2	1	—	14	14	14	15	5	32
Hypericum humifusum	L	2	1	—	208	208	208	5	5	69
Hypericum humifusum	L	3	1	—	29	29	29	12.5	5	126
Hypericum humifusum	L	3	1	>100	100	100	100	32	5	169
Hypericum humifusum	L	3	6	—	32	183	117	30	5	45
Hypericum humifusum	L	3	1	—	172	172	172	20	5	67
Hypericum humifusum	L	3	1	>18	467	467	467	32	5	169
Hypericum humifusum	L	4	3	—	227	706	487	17.8	5	155
Hypericum humifusum	L	4	1	—	1567	1567	1567	3	5	96
Hypericum humifusum	L	4	2	—	560	8736	4648	5	5	69
Hypericum maculatum	Crantz	1	2	—	0	0	0	12	5	171
Hypericum maculatum	Crantz	1	6	—	0	0	0	3	5	208
Hypericum maculatum	Crantz	1	1	—	0	0	0	20	4	225
Hypericum maculatum	Crantz	1	6	—	0	0	0	10	5	255
Hypericum maculatum	Crantz	2	4	—	1	10	4	5	6	170
Hypericum maculatum	Crantz	2	8	—	38	600	249	10	5	255
Hypericum maculatum	Crantz	2	1	—	5904	5904	5904	10	5	150
Hypericum maculatum	Crantz	3	1	—	123	123	123	12.5	5	126
Hypericum maculatum	Crantz	3	1	—	5933	5933	5933	10	5	150
Hypericum maculatum	Crantz	4	13	—	38	7800	1164	10	5	255
Hypericum maculatum	Crantz	4	10	—	1052	31944	7806	10	5	150
Hypericum montanum	L	2	1	—	184	184	184	20	5	219
Hypericum perforatum	L	1	1	—	0	0	0	10	5	152
Hypericum perforatum	L	1	1	—	0	0	0	6	5	181
Hypericum perforatum	L	1	1	—	0	0	0	6.5	7	182
Hypericum perforatum	L	1	2	—	0	0	0	10	5	184
Hypericum perforatum	L	1	1	—	0	0	0	10	5	255
Hypericum perforatum	L	1	1	—	53	53	53	8	7	275
Hypericum perforatum	L	2	3	—	6	11	8	3	7	235
Hypericum perforatum	L	2	1	—	43	43	43	10	5	126
Hypericum perforatum	L	2	3	—	45	189	108	6	5	49
Hypericum perforatum	L	2	1	—	144	144	144	16	5	168

Species										
Hypericum perforatum	L	2	12	—	38	1088	245	10	5	255
Hypericum perforatum	L	2	4	—	12	889	252	12	5	174
Hypericum perforatum	L	2	1	—	256	256	256	50	5	219
Hypericum perforatum	L	2	1	—	408	408	408	8	7	275
Hypericum perforatum	L	2	7	—	50	1550	648	3	5	96
Hypericum perforatum	L	2	1	—	700	700	700	6	5	162
Hypericum perforatum	L	2	1	—	833	833	833	20	5	97
Hypericum perforatum	L	2	3	—	234	1794	1066	10	5	152
Hypericum perforatum	L	3	1	—	3400	3400	3400	12	5	81
Hypericum perforatum	L	3	1	11	0	0	0	30	1	35
Hypericum perforatum	L	3	1	>5	10	10	10	7.5	2	195
Hypericum perforatum	L	3	1	—	40	40	40	5	5	234
Hypericum perforatum	L	3	1	>30	43	43	43	10	5	181
Hypericum perforatum	L	3	1	—	56	56	56	12.5	5	126
Hypericum perforatum	L	3	1	>45	56	56	56	24	5	168
Hypericum perforatum	L	3	6	—	37	210	105	11.5	5	134
Hypericum perforatum	L	3	1	—	178	178	178	16	5	168
Hypericum perforatum	L	3	1	>22	122	478	300	24	5	168
Hypericum perforatum	L	3	1	>40	800	800	800	20	5	173
Hypericum perforatum	L	3	5	—	260	3960	1828	15	5	257
Hypericum perforatum	L	3	1	—	2708	2708	2708	20	5	97
Hypericum perforatum	L	4	1	—	4360	4360	4360	30	5	173
Hypericum perforatum	L	4	1	—	24	24	24	6	5	181
Hypericum perforatum	L	4	1	—	69	69	69	6	5	49
Hypericum perforatum	L	4	1	—	200	200	200	1	5	96
Hypericum perforatum	L	4	1	—	346	346	346	11.5	5	134
Hypericum perforatum	L	4	9	—	50	713	406	10	5	255
Hypericum perforatum	L	4	1	—	816	816	816	8.5	4	254
Hypericum perforatum	L	4	7	—	26	5887	965	30	4	13
Hypericum perforatum	L	4	10	—	283	6300	1970	3	5	96
Hypericum perforatum	L	4	2	—	1404	12090	6747	10	5	152
Hypericum pulchrum	L	2	1	—	8	8	8	15	5	32
Hypericum pulchrum	L	2	1	—	84	84	84	10	4	51
Hypericum pulchrum	L	2	1	—	288	288	288	5	5	69
Hypericum pulchrum	L	3	1	—	126	126	126	10	4	93
Hypericum pulchrum	L	3	1	—	163	163	163	12	4	51
Hypericum pulchrum	L	3	1	—	454	454	454	16	5	51
Hypericum pulchrum	L	3	7	—	60	1920	926	15	5	257
Hypericum tetrapterum	Fries	1	1	—	0	0	0	6	5	99
Hypericum tetrapterum	Fries	1	1	—	0	0	0	10	5	149
Hypericum tetrapterum	Fries	1	2	—	0	0	0	6	5	208
Hypericum tetrapterum	Fries	1	2	—	0	0	0	20	4	225
Hypericum tetrapterum	Fries	1	2	—	34	78	56	2	7	99
Hypericum tetrapterum	Fries	2	1	—	22	22	22	15	5	32
Hypericum tetrapterum	Fries	2	1	—	50	50	50	3	5	96
Hypericum tetrapterum	Fries	2	1	—	80	80	80	6.5	5	128

Species	Authority	Seed bank type	Number of records	Longevity (y)	Minimum density (seeds m^{-2})	Maximum density (seeds m^{-2})	Mean density (seeds m^{-2})	Depth (cm)	Method	Source code
Hypericum tetrapterum	Fries	3	1	—	142	142	142	6	5	208
Hypericum tetrapterum	Fries	3	1	—	160	160	160	15	5	257
Haloragaceae										
Myriophyllum spicatum	L.	1	3	—	0	0	0	5	5	116
Hydrocharitaceae										
Elodea nuttallii	(Planchon) H. St. John	1	1	—	0	0	0	3	5	96
Iridaceae										
Iris foetidissima	L.	1	2	—	0	0	0	15	5	257
Iris foetidissima	L.	3	1	—	80	80	80	15	5	257
Iris pseudacorus	L.	1	3	—	0	0	0	3	5	208
Iris pseudacorus	L.	1	3	—	0	0	0	6	5	208
Iris pseudacorus	L.	1	1	—	0	0	0	20	4	225
Iris pseudacorus	L.	1	1	—	0	0	0	10	5	255
Sisyrinchium bermudiana	L.	4	1	—	3120	3120	3120	6	5	99
Juncaceae										
Juncus acutiflorus	Ehrh. ex Hoffm.	1	2	—	0	0	0	10	5	27
Juncus acutiflorus	Ehrh. ex Hoffm.	1	3	—	0	0	0	3	5	208
Juncus acutiflorus	Ehrh. ex Hoffm.	1	3	—	0	832	277	2	7	99
Juncus acutiflorus	Ehrh. ex Hoffm.	2	4	—	41	163	105	5	5	222
Juncus acutiflorus	Ehrh. ex Hoffm.	2	6	—	38	863	200	10	5	255
Juncus acutiflorus	Ehrh. ex Hoffm.	2	1	—	1092	1092	1092	6	5	99
Juncus acutiflorus	Ehrh. ex Hoffm.	2	2	—	1986	5418	3702	2	7	99
Juncus acutiflorus	Ehrh. ex Hoffm.	3	1	—	1600	1600	1600	20	5	69
Juncus acutiflorus	Ehrh. ex Hoffm.	4	1	—	183	183	183	3	5	96
Juncus acutiflorus	Ehrh. ex Hoffm.	4	4	—	31	857	291	5	5	222
Juncus acutiflorus	Ehrh. ex Hoffm.	4	3	—	788	1100	984	10	5	255
Juncus acutiflorus	Ehrh. ex Hoffm.	4	7	—	1768	26000	8454	6	5	99
Juncus articulatus	L.	1	1	—	0	0	0	5	5	69
Juncus articulatus	L.	1	1	—	0	0	0	6	5	99
Juncus articulatus	L.	1	1	—	0	0	0	20	4	225
Juncus articulatus	L.	1	2	—	34	91	63	2	7	99
Juncus articulatus	L.	1	1	—	156	156	156	6	5	11
Juncus articulatus	L.	2	6	—	50	400	122	3	5	96
Juncus articulatus	L.	2	3	—	52	455	197	2	7	99
Juncus articulatus	L.	2	2	—	234	234	234	10	5	152
Juncus articulatus	L.	2	7	—	12	1065	338	12	5	174
Juncus articulatus	L.	2	3	—	6	1240	433	3	7	235

Species										
Juncus articulatus	L.	2	3	—	360	600	453	6.5	5	128
Juncus articulatus	L.	2	2	—	97	1970	1034	30	5	45
Juncus articulatus	L.	2	1	—	1144	1144	1144	6	5	99
Juncus articulatus	L.	3	1	>35	64	64	64	10	5	84
Juncus articulatus	L.	3	1	—	85	85	85	6	5	208
Juncus articulatus	L.	3	1	—	96	96	96	9	5	84
Juncus articulatus	L.	3	1	—	208	208	208	6	5	11
Juncus articulatus	L.	3	4	—	156	312	247	6	5	99
Juncus articulatus	L.	3	1	—	342	342	342	25	5	107
Juncus articulatus	L.	3	3	—	32	980	395	30	5	45
Juncus articulatus	L.	3	5	—	200	1000	500	20	5	4
Juncus articulatus	L.	3	1	—	520	520	520	12.5	5	126
Juncus articulatus	L.	3	1	—	800	800	800	20	5	69
Juncus articulatus	L.	3	2	—	1488	2416	1952	20	5	67
Juncus articulatus	L.	3	2	—	912	3477	2195	30	5	107
Juncus articulatus	L.	3	2	—	2057	5367	3712	16	5	216
Juncus articulatus	L.	4	4	—	75	1108	438	5	5	98
Juncus articulatus	L.	4	5	—	156	1872	874	6	5	99
Juncus articulatus	L.	4	1	—	1008	1008	1008	5	5	69
Juncus articulatus	L.	4	2	—	1312	1422	1367	12	5	174
Juncus articulatus	L.	4	4	—	460	12660	5180	6.5	5	128
Juncus articulatus	L.	4	6	—	1559	45998	11868	50	5	216
Juncus balticus	Willd.	2	6	—	40	333	125	5	5	177
Juncus balticus	Willd.	4	1	—	962	962	962	5	5	177
Juncus bufonius	L.	1	1	—	0	0	0	6	5	11
Juncus bufonius	L.	1	3	—	0	0	0	25	4	36
Juncus bufonius	L.	1	1	—	0	0	0	20	4	225
Juncus bufonius	L.	1	3	—	26	78	54	2	7	99
Juncus bufonius	L.	1	2	—	78	130	104	6	5	28
Juncus bufonius	L.	1	4	—	53	132	106	8	7	275
Juncus bufonius	L.	1	4	—	208	3380	1092	10	5	27
Juncus bufonius	L.	2	3	—	28	47	35	3	7	235
Juncus bufonius	L.	2	1	—	52	52	52	25	5	2
Juncus bufonius	L.	2	1	—	57	57	57	15	5	32
Juncus bufonius	L.	2	2	—	83	133	108	3	5	96
Juncus bufonius	L.	2	2	—	100	125	113	2	7	99
Juncus bufonius	L.	2	1	—	179	179	179	16	5	216
Juncus bufonius	L.	2	3	—	151	227	184	17.8	5	155
Juncus bufonius	L.	2	1	—	187	187	187	12	5	162
Juncus bufonius	L.	2	3	—	208	832	485	6	5	99
Juncus bufonius	L.	2	1	—	546	546	546	6	5	28
Juncus bufonius	L.	2	1	—	570	570	570	30	5	107
Juncus bufonius	L.	2	6	—	48	2160	608	5	5	69
Juncus bufonius	L.	2	1	—	832	832	832	40	5	11
Juncus bufonius	L.	2	1	—	850	850	850	3	5	132
Juncus bufonius	L.	2	1	—	858	858	858	15	5	267
Juncus bufonius	L.	2	1	—	869	869	869	5	5	244

Species	Authority	Seed bank type	Number of records	Longevity (y)	Minimum density (seeds m⁻²)	Maximum density (seeds m⁻²)	Mean density (seeds m⁻²)	Depth (cm)	Method	Source code
Juncus bufonius	L.	2	1	—	934	934	934	3	5	208
Juncus bufonius	L.	2	3	—	552	1604	969	8	7	275
Juncus bufonius	L.	2	7	—	208	3172	1114	6	5	11
Juncus bufonius	L.	2	1	—	1438	1438	1438	17	5	162
Juncus bufonius	L.	2	12	—	122	4898	1672	5	5	222
Juncus bufonius	L.	2	1	—	3229	3229	3229	30	5	45
Juncus bufonius	L.	2	23	—	38	89063	5083	10	5	255
Juncus bufonius	L.	2	3	—	6237	15593	9616	50	5	216
Juncus bufonius	L.	2	1	—	25000	25000	25000	20	5	210
Juncus bufonius	L.	2	1	—	28940	28940	28940	25	4	3
Juncus bufonius	L.	3	1	>4	0	0	0	100	1	54
Juncus bufonius	L.	3	1	>4	0	0	0	200	1	54
Juncus bufonius	L.	3	2	>4	0	0	0	10	3	216
Juncus bufonius	L.	3	1	—	59	59	59	5	5	234
Juncus bufonius	L.	3	1	—	60	60	60	13	5	68
Juncus bufonius	L.	3	1	>30	74	74	74	9.5	5	84
Juncus bufonius	L.	3	1	>35	111	111	111	32	5	169
Juncus bufonius	L.	3	1	—	266	266	266	9	5	84
Juncus bufonius	L.	3	1	—	312	312	312	6	5	99
Juncus bufonius	L.	3	2	—	228	684	456	30	5	107
Juncus bufonius	L.	3	1	—	468	468	468	40	5	11
Juncus bufonius	L.	3	2	—	416	780	598	6	5	11
Juncus bufonius	L.	3	1	—	753	753	753	22.5	5	45
Juncus bufonius	L.	3	1	—	948	948	948	8	5	108
Juncus bufonius	L.	3	2	>18	944	978	961	32	5	169
Juncus bufonius	L.	3	1	>100	1267	1267	1267	16	5	169
Juncus bufonius	L.	3	9	—	32	7653	1338	30	5	45
Juncus bufonius	L.	3	1	—	1938	1938	1938	50	5	107
Juncus bufonius	L.	3	3	>100	1033	3767	2007	32	5	169
Juncus bufonius	L.	3	1	>100	2167	2167	2167	24	5	169
Juncus bufonius	L.	3	2	—	513	3876	2195	20	5	107
Juncus bufonius	L.	3	1	—	6000	6000	6000	6	5	208
Juncus bufonius	L.	3	2	—	1300	19300	10300	20	5	69
Juncus bufonius	L.	3	1	—	50000	50000	50000	20	5	210
Juncus bufonius	L.	4	3	—	24	70	49	25	4	36
Juncus bufonius	L.	4	4	—	25	117	65	5	5	98
Juncus bufonius	L.	4	2	—	65	296	181	12	4	230
Juncus bufonius	L.	4	2	—	10	419	215	15	4	31
Juncus bufonius	L.	4	1	—	1100	1100	1100	10	5	147
Juncus bufonius	L.	4	3	—	745	4857	2180	5	5	222
Juncus bufonius	L.	4	3	—	64	6368	2645	5	5	69
Juncus bufonius	L.	4	8	—	500	7667	2792	20	5	74

Species	Authority									
Juncus bufonius	L.	4	11	—	40	47539	5567	30	4	13
Juncus bufonius	L.	4	1	—	5755	5755	5755	15	5	207
Juncus bufonius	L.	4	1	—	26539	26539	26539	20	4	111
Juncus bulbosus	L.	1	3	—	0	26	17	2	7	99
Juncus bulbosus	L.	2	1	—	156	156	156	6	5	11
Juncus bulbosus	L.	2	1	—	32	32	32	22.5	5	45
Juncus bulbosus	L.	2	9	—	31	1837	388	5	5	222
Juncus bulbosus	L.	2	4	—	156	1352	572	6	5	11
Juncus bulbosus	L.	2	1	—	725	725	725	10	5	255
Juncus bulbosus	L.	3	1	—	64	64	64	9	5	84
Juncus bulbosus	L.	3	1	—	65	65	65	15	5	45
Juncus bulbosus	L.	3	1	>30	96	96	96	9.5	5	84
Juncus bulbosus	L.	3	2	—	156	260	208	40	5	11
Juncus bulbosus	L.	3	1	—	260	260	260	6	5	11
Juncus bulbosus	L.	3	2	—	208	364	286	6	5	99
Juncus bulbosus	L.	3	3	—	65	1518	560	30	5	45
Juncus bulbosus	L.	3	1	—	741	741	741	20	5	107
Juncus bulbosus	L.	3	1	—	1572	1572	1572	22.5	5	45
Juncus bulbosus	L.	3	2	—	848	7525	4187	16	5	216
Juncus bulbosus	L.	3	2	—	1596	8436	5016	30	5	107
Juncus bulbosus	L.	4	2	—	42	57	50	2	7	99
Juncus bulbosus	L.	4	2	—	122	214	168	5	5	222
Juncus bulbosus	L.	4	4	—	101	455	240	17.8	5	156
Juncus bulbosus	L.	4	7	—	299	4614	1652	17.8	5	155
Juncus canadensis	Gay	2	1	—	669	669	669	0.538	5	269
Juncus capitatus	Weigel	2	1	—	350	350	350	3	5	132
Juncus compressus	Jacq.	3	1	—	400	400	400	20	5	69
Juncus conglomeratus	L.	1	3	—	0	0	0	6	5	28
Juncus conglomeratus	L.	1	6	—	0	0	0	3	5	208
Juncus conglomeratus	L.	1	3	—	0	0	0	6	5	208
Juncus conglomeratus	L.	1	1	—	0	0	0	20	4	225
Juncus conglomeratus	L.	2	5	—	0	0	0	10	5	255
Juncus conglomeratus	L.	2	6	—	63	2763	594	10	5	255
Juncus conglomeratus	L.	2	1	—	624	624	624	10	5	27
Juncus conglomeratus	L.	3	1	—	1200	1200	1200	20	4	225
Juncus conglomeratus	L.	3	3	>100	211	211	211	16	5	169
Juncus conglomeratus	L.	3	3	>100	400	1456	848	32	5	169
Juncus conglomeratus	L.	3	1	>100	1111	1111	1111	24	5	169
Juncus conglomeratus	L.	3	3	—	572	2600	1508	10	5	27
Juncus conglomeratus	L.	3	3	—	3324	9624	5428	20	5	67
Juncus conglomeratus	L.	4	6	—	88	16350	3196	10	5	255
Juncus conglomeratus	L.	4	1	—	3200	3200	3200	20	4	225
Juncus effusus	L.	1	1	—	0	0	0	5	4	88
Juncus effusus	L.	1	1	—	0	0	0	6.5	5	128
Juncus effusus	L.	1	1	—	0	0	0	3	5	208
Juncus effusus	L.	1	2	—	0	0	0	20	4	225
Juncus effusus	L.	1	2	—	0	0	0	10	5	255

Species	Authority	Seed bank type	Number of records	Longevity (y)	Minimum density (seeds m^{-2})	Maximum density (seeds m^{-2})	Mean density (seeds m^{-2})	Depth (cm)	Method	Source code
Juncus effusus	L.	1	1	—	53	53	53	8	7	275
Juncus effusus	L.	1	1	—	80	80	80	30	5	180
Juncus effusus	L.	1	1	—	624	624	624	6	5	11
Juncus effusus	L.	1	1	—	936	936	936	6	5	28
Juncus effusus	L.	2	1	—	52	52	52	10	5	119
Juncus effusus	L.	2	1	—	85	85	85	3	5	208
Juncus effusus	L.	2	1	—	86	86	86	15	5	45
Juncus effusus	L.	2	1	—	97	97	97	22.5	5	45
Juncus effusus	L.	2	2	—	43	289	166	5	5	220
Juncus effusus	L.	2	2	—	96	288	192	5	5	69
Juncus effusus	L.	2	1	—	198	198	198	6	5	208
Juncus effusus	L.	2	6	—	127	906	406	5	5	92
Juncus effusus	L.	2	1	—	430	430	430	10	5	143
Juncus effusus	L.	2	6	—	28	3746	751	3	7	235
Juncus effusus	L.	2	20	—	38	4200	977	10	5	255
Juncus effusus	L.	2	26	—	50	9400	1074	3	5	96
Juncus effusus	L.	2	3	—	13	2299	1178	7	5	232
Juncus effusus	L.	2	14	—	82	4959	1525	5	5	222
Juncus effusus	L.	2	2	—	48	3360	1704	13	5	68
Juncus effusus	L.	2	1	—	1730	1730	1730	15	5	32
Juncus effusus	L.	2	1	—	2075	2075	2075	6.5	5	68
Juncus effusus	L.	2	2	—	360	4200	2280	15	5	257
Juncus effusus	L.	2	1	—	2288	2288	2288	40	5	11
Juncus effusus	L.	2	7	—	936	6292	2782	6	5	28
Juncus effusus	L.	2	1	—	2815	2815	2815	10	5	126
Juncus effusus	L.	2	5	—	270	4680	2932	2	7	99
Juncus effusus	L.	2	3	—	108	5845	3219	30	5	45
Juncus effusus	L.	2	1	—	4700	4700	4700	5	5	251
Juncus effusus	L.	2	1	—	5351	5351	5351	21	5	160
Juncus effusus	L.	2	2	—	3484	10296	6890	6	5	99
Juncus effusus	L.	2	9	—	1976	52936	11411	6	5	11
Juncus effusus	L.	2	1	—	680000	680000	680000	20	4	225
Juncus effusus	L.	3	1	>50	0	0	0	50	3	115
Juncus effusus	L.	3	1	—	29	29	29	10	5	93
Juncus effusus	L.	3	1	>35	32	32	32	10	5	84
Juncus effusus	L.	3	1	—	64	64	64	12	5	84
Juncus effusus	L.	3	1	>73	74	74	74	7.5	5	84
Juncus effusus	L.	3	3	>64	111	186	140	11.5	5	134
Juncus effusus	L.	3	1	—	383	383	383	9.5	5	84
Juncus effusus	L.	3	1	>30	438	438	438	8	5	108
Juncus effusus	L.	3	1	—	456	456	456	25	5	107
Juncus effusus	L.	3	1	—	564	564	564	9	5	84

Species	Author									
Juncus effusus	L.	3	2	—	456	1311	884	20	5	107
Juncus effusus	L.	3	7	—	0	5457	1024	30	5	45
Juncus effusus	L.	3	1	—	1044	1044	1044	22.5	5	45
Juncus effusus	L.	3	1	—	1280	1280	1280	32	5	131
Juncus effusus	L.	3	2	—	85	3255	1670	6	5	208
Juncus effusus	L.	3	2	—	504	3024	1764	20	5	67
Juncus effusus	L.	3	2	—	2280	2793	2537	50	5	107
Juncus effusus	L.	3	1	—	2562	2562	2562	17.5	5	126
Juncus effusus	L.	3	2	—	2309	2844	2577	12.5	5	126
Juncus effusus	L.	3	1	—	2600	2600	2600	20	5	69
Juncus effusus	L.	3	3	—	2264	3198	2858	3	5	208
Juncus effusus	L.	3	15	—	1600	4900	3547	5	5	251
Juncus effusus	L.	3	3	—	312	12116	5512	6	5	99
Juncus effusus	L.	3	2	—	9177	14250	11714	30	5	107
Juncus effusus	L.	3	5	—	7750	26650	15960	20	5	4
Juncus effusus	L.	3	1	—	17420	17420	17420	40	5	11
Juncus effusus	L.	3	11	—	60	105600	23293	15	5	257
Juncus effusus	L.	3	4	—	23750	29000	27313	10	4	88
Juncus effusus	L.	3	8	—	2444	97032	41522	10	5	27
Juncus effusus	L.	3	5	—	219514	219514	219514	45	5	145
Juncus effusus	L.	4	1	—	19	19	19	0.538	5	269
Juncus effusus	L.	4	1	—	36	36	36	13.5	5	163
Juncus effusus	L.	4	2	—	29	628	329	5	5	220
Juncus effusus	L.	4	4	—	170	764	411	3	5	208
Juncus effusus	L.	4	3	—	300	600	433	15	5	249
Juncus effusus	L.	4	2	—	48	1552	800	5	5	69
Juncus effusus	L.	4	1	—	975	975	975	5	5	117
Juncus effusus	L.	4	7	—	101	3006	1144	17.8	5	156
Juncus effusus	L.	4	4	—	983	1392	1156	5	5	98
Juncus effusus	L.	4	7	—	504	2974	1252	17.8	5	155
Juncus effusus	L.	4	3	—	238	3675	1704	10	5	255
Juncus effusus	L.	4	3	—	1010	2898	1813	5	5	222
Juncus effusus	L.	4	1	—	3833	3833	3833	3	5	96
Juncus effusus	L.	4	2	—	2079	11435	6757	50	5	216
Juncus effusus	L.	4	1	—	7380	7380	7380	15	5	257
Juncus effusus	L.	4	1	—	10813	10813	10813	10	5	126
Juncus effusus	L.	4	14	—	1404	40664	15771	6	5	99
Juncus effusus	L.	4	2	—	2000	520000	261000	20	4	225
Juncus effusus	L.	1	9	—	0	0	0	3	5	208
Juncus filiformis	L.	1	3	—	0	0	0	6	5	208
Juncus filiformis	L.	1	2	—	0	0	0	10	5	255
Juncus filiformis	L.	2	5	—	138	1163	563	10	5	255
Juncus filiformis	L.	2	1	—	3739	3739	3739	0.538	5	269
Juncus filiformis	L.	3	2	—	3018	40048	21533	16	5	216
Juncus filiformis	L.	4	1	—	50	50	50	10	5	255
Juncus gerardii	Lois.	1	3	—	0	0	0	1	5	57
Juncus gerardii	Lois.	2	2	—	600	600	600	10	4	20

Species	Authority	Seed bank type	Number of records	Longevity (y)	Minimum density (seeds m^{-2})	Maximum density (seeds m^{-2})	Mean density (seeds m^{-2})	Depth (cm)	Method	Source code
Juncus gerardii	Lois.	2	3	—	300	1200	833	1	5	57
Juncus gerardii	Lois.	2	8	—	13353	128353	66272	7	5	112
Juncus gerardii	Lois.	3	2	—	10000	12529	11265	7	5	112
Juncus gerardii	Lois.	4	1	—	3	3	3	15	4	21
Juncus gerardii	Lois.	4	1	—	400	400	400	1	5	57
Juncus inflexus	L.	1	3	—	53	105	70	8	7	275
Juncus inflexus	L.	2	3	—	6	103	47	3	7	235
Juncus inflexus	L.	2	1	—	167	167	167	3	5	96
Juncus inflexus	L.	2	1	—	320	320	320	5	5	69
Juncus inflexus	L.	2	3	—	33	656	348	16	5	168
Juncus inflexus	L.	2	1	—	5842	5842	5842	15	5	267
Juncus inflexus	L.	3	1	>30	56	56	56	24	5	168
Juncus inflexus	L.	3	1	—	300	300	300	20	5	69
Juncus inflexus	L.	3	2	>40	44	1133	589	24	5	168
Juncus inflexus	L.	4	1	—	17	17	17	7	5	58
Juncus inflexus	L.	4	1	—	565	565	565	8	7	275
Juncus inflexus	L.	4	1	—	3350	3350	3350	3	5	96
Juncus maritimus	Lam.	2	1	—	300	300	300	1	5	57
Juncus maritimus	Lam.	4	2	—	152	404	278	17.8	5	156
Juncus squarrosus	L.	1	2	—	0	32	16	15	5	45
Juncus squarrosus	L.	2	1	—	43	43	43	22.5	5	45
Juncus squarrosus	L.	2	2	—	156	225	191	7	5	232
Juncus squarrosus	L.	2	6	—	87	367	216	5	5	92
Juncus squarrosus	L.	3	1	—	32	32	32	30	5	45
Juncus squarrosus	L.	3	1	>35	32	32	32	12	5	84
Juncus squarrosus	L.	3	1	—	420	420	420	22.5	5	45
Juncus squarrosus	L.	4	1	—	63	63	63	2	5	153
Juncus squarrosus	L.	4	11	—	101	2703	1052	17.8	5	156
Juncus squarrosus	L.	4	6	—	203	3878	1630	17.8	5	155
Juncus squarrosus	L.	4	3	—	2078	12276	6957	7	5	232
Juncus subnodulosus	Schrank	1	3	—	0	0	0	6.5	5	128
Juncus subnodulosus	Schrank	1	1	—	0	0	0	5	5	248
Juncus tenuis	Willd.	1	1	—	0	0	0	5	5	69
Juncus tenuis	Willd.	2	1	—	48	48	48	5	5	69
Juncus tenuis	Willd.	2	2	—	463	475	469	10	5	255
Juncus tenuis	Willd.	2	5	—	37	1274	750	11.5	5	134
Juncus tenuis	Willd.	2	1	—	14245	14245	14245	16	5	23
Juncus tenuis	Willd.	3	10	—	235	2079	549	11.5	5	134
Juncus tenuis	Willd.	4	1	—	458	458	458	11.5	5	134
Juncus trifidus	L.	4	1	—	3014	3014	3014	12	5	187
Juncus trifidus	L.	1	1	—	0	0	0	5	5	55
Juncus triglumis	L.	3	2	—	32	657	345	30	5	45

Juncus triglumis	L.	4	1	—	301	301	301	17.8	5	155
Luzula alpinopilosa	(Chaix) Breistr.	1	1	—	0	0	0	5	5	55
Luzula arctica	Blytt	1	2	—	0	0	0	20	5	191
Luzula campestris	(L.) DC.	1	1	—	0	0	0	40	5	11
Luzula campestris	(L.) DC.	1	1	—	0	0	0	15	5	45
Luzula campestris	(L.) DC.	1	1	—	0	0	0	6	5	67
Luzula campestris	(L.) DC.	1	1	—	0	0	0	6.5	5	68
Luzula campestris	(L.) DC.	1	1	—	0	0	0	13	5	68
Luzula campestris	(L.) DC.	1	2	—	0	0	0	5	5	69
Luzula campestris	(L.) DC.	1	1	—	0	0	0	3	5	96
Luzula campestris	(L.) DC.	1	5	—	0	0	0	2	7	99
Luzula campestris	(L.) DC.	1	1	—	0	0	0	5	5	123
Luzula campestris	(L.) DC.	1	4	—	0	0	0	10	5	149
Luzula campestris	(L.) DC.	1	1	—	0	0	0	7	5	232
Luzula campestris	(L.) DC.	1	1	—	0	0	0	5	5	244
Luzula campestris	(L.) DC.	1	1	—	0	0	0	5	5	250
Luzula campestris	(L.) DC.	1	12	—	0	0	0	10	5	255
Luzula campestris	(L.) DC.	1	3	—	3	5	4	3	7	235
Luzula campestris	(L.) DC.	2	1	—	54	54	54	30	5	29
Luzula campestris	(L.) DC.	2	3	—	9	99	55	12	5	174
Luzula campestris	(L.) DC.	2	3	—	38	113	84	10	5	255
Luzula campestris	(L.) DC.	2	1	—	200	200	200	20	5	67
Luzula campestris	(L.) DC.	2	1	—	592	592	592	30	5	45
Luzula campestris	(L.) DC.	3	1	>35	33	33	33	32	5	169
Luzula campestris	(L.) DC.	3	3	—	32	54	39	30	5	45
Luzula campestris	(L.) DC.	3	1	>100	44	44	44	24	5	169
Luzula campestris	(L.) DC.	3	2	>30	56	111	84	24	5	168
Luzula campestris	(L.) DC.	3	3	>30	78	133	104	24	5	168
Luzula campestris	(L.) DC.	3	1	—	108	108	108	8	5	108
Luzula campestris	(L.) DC.	3	1	—	280	280	280	30	5	29
Luzula campestris	(L.) DC.	4	1	—	17	17	17	7	5	232
Luzula campestris	(L.) DC.	4	1	—	18	18	18	5	5	233
Luzula campestris	(L.) DC.	4	1	—	46	46	46	5	5	123
Luzula campestris	(L.) DC.	4	2	—	36	62	49	16	5	216
Luzula campestris	(L.) DC.	4	1	—	87	87	87	15	5	267
Luzula campestris	(L.) DC.	4	7	—	38	225	95	10	5	255
Luzula campestris	(L.) DC.	4	5	—	12	225	96	12	5	174
Luzula campestris	(L.) DC.	4	1	—	101	101	101	17.8	5	156
Luzula campestris	(L.) DC.	4	3	—	64	384	187	5	5	69
Luzula campestris	(L.) DC.	4	1	—	191	191	191	9	5	84
Luzula campestris	(L.) DC.	4	6	—	126	655	370	17.8	5	155
Luzula campestris	(L.) DC.	4	2	—	75	840	458	5	5	250
Luzula campestris	(L.) DC.	4	4	—	300	700	500	15	5	249
Luzula campestris	(L.) DC.	4	1	—	1837	1837	1837	8.5	4	254
Luzula campestris	(L.) DC.	4	2	—	3600	9600	6600	20	4	225
Luzula luzuloides	(Lam.) Dandy & Wilm.	1	1	—	0	0	0	20	5	67

Species	Authority	Seed bank type	Number of records	Longevity (y)	Minimum density (seeds m^{-2})	Maximum density (seeds m^{-2})	Mean density (seeds m^{-2})	Depth (cm)	Method	Source code
Luzula luzuloides	(Lam.) Dandy & Wilm.	2	4	—	38	763	269	10	5	255
Luzula luzuloides	(Lam.) Dandy & Wilm.	4	3	—	50	88	75	10	5	255
Luzula multiflora	(Ehrh.) Lej.	1	1	—	0	0	0	40	5	11
Luzula multiflora	(Ehrh.) Lej.	1	1	—	0	0	0	5	5	55
Luzula multiflora	(Ehrh.) Lej.	1	1	—	0	0	0	2	5	153
Luzula multiflora	(Ehrh.) Lej.	1	4	—	0	0	0	3	5	208
Luzula multiflora	(Ehrh.) Lej.	2	1	—	49	49	49	11.5	5	134
Luzula multiflora	(Ehrh.) Lej.	2	1	—	85	85	85	3	5	208
Luzula multiflora	(Ehrh.) Lej.	2	1	—	184	184	184	5	5	222
Luzula multiflora	(Ehrh.) Lej.	3	1	>20	41	41	41	5	5	222
Luzula multiflora	(Ehrh.) Lej.	4	1	—	27	27	27	2	5	153
Luzula multiflora	(Ehrh.) Lej.	4	1	—	113	113	113	3	5	208
Luzula multiflora	(Ehrh.) Lej.	4	2	—	122	184	153	5	5	222
Luzula pallidula	Kirschner	3	2	—	69	69	69	12	5	171
Luzula parviflora	(Ehrh.) Desv.	1	3	—	0	0	0	6	5	39
Luzula parviflora	(Ehrh.) Desv.	3	1	200	160	160	160	15	5	146
Luzula pilosa	(L.) Willd.	1	1	—	0	0	0	20	5	67
Luzula pilosa	(L.) Willd.	1	1	—	0	0	0	3	7	235
Luzula pilosa	(L.) Willd.	1	1	—	42	42	42	12	5	171
Luzula pilosa	(L.) Willd.	2	1	—	8	8	8	15	5	32
Luzula pilosa	(L.) Willd.	2	1	—	60	60	60	15	5	257
Luzula pilosa	(L.) Willd.	2	1	—	78	78	78	32	5	169
Luzula pilosa	(L.) Willd.	2	1	—	83	83	83	12	5	171
Luzula pilosa	(L.) Willd.	3	1	—	44	44	44	24	5	169
Luzula pilosa	(L.) Willd.	3	1	—	65	65	65	17.5	5	126
Luzula pilosa	(L.) Willd.	3	1	—	67	67	67	16	5	168
Luzula pilosa	(L.) Willd.	3	2	—	80	220	150	15	5	257
Luzula pilosa	(L.) Willd.	4	1	—	25	25	25	10	5	126
Luzula pilosa	(L.) Willd.	4	1	—	32	32	32	9	5	84
Luzula pilosa	(L.) Willd.	4	1	—	107	107	107	3	5	82
Luzula pilosa	(L.) Willd.	4	1	—	117	117	117	2	5	82
Luzula pilosa	(L.) Willd.	4	3	—	46	224	124	5	5	82
Luzula spicata	(L.) DC.	1	2	—	0	0	0	5	4	91
Luzula spicata	(L.) DC.	4	4	—	13	46	27	6	5	39
Luzula spicata	(L.) DC.	4	1	—	36	36	36	5	5	55
Luzula sylvatica	(Hudson) Gaudin	1	1	—	0	0	0	2	5	153
Luzula sylvatica	(Hudson) Gaudin	1	1	—	0	0	0	17.8	5	156
Luzula sylvatica	(Hudson) Gaudin	2	1	—	6	6	6	3	7	235
Luzula sylvatica	(Hudson) Gaudin	2	1	—	63	63	63	12	4	51
Luzula sylvatica	(Hudson) Gaudin	4	1	—	583	583	583	3	5	96

Species	Author									
Juncaginaceae	—	—	—	—	—	—	—	—	—	—
Triglochin maritimum	L.	1	4	—	0	0	0	1	5	57
Triglochin maritimum	L.	1	1	—	0	0	0	10	4	103
Triglochin maritimum	L.	2	10	—	176	15706	5088	7	5	112
Labiatae (Lamiaceae)	—	—	—	—	—	—	—	—	—	—
Ajuga chamaepitys	(L.) Schreber	4	8	—	20	173	51	30	4	13
Ajuga genevensis	L.	1	1	—	0	0	0	6.5	7	182
Ajuga genevensis	L.	2	1	—	914	914	914	15	5	272
Ajuga pyramidalis	L.	3	2	—	88	88	88	10	5	150
Ajuga pyramidalis	L.	4	2	—	205	263	234	10	5	150
Ajuga reptans	L.	1	3	—	0	0	0	6	5	49
Ajuga reptans	L.	1	2	—	0	0	0	3	5	96
Ajuga reptans	L.	1	1	—	0	0	0	12.5	5	126
Ajuga reptans	L.	1	3	—	0	0	0	6.5	5	128
Ajuga reptans	L.	1	3	—	0	0	0	10	5	149
Ajuga reptans	L.	1	2	—	0	0	0	12	5	171
Ajuga reptans	L.	1	1	—	0	0	0	30	5	173
Ajuga reptans	L.	1	4	—	0	0	0	12	5	174
Ajuga reptans	L.	1	3	—	0	0	0	6	5	208
Ajuga reptans	L.	1	2	—	0	0	0	15	5	257
Ajuga reptans	L.	2	3	—	0	39	22	2	7	99
Ajuga reptans	L.	2	2	—	1	7	4	5	6	170
Ajuga reptans	L.	2	1	—	12	12	12	25	4	36
Ajuga reptans	L.	2	7	—	38	113	61	10	5	255
Ajuga reptans	L.	2	1	—	100	100	100	20	5	67
Ajuga reptans	L.	3	1	—	216	216	216	13	5	68
Ajuga reptans	L.	3	1	—	48	48	48	13	5	68
Ajuga reptans	L.	3	3	—	100	100	100	6.5	5	68
Ajuga reptans	L.	3	3	—	170	764	387	3	5	208
Ajuga reptans	L.	3	4	—	200	600	438	20	5	4
Ajuga reptans	L.	4	1	—	1593	1593	1593	30	5	29
Ajuga reptans	L.	4	1	—	13	13	13	15	5	32
Ajuga reptans	L.	4	3	—	33	33	33	10	5	126
Ajuga reptans	L.	4	3	—	50	50	50	10	5	255
Ajuga reptans	L.	4	4	—	170	425	264	3	5	208
Ballota nigra	L.	1	2	—	0	0	0	3	5	96
Ballota nigra	L.	2	3	—	135	135	135	11	5	162
Ballota nigra	L.	2	2	—	438	438	438	17	5	162
Ballota nigra	L.	2	1	—	1082	1082	1082	8	7	275
Ballota nigra	L.	3	1	>5	0	0	0	7.5	2	195
Ballota nigra	L.	3	1	>35	538	538	538	65	5	161
Clinopodium acinos	(L.) Kuntze	1	1	—	0	0	0	1	5	96
Clinopodium acinos	(L.) Kuntze	2	1	—	15	15	15	6	5	211
Clinopodium acinos	(L.) Kuntze	2	1	—	648	648	648	10	5	152
Clinopodium acinos	(L.) Kuntze	3	1	>35	33	33	33	24	5	168

Species	Authority	Seed bank type	Number of records	Longevity (y)	Minimum density (seeds m⁻²)	Maximum density (seeds m⁻²)	Mean density (seeds m⁻²)	Depth (cm)	Method	Source code
Clinopodium acinos	(L.) Kuntze	3	1	>40	56	56	56	24	5	168
Clinopodium acinos	(L.) Kuntze	4	1	—	169	169	169	10	5	110
Clinopodium acinos	(L.) Kuntze	4	1	—	702	702	702	10	5	152
Clinopodium vulgare	L.	1	1	—	0	0	0	20	5	97
Clinopodium vulgare	L.	2	1	—	10	10	10	6	5	49
Clinopodium vulgare		3	1	—	144	144	144	70	5	219
Galeopsis speciosa	Miller	1	2	—	0	0	0	6	5	49
Galeopsis speciosa	Miller	1	1	—	0	0	0	10	5	255
Galeopsis speciosa	Miller	4	1	—	188	188	188	10	5	255
Galeopsis tetrahit	L.	1	2	—	0	0	0	6	5	11
Galeopsis tetrahit	L.	1	1	—	0	0	0	4	5	176
Galeopsis tetrahit	L.	1	1	—	0	0	0	10	5	181
Galeopsis tetrahit	L.	1	3	—	0	0	0	3	5	208
Galeopsis tetrahit	L.	1	1	—	0	0	0	20	4	225
Galeopsis tetrahit	L.	1	1	—	0	0	0	5	5	248
Galeopsis tetrahit	L.	1	9	—	0	0	0	10	5	255
Galeopsis tetrahit	L.	1	1	—	123	123	123	8	7	275
Galeopsis tetrahit	L.	2	1	>2	0	0	0	2	1	47
Galeopsis tetrahit	L.	2	1	>2	0	0	0	15	1	47
Galeopsis tetrahit	L.	2	1	—	0	0	0	5	6	170
Galeopsis tetrahit	L.	2	1	—	52	52	52	25	4	36
Galeopsis tetrahit	L.	2	1	—	153	153	153	15	5	142
Galeopsis tetrahit	L.	2	2	—	125	250	188	10	5	255
Galeopsis tetrahit	L.	2	3	—	353	3075	1445	17.8	5	155
Galeopsis tetrahit	L.	3	1	>5	0	0	0	7.5	2	195
Galeopsis tetrahit	L.	3	1	—	32	32	32	30	5	45
Galeopsis tetrahit	L.	3	1	>24	32	32	32	30	5	45
Galeopsis tetrahit	L.	3	1	>6	237	237	237	30	5	45
Galeopsis tetrahit	L.	3	1	>8	237	237	237	30	5	45
Galeopsis tetrahit	L.	3	1	—	403	403	403	17.8	5	155
Galeopsis tetrahit	L.	4	10	—	21	21	21	20	5	111
Galeopsis tetrahit	L.	4	1	—	50	288	133	10	5	255
Galeopsis tetrahit	L.	4	4	—	542	542	542	7.5	5	10
Galeopsis tetrahit	L.	4	4	—	850	1440	1113	15	5	94
Glechoma hederacea	L.	1	2	—	0	0	0	25	4	36
Glechoma hederacea	L.	1	1	—	0	0	0	13	5	68
Glechoma hederacea	L.	1	1	—	0	0	0	5	5	69
Glechoma hederacea	L.	1	1	—	0	0	0	5	4	88
Glechoma hederacea	L.	1	2	—	0	0	0	3	5	96
Glechoma hederacea	L.	1	2	—	0	0	0	20	5	97
Glechoma hederacea	L.	1	2	—	0	0	0	6.5	5	128
Glechoma hederacea	L.	1	6	—	0	0	0	3	5	208

Species	Authority									
Glechoma hederacea	L.	1	1	—	0	0	0	20	4	225
Glechoma hederacea	L.	1	1	—	0	0	0	5	5	250
Glechoma hederacea	L.	1	1	—	0	0	0	10	5	255
Glechoma hederacea	L.	1	4	—	0	0	0	15	5	257
Glechoma hederacea	L.	1	1		50	50	50	6.5	5	68
Glechoma hederacea	L.	5	5	—	53	210	102	8	7	275
Glechoma hederacea	L.	1	2	—	112	112	112	12	5	162
Glechoma hederacea	L.	1	2	—	280	280	280	15	5	257
Glechoma hederacea	L.	1	2	—	281	281	281	8	7	275
Glechoma hederacea	L.	1	3	>460	53	53	53	62	5	161
Glechoma hederacea	L.	1	3	>40	60	60	60	20	5	173
Glechoma hederacea	L.	1	3	>22	111	111	111	24	5	168
Glechoma hederacea	L.	2	3	—	0	0	0	20	5	67
Lamiastrum galeobdolon	(L.) Ehrend. & Polatschek	1	4	—	0	0	0	3	5	96
Lamiastrum galeobdolon	(L.) Ehrend. & Polatschek	2	1	—	0	0	0	12.5	5	126
Lamiastrum galeobdolon	(L.) Ehrend. & Polatschek	1	2	—	0	0	0	5	6	170
Lamiastrum galeobdolon	(L.) Ehrend. & Polatschek	1	1	—	0	0	0	12	5	171
Lamiastrum galeobdolon	(L.) Ehrend. & Polatschek	1	1	—	0	0	0	30	5	173
Lamiastrum galeobdolon	(L.) Ehrend. & Polatschek	1	1	—	0	0	0	4	5	176
Lamiastrum galeobdolon	(L.) Ehrend. & Polatschek	1	1	—	0	0	0	13	5	183
Lamiastrum galeobdolon	(L.) Ehrend. & Polatschek	1	4	—	0	0	0	5	5	220
Lamiastrum galeobdolon	(L.) Ehrend. & Polatschek	1	1	—	0	0	0	20	4	225
Lamiastrum galeobdolon	(L.) Ehrend. & Polatschek	1	1	—	0	0	0	3	7	235
Lamiastrum galeobdolon	(L.) Ehrend. & Polatschek	1	1	—	44	44	44	15	5	257
Lamiastrum galeobdolon	(L.) Ehrend. & Polatschek	2	1	—	44	44	44	16	5	168
Lamiastrum galeobdolon	(L.) Ehrend. & Polatschek	4	1	—	1	1	1	4	5	176
Lamiastrum galeobdolon	(L.) Ehrend. & Polatschek	1	3	—	0	0	0	3	5	96
Lamium album	L.	1	3	—	53	70	59	8	7	275
Lamium album	L.	2	4	—	113	361	230	8	7	275
Lamium album	L.	2	1	—	700	700	700	6	5	162
Lamium album	L.	2	1	—	875	875	875	10	5	162
Lamium album	L.	3	3	>5	0	0	0	7.5	2	198
Lamium album	L.	3	1	—	68	68	68	40	5	219
Lamium album	L.	3	1	>660	90	90	90	50	5	161
Lamium album	L.	4	1	—	50	50	50	3	5	96
Lamium amplexicaule	L.	1	2	—	0	0	0	10	4	22
Lamium amplexicaule	L.	1	1	—	0	0	0	3	5	132
Lamium amplexicaule	L.	2	1	—	235	235	235	50	5	260
Lamium amplexicaule	L.	2	2	—	67	67	67	10	5	184
Lamium amplexicaule	L.	3	2	>20	188	375	282	17	5	162
Lamium amplexicaule	L.	3	1	—	7	7	7	20	5	40
Lamium amplexicaule	L.	3	1	>25	188	188	188	17	5	162
Lamium amplexicaule	L.	4	1	—	60	60	60	20	5	111
Lamium amplexicaule	L.	4	2	—	87	229	158	15	4	31
Lamium amplexicaule	L.	4	2	—	254	254	254	25	5	2
Lamium amplexicaule	L.	4	2	—	303	1294	799	25	5	260
Lamium amplexicaule	L.	4	1	—	2025	2025	2025	15	5	207

Species	Authority	Seed bank type	Number of records	Longevity (y)	Minimum density (seeds m⁻²)	Maximum density (seeds m⁻²)	Mean density (seeds m⁻²)	Depth (cm)	Method	Source code
Lamium purpureum	L.	1	2	—	0	0	0	25	4	36
Lamium purpureum	L.	1	1	—	0	0	0	30	5	66
Lamium purpureum	L.	1	1	—	0	0	0	50	5	66
Lamium purpureum	L.	2	1	>3	0	0	0	20	3	268
Lamium purpureum	L.	2	1	—	20	20	20	15	5	272
Lamium purpureum	L.	2	1	—	28	28	28	40	5	219
Lamium purpureum	L.	2	1	—	240	240	240	18	5	162
Lamium purpureum	L.	2	8	—	38	1100	478	10	5	255
Lamium purpureum	L.	2	1	—	688	688	688	17	5	162
Lamium purpureum	L.	3	1	>160	70	70	70	57	5	161
Lamium purpureum	L.	3	1	>660	84	84	84	74	5	161
Lamium purpureum	L.	3	1	>660	103	103	103	55	5	161
Lamium purpureum	L.	3	1	>660	110	110	110	50	5	111
Lamium purpureum	L.	4	1	—	14	14	14	20	5	272
Lamium purpureum	L.	4	2	—	21	104	63	15	5	255
Lamium purpureum	L.	4	2	—	263	300	282	10	5	13
Lamium purpureum	L.	4	7	—	60	2437	529	30	4	207
Lamium purpureum	L.	4	1	—	618	618	618	15	5	185
Lamium purpureum	L.	4	2	—	366	896	631	25	4	255
Lycopus europaeus	L.	1	1	—	3	3	3	10	5	235
Lycopus europaeus	L.	1	1	—	3	3	3	3	7	98
Melittis melissophyllum	L.	2	1	—	1427	1427	1427	8	7	275
Mentha aquatica	L.	1	1	—	0	0	0	3	5	96
Mentha aquatica	L.	1	1	—	0	0	0	3	5	208
Mentha aquatica	L.	1	2	—	0	0	0	20	4	225
Mentha aquatica	L.	1	1	—	0	0	0	3	7	235
Mentha aquatica	L.	1	5	—	0	26	5	2	7	99
Mentha aquatica	L.	1	1	—	75	75	75	30	5	45
Mentha aquatica	L.	1	2	—	53	105	79	8	7	275
Mentha aquatica	L.	1	1	—	156	156	156	10	5	27
Mentha aquatica	L.	2	1	—	215	215	215	30	5	45
Mentha aquatica	L.	2	3	—	500	660	560	6.5	5	128
Mentha aquatica	L.	3	2	—	54	65	60	30	5	45
Mentha aquatica	L.	3	1	—	208	260	234	6	5	99
Mentha aquatica	L.	3	1	—	520	520	520	10	5	27
Mentha aquatica	L.	4	1	—	42	42	42	5	5	98
Mentha aquatica	L.	4	5	—	100	5360	1620	6.5	5	128
Mentha aquatica	L.	4	1	—	2517	2517	2517	3	5	96
Mentha arvensis	L.	1	1	—	0	0	0	25	5	2
Mentha arvensis	L.	1	1	—	0	0	0	25	4	36
Mentha arvensis	L.	1	3	—	0	0	0	3	5	208

Species	Authority									
Mentha arvensis	L.	1	6	—	0	0	0	10	5	255
Mentha arvensis	L.	1	2	—	142	142	142	6	5	208
Mentha arvensis	L.	1	1	—	278	278	278	50	5	260
Mentha arvensis	L.	2	3	—	38	188	88	10	5	255
Mentha arvensis	L.	2	1	—	100	100	100	3	5	96
Mentha arvensis	L.	2	1	—	124	124	124	16	5	216
Mentha arvensis	L.	2	1	—	142	142	142	6	5	208
Mentha arvensis	L.	3	1	>30	15	15	15	6	5	211
Mentha arvensis	L.	3	3	—	85	283	170	6	5	208
Mentha arvensis	L.	3	1	—	541	541	541	50	5	260
Mentha arvensis	L.	4	1	—	14	14	14	10	5	8
Mentha arvensis	L.	4	1	—	21	21	21	5	5	166
Mentha arvensis	L.	4	1	—	34	34	34	25	4	36
Mentha arvensis	L.	4	1	—	40	40	40	30	4	13
Mentha arvensis	L.	4	1	—	50	50	50	10	5	255
Mentha arvensis	L.	4	1	—	96	96	96	20	4	111
Mentha arvensis	L.	4	5	—	85	538	204	3	5	208
Mentha arvensis	L.	4	4	—	780	3119	1689	50	5	216
Mentha arvensis	L.	4	1	—	3783	3783	3783	3	5	96
Mentha pulegium	L.	4	4	—	13800	13800	13800	6.5	4	1
Mentha pulegium	L.	4	14	—	800	176000	46421	6.5	4	164
Mentha × verticillata	L.	4	1	—	77	77	77	17.8	5	155
Origanum vulgare	L.	1	1	—	0	0	0	3	5	96
Origanum vulgare	L.	1	1	—	0	0	0	6	5	181
Origanum vulgare	L.	1	1	>3	299	299	299	6	5	48
Origanum vulgare	L.	2	2	—	0	0	0	6	3	179
Origanum vulgare	L.	2	2	—	62	351	207	3	7	235
Origanum vulgare	L.	2	2	—	234	234	234	10	5	152
Origanum vulgare	L.	2	1	—	712	712	712	13	5	183
Origanum vulgare	L.	2	1	—	1771	1771	1771	20	5	97
Origanum vulgare	L.	3	1	—	521	521	521	20	5	97
Origanum vulgare	L.	3	1	>20	1152	1152	1152	13	5	183
Origanum vulgare	L.	3	1	—	4910	4910	4910	6	5	48
Origanum vulgare	L.	4	1	—	128	128	128	6	5	181
Origanum vulgare	L.	4	1	—	183	183	183	3	5	96
Origanum vulgare	L.	4	1	—	2340	2340	2340	10	5	152
Prunella grandiflora	(L.) Scholler	1	2	—	0	0	0	6	5	181
Prunella grandiflora	(L.) Scholler	1	1	—	0	0	0	10	5	255
Prunella grandiflora	(L.) Scholler	2	2	—	25	79	52	6.5	7	182
Prunella grandiflora	(L.) Scholler	4	1	—	64	64	64	6	5	181
Prunella laciniata	(L.) L.	4	3	—	408	408	408	8.5	4	254
Prunella vulgaris	L.	1	1	—	0	0	0	25	4	36
Prunella vulgaris	L.	1	1	—	0	0	0	30	5	45
Prunella vulgaris	L.	1	2	—	0	0	0	5	4	53
Prunella vulgaris	L.	1	1	—	0	0	0	3	5	59
Prunella vulgaris	L.	1	1	—	0	0	0	5	5	69
Prunella vulgaris	L.	1	1	—	0	0	0	5	4	88

Species	Authority	Seed bank type	Number of records	Longevity (y)	Minimum density (seeds m⁻²)	Maximum density (seeds m⁻²)	Mean density (seeds m⁻²)	Depth (cm)	Method	Source code
Prunella vulgaris	L	1	1	—	0	0	0	10	4	88
Prunella vulgaris	L	1	1	—	0	0	0	3	5	96
Prunella vulgaris	L	1	5	—	0	0	0	20	5	97
Prunella vulgaris	L	1	1	—	0	0	0	6.5	5	128
Prunella vulgaris	L	1	1	—	0	0	0	10	5	150
Prunella vulgaris	L	1	5	—	0	0	0	10	5	152
Prunella vulgaris	L	1	1	—	0	0	0	12	5	174
Prunella vulgaris	L	1	2	—	0	0	0	10	5	184
Prunella vulgaris	L	1	1	—	0	0	0	20	4	225
Prunella vulgaris	L	1	2	—	0	0	0	5	5	251
Prunella vulgaris	L	1	1	—	0	0	0	10	5	252
Prunella vulgaris	L	1	1	—	0	0	0	10	5	255
Prunella vulgaris	L	1	1	—	0	0	0	15	5	257
Prunella vulgaris	L	1	5	—	22	22	22	16	5	23
Prunella vulgaris	L	1	5	—	78	104	87	6	5	28
Prunella vulgaris	L	2	5	>1	0	0	0	—	1	275
Prunella vulgaris	L	2	2	—	10	12	11	25	4	36
Prunella vulgaris	L	2	1	—	25	25	25	12	5	174
Prunella vulgaris	L	2	1	—	43	43	43	30	5	29
Prunella vulgaris	L	2	3	—	60	240	153	10	5	149
Prunella vulgaris	L	2	4	—	54	635	353	30	5	45
Prunella vulgaris	L	3	1	>5	0	0	0	7.5	2	195
Prunella vulgaris	L	3	1	>45	67	67	67	24	5	168
Prunella vulgaris	L	3	2	—	258	484	371	30	5	45
Prunella vulgaris	L	4	1	—	7	7	7	10	5	33
Prunella vulgaris	L	4	1	—	24	24	24	25	4	36
Prunella vulgaris	L	4	2	—	24	37	31	5	5	122
Prunella vulgaris	L	4	1	—	39	39	39	12	5	174
Prunella vulgaris	L	4	1	—	43	43	43	3	5	59
Prunella vulgaris	L	4	2	—	60	120	90	6.5	5	128
Prunella vulgaris	L	4	3	—	29	277	114	7	5	58
Prunella vulgaris	L	4	1	—	144	144	144	5	5	69
Prunella vulgaris	L	4	2	—	50	283	167	3	5	96
Prunella vulgaris	L	4	2	—	160	300	230	10	5	149
Prunella vulgaris	L	4	3	—	66	477	240	10	5	110
Prunella vulgaris	L	4	2	—	312	312	312	10	5	152
Prunella vulgaris	L	4	1	—	409	409	409	10	5	150
Prunella vulgaris	L	4	3	—	605	8502	4323	17.8	5	155
Salvia pratensis	L	1	3	—	0	0	0	6	5	181
Salvia pratensis	L	1	1	—	0	0	0	6.5	7	182
Salvia pratensis	L	4	1	—	2245	2245	2245	8.5	4	254
Scutellaria galericulata	L	1	1	—	0	0	0	10	5	255

Species	Authority									
Scutellaria galericulata	L.	4	1	—	0	0	0	5	5	166
Scutellaria minor	Hudson	1	2	—	86	172	129	30	5	45
Scutellaria minor	Hudson	4	1	—	252	252	252	17.8	5	155
Stachys annua	(L.) L.	4	1	—	97	97	97	25	4	136
Stachys arvensis	(L.) L.	2	1	>8	227	227	227	17.8	5	155
Stachys arvensis	(L.) L.	3	1	>20	75	75	75	30	5	45
Stachys arvensis	(L.) L.	3	1	—	89	89	89	24	5	168
Stachys arvensis	(L.) L.	3	1	—	277	277	277	17.8	5	155
Stachys arvensis	(L.) L.	4	1	—	4	4	4	20	4	111
Stachys arvensis	(L.) L.	4	10	—	20	413	126	30	4	13
Stachys officinalis	(L.) Trev. St. Leon	1	1	—	0	0	0	6.5	7	182
Stachys officinalis	(L.) Trev. St. Leon	1	3	—	0	0	0	10	5	255
Stachys officinalis	(L.) Trev. St. Leon	1	1	—	3	3	3	3	7	235
Stachys officinalis	(L.) Trev. St. Leon	2	1	4	0	0	0	7.5	2	195
Stachys officinalis	(L.) Trev. St. Leon	2	1	—	80	80	80	10	5	149
Stachys palustris	L.	1	1	—	0	0	0	20	4	225
Stachys palustris	L.	1	10	>20	0	0	0	10	5	255
Stachys palustris	L.	3	1	—	33	33	33	24	5	168
Stachys palustris	L.	4	1	—	4	4	4	5	5	166
Stachys palustris	L.	4	1	—	30	30	30	20	5	111
Stachys recta	L.	1	1	—	0	0	0	6	5	181
Stachys recta	L.	1	1	—	0	0	0	13	5	183
Stachys sylvatica	L.	1	1	—	0	0	0	6	5	49
Stachys sylvatica	L.	1	2	—	0	0	0	20	5	67
Stachys sylvatica	L.	1	2	—	0	0	0	12.5	5	126
Stachys sylvatica	L.	1	1	—	0	0	0	20	4	225
Stachys sylvatica	L.	1	1	—	0	0	0	3	7	235
Stachys sylvatica	L.	1	1	—	120	120	120	15	5	257
Stachys sylvatica	L.	2	1	—	9	9	9	3	7	235
Stachys sylvatica	L.	2	1	—	2188	2188	2188	17	5	162
Stachys sylvatica	L.	3	2	>5	0	0	0	7.5	2	198
Stachys sylvatica	L.	3	1	—	80	80	80	15	5	257
Stachys sylvatica	L.	4	1	—	10	10	10	6	5	49
Stachys sylvatica	L.	4	1	—	17	17	17	7	5	58
Stachys sylvatica	L.	4	1	—	50	50	50	3	5	96
Stachys sylvatica	L.	4	1	—	58	58	58	5	5	220
Stachys sylvatica	L.	4	1	—	72	72	72	10	5	126
Teucrium botrys	L.	3	1	—	10	10	10	6	5	211
Teucrium chamaedrys	L.	1	1	—	0	0	0	8.5	4	254
Teucrium chamaedrys	L.	3	1	>20	40	40	40	13	5	183
Teucrium scorodonia	L.	1	1	—	0	0	0	10	5	255
Teucrium scorodonia	L.	1	3	—	0	0	0	15	5	257
Teucrium scorodonia	L.	2	2	—	0	3	2	3	7	235
Teucrium scorodonia	L.	2	5	>1	0	0	0	—	1	275
Teucrium scorodonia	L.	2	3	—	100	100	100	10	5	255
Teucrium scorodonia	L.	3	3	—	60	200	113	15	5	257

Species	Authority	Seed bank type	Number of records	Longevity (y)	Minimum density (seeds m⁻²)	Maximum density (seeds m⁻²)	Mean density (seeds m⁻²)	Depth (cm)	Method	Source code
Teucrium scorodonia	L.	4	3	—	38	100	67	10	5	255
Teucrium scorodonia	L.	4	1	—	617	617	617	3	5	96
Thymus polytrichus	A. Kerner ex Borbas	1	1	—	0	0	0	1	5	96
Thymus polytrichus	A. Kerner ex Borbas	1	3	—	0	0	0	3	5	96
Thymus polytrichus	A. Kerner ex Borbas	1	1	—	0	0	0	2	5	153
Thymus polytrichus	A. Kerner ex Borbas	1	1	—	4	4	4	3	7	235
Thymus polytrichus	A. Kerner ex Borbas	2	1	—	25	25	25	3	7	235
Thymus polytrichus	A. Kerner ex Borbas	4	2	—	8	29	19	5	5	122
Thymus polytrichus	A. Kerner ex Borbas	4	1	—	106	106	106	5	5	55
Thymus polytrichus	A. Kerner ex Borbas	4	1	—	160	160	160	5	5	69
Thymus polytrichus	A. Kerner ex Borbas	4	2	—	198	505	352	10	5	110
Thymus pulegioides	L.	1	4	—	0	0	0	10	5	255
Thymus pulegioides	L.	1	1	—	17	17	17	6.5	7	182
Thymus pulegioides	L.	2	1	—	43	43	43	6.5	7	182
Thymus pulegioides	L.	2	1	—	56	56	56	13	5	183
Thymus pulegioides	L.	3	1	—	24	24	24	40	5	219
Thymus pulegioides	L.	3	1	—	68	68	68	70	5	219
Thymus pulegioides	L.	4	2	—	88	100	94	10	5	255
Thymus pulegioides	L.	4	2	—	264	576	420	6	5	181
Thymus serpyllum	L.	4	1	—	28	28	28	5	5	55
Leguminosae (Fabaceae)	—									
Anthyllis vulneraria	L.	1	1	—	0	0	0	5	5	55
Anthyllis vulneraria	L.	1	1	—	0	0	0	6	5	67
Anthyllis vulneraria	L.	1	1	—	0	0	0	3	5	96
Anthyllis vulneraria	L.	1	3	—	0	0	0	10	5	152
Anthyllis vulneraria	L.	1	2	—	0	0	0	10	5	255
Astragalus glycyphyllos	L.	2	1	>3	0	0	0	0	3	221
Astragalus norvegicus	Weber	1	1	—	0	0	0	10	5	255
Cytisus scoparius	(L.) Link	1	1	—	0	0	0	10	5	255
Cytisus scoparius	(L.) Link	2	2	—	63	250	157	10	5	255
Cytisus scoparius	(L.) Link	4	1	—	47	47	47	2	5	153
Cytisus scoparius	(L.) Link	4	6	—	225	738	415	10	5	255
Cytisus scoparius	(L.) Link	4	1	—	56834	56834	56834	5	4	104
Galega officinalis	L.	4	1	—	5167	5167	5167	5	4	104
Genista anglica	L.	1	1	—	0	0	0	5	5	92
Genista anglica	L.	1	1	—	0	0	0	2	5	153
Genista germanica	L.	1	1	—	0	0	0	10	5	255
Genista pilosa	L.	1	1	—	0	0	0	5	5	69
Genista tinctoria	L.	1	1	—	0	0	0	8.5	4	254
Genista tinctoria	L.	1	2	—	0	0	0	10	5	255

Taxon	Authority									
Hippocrepis comosa	L.	1	1	—	0	0	0	5	5	55
Hippocrepis comosa	L.	1	1	—	0	0	0	6	5	67
Hippocrepis comosa	L.	1	3	—	0	0	0	6	5	181
Hippocrepis comosa	L.	1	1	—	0	0	0	6.5	7	182
Hippocrepis comosa	L.	2	1	—	117	117	117	6.5	7	182
Lathyrus japonicus	Willd.	1	1	—	0	0	0	6.5	7	67
Lathyrus linifolius	(Reichard) Bassler	1	2	—	0	0	0	20	5	92
Lathyrus linifolius	(Reichard) Bassler	1	2	—	0	0	0	5	5	150
Lathyrus linifolius	(Reichard) Bassler	1	1	—	0	0	0	10	5	153
Lathyrus linifolius	(Reichard) Bassler	1	2	—	0	0	0	2	7	235
Lathyrus linifolius	(Reichard) Bassler	1	4	—	0	0	0	3	5	255
Lathyrus linifolius	(Reichard) Bassler	4	1	—	100	100	100	10	5	255
Lathyrus niger	(L.) Bernh.	1	4	—	0	0	0	10	5	49
Lathyrus palustris	L.	1	1	—	0	0	0	6	4	128
Lathyrus palustris	L.	1	1	—	0	0	0	6.5	5	148
Lathyrus palustris	L.	1	1	—	0	0	0	15	5	216
Lathyrus pratensis	L.	1	1	—	0	0	0	16	5	68
Lathyrus pratensis	L.	1	3	—	0	0	0	6.5	4	68
Lathyrus pratensis	L.	1	4	—	0	0	0	13	4	69
Lathyrus pratensis	L.	1	4	—	0	0	0	5	5	88
Lathyrus pratensis	L.	1	2	—	0	0	0	5	5	88
Lathyrus pratensis	L.	1	10	—	0	0	0	10	5	96
Lathyrus pratensis	L.	1	1	—	0	0	0	3	5	97
Lathyrus pratensis	L.	1	5	—	0	0	0	20	5	110
Lathyrus pratensis	L.	1	1	—	0	0	0	10	5	150
Lathyrus pratensis	L.	1	4	—	0	0	0	10	5	151
Lathyrus pratensis	L.	1	1	—	0	0	0	6	4	152
Lathyrus pratensis	L.	1	8	—	0	0	0	10	7	181
Lathyrus pratensis	L.	1	1	—	0	0	0	6	5	225
Lathyrus pratensis	L.	1	13	—	0	0	0	20	5	235
Lathyrus pratensis	L.	1	1	—	0	0	0	3	5	251
Lathyrus pratensis	L.	1	2	—	0	0	0	5	7	252
Lathyrus pratensis	L.	1	5	—	0	0	0	10	1	255
Lathyrus pratensis	L.	1	1	—	32	32	32	10	5	45
Lathyrus pratensis	L.	2	2	>1	53	79	66	30	2	275
Lathyrus pratensis	L.	2	5	—	0	0	0	8	5	275
Lathyrus pratensis	L.	3	1	—	17	17	17	6	5	49
Lathyrus pratensis	L.	4	3	>5	0	0	0	7.5	2	200
Lathyrus pratensis	L.	4	5	—	60	120	84	10	5	149
Lathyrus pratensis	L.	4	4	—	101	101	101	17.8	5	155
Lathyrus tuberosus	L.	4	4	—	26	280	115	30	4	13
Lathyrus vernus	(L.) Bernh.	1	1	—	0	0	0	16	5	216
Lotus corniculatus	L.	1	3	—	0	0	0	5	4	53
Lotus corniculatus	L.	1	2	—	0	0	0	6	5	67
Lotus corniculatus	L.	1	1	—	0	0	0	20	5	67

Species	Authority	Seed bank type	Number of records	Longevity (y)	Minimum density (seeds m⁻²)	Maximum density (seeds m⁻²)	Mean density (seeds m⁻²)	Depth (cm)	Method	Source code
Lotus corniculatus	L.	1	3	—	0	0	0	5	5	69
Lotus corniculatus	L.	1	2	—	0	0	0	10	4	88
Lotus corniculatus	L.	1	7	—	0	0	0	3	5	96
Lotus corniculatus	L.	1	4	—	0	0	0	20	5	97
Lotus corniculatus	L.	1	2	—	0	0	0	10	5	110
Lotus corniculatus	L.	1	1	—	0	0	0	5	5	122
Lotus corniculatus	L.	1	1	—	0	0	0	10	5	149
Lotus corniculatus	L.	1	1	—	0	0	0	10	5	150
Lotus corniculatus	L.	1	5	—	0	0	0	10	5	152
Lotus corniculatus	L.	1	1	—	0	0	0	2	5	153
Lotus corniculatus	L.	1	2	—	0	0	0	6	5	181
Lotus corniculatus	L.	1	1	—	0	0	0	5	5	244
Lotus corniculatus	L.	1	4	—	0	0	0	5	5	251
Lotus corniculatus	L.	1	1	—	0	0	0	10	5	252
Lotus corniculatus	L.	1	7	—	0	0	0	10	5	255
Lotus corniculatus	L.	1	4	—	0	5	3	3	7	235
Lotus corniculatus	L.	2	1	4	0	0	0	8	1	61
Lotus corniculatus	L.	2	1	—	4	4	4	3	7	235
Lotus corniculatus	L.	2	1	—	6	6	6	15	5	32
Lotus corniculatus	L.	2	1	—	48	48	48	13	5	183
Lotus corniculatus	L.	2	2	—	62	86	74	6.5	7	182
Lotus corniculatus	L.	2	1	—	2056	2056	2056	30	5	45
Lotus corniculatus	L.	3	1	>6	0	0	0	20	1	61
Lotus corniculatus	L.	3	1	5	0	0	0	30	1	61
Lotus corniculatus	L.	3	1	>5	0	0	0	7.5	2	200
Lotus corniculatus	L.	3	1	5	0	0	0	7.5	2	200
Lotus corniculatus	L.	3	1	>10	3	3	3	10	4	241
Lotus corniculatus	L.	3	2	>100	44	56	50	32	5	169
Lotus corniculatus	L.	3	1	>20	64	64	64	13	5	183
Lotus corniculatus	L.	3	1	—	65	65	65	30	5	29
Lotus corniculatus	L.	3	2	—	180	980	580	15	5	257
Lotus corniculatus	L.	4	1	—	6	6	6	5	5	122
Lotus corniculatus	L.	4	1	—	7	7	7	5	5	19
Lotus corniculatus	L.	4	3	—	13	44	31	10	5	110
Lotus corniculatus	L.	4	1	—	32	32	32	9	5	84
Lotus corniculatus	L.	4	1	—	48	48	48	6	5	181
Lotus corniculatus	L.	4	1	—	80	80	80	10	5	149
Lotus corniculatus	L.	4	2	—	17	237	127	7	5	58
Lotus corniculatus	L.	4	2	—	101	252	177	17.8	5	155
Lotus pedunculatus	Cav.	1	4	—	0	0	0	10	5	27
Lotus pedunculatus	Cav.	1	1	—	0	0	0	20	5	67
Lotus pedunculatus	Cav.	1	1	—	0	0	0	5	5	69

Species	Author										
Lotus pedunculatus	Cav.	1	1	—	0	0	0	0	5	4	88
Lotus pedunculatus	Cav.	1	1	—	0	0	0	0	3	5	96
Lotus pedunculatus	Cav.	1	7	—	0	0	0	0	6	5	99
Lotus pedunculatus	Cav.	1	1	—	0	0	0	0	12	5	174
Lotus pedunculatus	Cav.	1	5	—	0	0	0	0	3	5	208
Lotus pedunculatus	Cav.	1	7	—	0	0	0	0	20	4	225
Lotus pedunculatus	Cav.	1	2	—	0	0	0	0	3	7	235
Lotus pedunculatus	Cav.	1	11	—	0	0	0	0	10	5	255
Lotus pedunculatus	Cav.	1	1	—	0	0	0	0	15	5	257
Lotus pedunculatus	Cav.	1	5	—	0	26	26	16	2	7	99
Lotus pedunculatus	Cav.	2	4	>1	0	0	0	0	—	1	275
Lotus pedunculatus	Cav.	2	1	1	0	0	0	0	—	1	275
Lotus pedunculatus	Cav.	2	3	—	12	37	37	21	12	5	174
Lotus pedunculatus	Cav.	3	1	—	150	150	150	150	20	5	4
Lotus pedunculatus	Cav.	3	1	—	1300	1300	1300	1300	20	5	69
Lotus pedunculatus	Cav.	3	1	—	1482	1482	1482	1482	30	5	107
Lotus pedunculatus	Cav.	4	1	—	9	9	9	9	8	4	42
Lotus pedunculatus	Cav.	4	1	—	48	48	48	48	5	5	69
Lotus pedunculatus	Cav.	4	5	—	58	58	58	58	5	5	98
Lotus pedunculatus	Cav.	4	1	—	38	175	175	103	10	5	255
Lotus pedunculatus	Cav.	4	2	—	113	113	113	113	3	5	208
Lotus pedunculatus	Cav.	4	2	—	92	143	143	118	5	5	222
Lotus pedunculatus	Cav.	4	2	—	9	385	385	197	12	5	174
Lotus pedunculatus	Cav.	4	13	—	161	5167	5167	1167	5	4	104
Lotus subbiflorus	Lagasca	4	12	—	161	12702	12702	3601	5	4	104
Lotus subbiflorus	L.	1	2	—	0	0	0	0	10	5	255
Lupinus angustifolius	(L.) Hudson	4	1	—	11787	11787	11787	11787	5	4	104
Medicago arabica	L.	1	1	—	0	0	0	0	5	5	69
Medicago lupulina	L.	1	2	—	0	0	0	0	1	5	96
Medicago lupulina	L.	1	2	—	0	0	0	0	3	5	96
Medicago lupulina	L.	1	5	—	0	0	0	0	20	5	97
Medicago lupulina	L.	1	3	—	0	0	0	0	6.5	5	128
Medicago lupulina	L.	1	1	—	0	0	0	0	10	5	149
Medicago lupulina	L.	1	4	—	0	0	0	0	10	5	152
Medicago lupulina	L.	1	1	4	13	13	13	13	8	5	108
Medicago lupulina	L.	2	1	4	0	0	0	0	13	1	133
Medicago lupulina	L.	2	1	>1	0	0	0	0	26	1	133
Medicago lupulina	L.	2	5	>1	0	0	0	0	—	1	275
Medicago lupulina	L.	2	3	—	63	88	88	75	10	5	255
Medicago lupulina	L.	2	2	—	6	388	388	197	6.5	7	182
Medicago lupulina	L.	2	1	—	217	217	217	217	12	5	81
Medicago lupulina	L.	2	1	—	234	234	234	234	10	5	152
Medicago lupulina	L.	2	1	—	592	592	592	592	13	5	183
Medicago lupulina	L.	2	1	—	700	700	700	700	6	5	162
Medicago lupulina	L.	3	3	>5	0	0	0	0	7	2	44
Medicago lupulina	L.	3	1	>26	0	0	0	0	25	1	125
Medicago lupulina	L.	3	1	>20	0	0	0	0	26	1	133

Species	Authority	Seed bank type	Number of records	Longevity (y)	Minimum density (seeds m⁻²)	Maximum density (seeds m⁻²)	Mean density (seeds m⁻²)	Depth (cm)	Method	Source code
Medicago lupulina	L.	3	1	>20	0	0	0	39	1	133
Medicago lupulina	L.	3	3	>5	0	0	0	7.5	2	200
Medicago lupulina	L.	3	1	>5	0	0	0	2.5	1	203
Medicago lupulina	L.	3	1	>5	0	0	0	2.5	2	203
Medicago lupulina	L.	3	1	>5	0	0	0	7.5	1	203
Medicago lupulina	L.	3	1	>5	0	0	0	7.5	2	203
Medicago lupulina	L.	3	1	>5	0	0	0	15	1	203
Medicago lupulina	L.	3	1	>5	0	0	0	15	2	203
Medicago lupulina	L.	3	1	>20	48	48	48	13	5	183
Medicago lupulina	L.	3	1	>30	56	56	56	10	5	181
Medicago lupulina	L.	3	1	>80	75	75	75	90	5	162
Medicago lupulina	L.	4	1	—	8	8	8	5	5	122
Medicago lupulina	L.	4	1	—	35	35	35	7	5	58
Medicago lupulina	L.	4	3	—	32	48	40	6	5	181
Medicago lupulina	L.	4	3	—	16	85	41	10	5	110
Medicago lupulina	L.	4	1	—	70	70	70	20	4	111
Medicago lupulina	L.	4	1	—	100	100	100	10	5	86
Medicago lupulina	L.	4	1	—	100	100	100	6.5	5	128
Medicago lupulina	L.	4	3	—	50	175	100	10	5	255
Medicago lupulina	L.	4	5	—	11	310	118	15	4	30
Medicago lupulina	L.	4	1	—	272	272	272	5	5	69
Medicago lupulina	L.	4	2	—	234	312	273	10	5	152
Medicago lupulina	L.	4	3	—	233	1783	761	3	5	96
Medicago lupulina	L.	4	6	—	600	2000	1317	10	4	165
Medicago minima	(L.) Bartal.	1	1	—	0	0	0	3	5	132
Medicago polymorpha	L.	3	1	>4	0	0	0	0	3	229
Medicago polymorpha	L.	3	1	>4	0	0	0	2	3	229
Medicago polymorpha	L.	3	1	>4	0	0	0	6	3	229
Medicago polymorpha	L.	3	1	>4	0	0	0	10	3	229
Medicago sativa	L.	1	1	—	0	0	0	10	5	255
Medicago sativa	L.	2	1	1	0	0	0	8	1	61
Medicago sativa	L.	2	1	1	0	0	0	20	1	61
Medicago sativa	L.	2	1	1	0	0	0	30	1	61
Medicago sativa	L.	2	1	3	0	0	0	100	1	240
Medicago sativa	L.	3	1	>20	0	0	0	13	1	133
Medicago sativa	L.	3	2	>20	0	0	0	26	1	133
Medicago sativa	L.	3	1	>20	0	0	0	39	1	133
Medicago sativa	L.	3	1	6	0	0	0	15	1	240
Medicago sativa	L.	3	1	6	0	0	0	50	1	240
Medicago sativa subsp. *falcata*	(L.) Arcang.	1	1	—	0	0	0	10	5	255
Melilotus altissimus	Thuill.	3	2	>5	0	0	0	7.5	2	200
Melilotus officinalis	(L.) Lam.	1	1	—	0	0	0	10	5	255

Species	Authority									
Melilotus officinalis	(L.) Lam.	3	1	>13	1292	1292	1292	15	4	223
Onobrychis viciifolia	Scop.	1	1	—	0	0	0	5	5	69
Onobrychis viciifolia	Scop.	1	1	—	0	0	0	6.5	7	182
Onobrychis viciifolia	Scop.	1	1	—	0	0	0	10	5	255
Ononis repens	L.	1	1	—	0	0	0	6	5	67
Ononis repens	L.	1	2	—	0	0	0	20	5	97
Ononis repens	L.	1	2	—	0	0	0	6.5	7	182
Ononis repens	L.	1	1	—	0	0	0	13	5	183
Ononis repens	L.	4	1	—	469	469	469	10	5	110
Ononis spinosa	L.	1	2	—	0	0	0	6	5	181
Ononis spinosa	L.	1	3	—	0	0	0	10	5	255
Ornithopus perpusillus	L.	1	1	—	0	0	0	20	4	111
Ornithopus perpusillus	L.	1	1	—	0	0	0	20	5	111
Ornithopus perpusillus	L.	1	1	—	0	0	0	5	5	250
Ornithopus perpusillus	L.	4	1	—	112	112	112	5	5	69
Ornithopus perpusillus	L.	4	2	—	50	250	150	3	5	96
Oxytropis campestris	(L.) DC.	4	3	—	10	27	19	6	5	39
Robinia pseudoacacia	L.	3	1	>30	0	0	0	15	1	240
Robinia pseudoacacia	L.	3	1	>39	0	0	0	50	1	240
Robinia pseudoacacia	L.	3	1	>39	0	0	0	100	1	240
Securigera varia	(L.) Lassen	1	2	—	0	0	0	6	5	181
Securigera varia	(L.) Lassen	1	1	—	0	0	0	13	5	183
Trifolium arvense	L.	1	2	—	0	0	0	3	5	96
Trifolium arvense	L.	1	3	—	0	0	0	3	5	132
Trifolium arvense	L.	2	2	>1	0	0	0	—	1	275
Trifolium arvense	L.	2	3	1	0	0	0	—	1	275
Trifolium arvense	L.	2	1	—	165	165	165	16	5	23
Trifolium arvense	L.	3	2	>5	0	0	0	7.5	2	200
Trifolium arvense	L.	4	1	—	7	7	7	20	4	111
Trifolium arvense	L.	4	1	—	50	50	50	3	5	132
Trifolium arvense	L.	4	1	—	72	72	72	6	5	67
Trifolium arvense	L.	4	12	—	26	413	123	30	4	13
Trifolium arvense	L.	4	2	—	63	225	144	13.5	5	163
Trifolium arvense	L.	4	1	>5	726	726	726	25	5	2
Trifolium campestre	Schreber	1	3	—	0	0	0	3	5	132
Trifolium campestre	Schreber	1	1	—	0	0	0	6.5	7	182
Trifolium campestre	Schreber	1	1	—	0	0	0	3	7	235
Trifolium campestre	Schreber	3	2	>5	0	0	0	7.5	2	200
Trifolium campestre	Schreber	4	1	—	15	15	15	25	5	2
Trifolium campestre	Schreber	4	1	—	50	50	50	3	5	96
Trifolium campestre	Schreber	4	1	—	2449	2449	2449	8.5	4	254
Trifolium dubium	Sibth.	1	1	—	0	0	0	30	5	45
Trifolium dubium	Sibth.	1	3	—	0	0	0	5	5	69
Trifolium dubium	Sibth.	1	1	—	0	0	0	3	5	96
Trifolium dubium	Sibth.	1	1	—	53	53	53	5	5	244
Trifolium dubium	Sibth.	1	2	—	53	53	53	8	7	275
Trifolium dubium	Sibth.	2	2	—	38	150	94	10	5	255

Species	Authority	Seed bank type	Number of records	Longevity (y)	Minimum density (seeds m⁻²)	Maximum density (seeds m⁻²)	Mean density (seeds m⁻²)	Depth (cm)	Method	Source code
Trifolium dubium	Sibth.	2	2	—	140	1023	582	30	5	45
Trifolium dubium	Sibth.	2	2	—	753	1507	1130	5	4	104
Trifolium dubium	Sibth.	3	1	>20	0	0	0	13	1	133
Trifolium dubium	Sibth.	3	2	>20	0	0	0	26	1	133
Trifolium dubium	Sibth.	3	1	>20	0	0	0	39	1	133
Trifolium dubium	Sibth.	3	3	>5	0	0	0	7.5	2	200
Trifolium dubium	Sibth.	3	1	—	129	129	129	30	5	45
Trifolium dubium	Sibth.	3	1	—	500	500	500	20	5	69
Trifolium dubium	Sibth.	4	1	—	9	9	9	12	5	174
Trifolium dubium	Sibth.	4	1	—	38	38	38	3	5	132
Trifolium dubium	Sibth.	4	2	—	77	101	89	17.8	5	155
Trifolium dubium	Sibth.	4	1	—	92	92	92	8	7	275
Trifolium dubium	Sibth.	4	2	—	117	167	142	3	5	96
Trifolium dubium	Sibth.	4	4	—	80	992	320	5	5	69
Trifolium dubium	Sibth.	4	1	—	700	700	700	15	5	249
Trifolium dubium	Sibth.	4	1	—	2541	2541	2541	8	4	42
Trifolium dubium	Sibth.	4	81	—	161	38427	5689	5	4	104
Trifolium fragiferum	L.	1	1	—	0	0	0	5	5	69
Trifolium fragiferum	L.	4	4	—	215	1561	700	5	4	104
Trifolium glomeratum	L.	1	2	—	0	0	0	3	5	132
Trifolium glomeratum	L.	4	1	—	43	43	43	10	5	147
Trifolium hybridum	L.	1	1	—	0	0	0	25	4	36
Trifolium hybridum	L.	1	1	—	0	0	0	10	5	150
Trifolium hybridum	L.	1	3	—	0	0	0	10	5	255
Trifolium hybridum	L.	3	1	>20	0	0	0	13	1	133
Trifolium hybridum	L.	3	2	>20	0	0	0	26	1	133
Trifolium hybridum	L.	3	1	>20	0	0	0	39	1	133
Trifolium hybridum	L.	3	1	21	0	0	0	15	1	240
Trifolium hybridum	L.	3	1	30	0	0	0	50	1	240
Trifolium hybridum	L.	4	1	30	4	4	4	100	1	240
Trifolium hybridum	L.	4	1	—	4	4	4	20	4	111
Trifolium hybridum	L.	4	1	—	150	150	150	3	5	96
Trifolium incarnatum	L.	3	1	>7	0	0	0	0	3	188
Trifolium incarnatum	L.	4	1	—	27	27	27	13.5	5	163
Trifolium medium	L.	1	1	—	0	0	0	7	5	58
Trifolium medium	L.	1	1	—	0	0	0	6	5	67
Trifolium medium	L.	1	1	—	0	0	0	3	5	96
Trifolium medium	L.	1	1	—	0	0	0	3	7	235
Trifolium medium	L.	1	5	—	0	0	0	10	5	255
Trifolium montanum	L.	1	1	—	0	0	0	8.5	4	254
Trifolium pratense	L.	1	2	—	0	0	0	6	5	11
Trifolium pratense	L.	1	7	—	0	0	0	10	5	27

Species										
Trifolium pratense	L.	1	4	—	0	0	0	6	5	28
Trifolium pratense	L.	1	2	—	0	0	0	25	4	36
Trifolium pratense	L.	1	2	—	0	0	0	5	4	53
Trifolium pratense	L.	1	1	—	0	0	0	5	5	55
Trifolium pratense	L.	1	2	—	0	0	0	1	5	57
Trifolium pratense	L.	1	2	—	0	0	0	3	5	59
Trifolium pratense	L.	1	1	—	0	0	0	6.5	5	68
Trifolium pratense	L.	1	4	—	0	0	0	5	5	69
Trifolium pratense	L.	1	1	—	0	0	0	5	4	88
Trifolium pratense	L.	1	3	—	0	0	0	10	4	88
Trifolium pratense	L.	1	4	—	0	0	0	20	5	97
Trifolium pratense	L.	1	3	—	0	0	0	6.5	5	128
Trifolium pratense	L.	1	1	—	0	0	0	10	5	149
Trifolium pratense	L.	1	3	—	0	0	0	12	5	174
Trifolium pratense	L.	1	1	—	0	0	0	6	5	181
Trifolium pratense	L.	1	2	—	0	0	0	6.5	7	182
Trifolium pratense	L.	1	1´	—	0	0	0	5	5	244
Trifolium pratense	L.	1	5	—	0	0	0	5	5	251
Trifolium pratense	L.	1	1	—	0	0	0	10	5	252
Trifolium pratense	L.	1	1	—	0	0	0	8.5	4	254
Trifolium pratense	L.	1	10	—	0	0	0	10	5	255
Trifolium pratense	L.	1	3	—	0	7	0	3	7	235
Trifolium pratense	L.	1	4	>1	0	0	0	—	1	275
Trifolium pratense	L.	1	1	1	0	0	0	—	1	275
Trifolium pratense	L.	2	2	—	28	40	34	25	4	36
Trifolium pratense	L.	2	1	—	90	90	90	13	5	68
Trifolium pratense	L.	2	2	—	97	183	140	30	5	45
Trifolium pratense	L.	2	1	—	403	403	403	10	5	252
Trifolium pratense	L.	3	1	>4	0	0	0	8	1	60
Trifolium pratense	L.	3	1	>4	0	0	0	20	1	60
Trifolium pratense	L.	3	1	>6	0	0	0	30	1	61
Trifolium pratense	L.	3	1	5	0	0	0	20	1	61
Trifolium pratense	L.	3	1	5	0	0	0	8	1	61
Trifolium pratense	L.	3	1	19	0	0	0	25	1	125
Trifolium pratense	L.	3	1	>20	0	0	0	13	1	133
Trifolium pratense	L.	3	2	>20	0	0	0	26	1	133
Trifolium pratense	L.	3	1	>20	0	0	0	39	1	133
Trifolium pratense	L.	3	1	>7	0	0	0	0	3	188
Trifolium pratense	L.	3	1	>30	0	0	0	15	1	240
Trifolium pratense	L.	3	1	>39	0	0	0	50	1	240
Trifolium pratense	L.	3	1	>39	0	0	0	100	1	240
Trifolium pratense	L.	3	1	—	32	32	32	30	5	45
Trifolium pratense	L.	3	3	—	60	60	60	13	5	68
Trifolium pratense	L.	3	3	—	32	3014	1127	30	5	29
Trifolium pratense	L.	4	1	—	32	32	32	6	5	181
Trifolium pratense	L.	4	1	—	58	58	58	7	5	58

Species	Authority	Seed bank type	Number of records	Longevity (y)	Minimum density (seeds m⁻²)	Maximum density (seeds m⁻²)	Mean density (seeds m⁻²)	Depth (cm)	Method	Source code
Trifolium pratense	L.	4	1	—	60	60	60	6.5	5	128
Trifolium pratense	L.	4	1	—	80	80	80	5	5	69
Trifolium pratense	L.	4	1	—	100	100	100	3	5	96
Trifolium pratense	L.	4	3	—	43	235	113	12	5	174
Trifolium pratense	L.	4	1	—	153	153	153	5	4	53
Trifolium pratense	L.	4	14	—	215	14101	3475	5	4	104
Trifolium repens	L.	1	8	—	0	0	0	10	5	27
Trifolium repens	L.	1	1	—	0	0	0	25	4	36
Trifolium repens	L.	1	1	—	0	0	0	30	5	45
Trifolium repens	L.	1	4	—	0	0	0	5	4	53
Trifolium repens	L.	1	7	—	0	0	0	1	5	57
Trifolium repens	L.	1	1	—	0	0	0	3	5	59
Trifolium repens	L.	1	1	—	0	0	0	6	5	67
Trifolium repens	L.	1	4	—	0	0	0	5	5	69
Trifolium repens	L.	1	1	—	0	0	0	5	4	88
Trifolium repens	L.	1	1	—	0	0	0	10	4	88
Trifolium repens	L.	1	8	—	0	0	0	3	5	96
Trifolium repens	L.	1	4	—	0	0	0	20	5	97
Trifolium repens	L.	1	16	—	0	0	0	6	5	99
Trifolium repens	L.	1	1	—	0	0	0	10	5	110
Trifolium repens	L.	1	1	—	0	0	0	5	5	122
Trifolium repens	L.	1	1	—	0	0	0	6.5	5	128
Trifolium repens	L.	1	3	—	0	0	0	10	5	149
Trifolium repens	L.	1	6	—	0	0	0	10	5	152
Trifolium repens	L.	1	2	—	0	0	0	6	5	181
Trifolium repens	L.	1	6	—	0	0	0	3	5	208
Trifolium repens	L.	1	4	—	0	0	0	5	5	222
Trifolium repens	L.	1	3	—	0	0	0	3	7	235
Trifolium repens	L.	1	1	—	0	0	0	5	5	244
Trifolium repens	L.	1	6	—	0	0	0	5	5	251
Trifolium repens	L.	1	4	—	0	0	0	10	5	255
Trifolium repens	L.	1	1	—	20	20	20	10	5	252
Trifolium repens	L.	1	5	—	0	208	83	2	7	99
Trifolium repens	L.	1	6	—	85	85	85	6	5	208
Trifolium repens	L.	1	3	—	104	104	104	6	5	28
Trifolium repens	L.	1	5	—	53	649	225	8	7	275
Trifolium repens	L.	2	1	>2	0	0	0	0	3	41
Trifolium repens	L.	2	1	4	0	0	0	8	1	61
Trifolium repens	L.	2	2	>1	0	0	0	—	1	275
Trifolium repens	L.	2	2	1	0	0	0	—	1	275
Trifolium repens	L.	2	1	—	5	5	5	5	5	189
Trifolium repens	L.	2	2	—	11	19	15	3	7	235

Species	Authority									
Trifolium repens	L.	2	1	—	41	41	41	5	5	222
Trifolium repens	L.	2	1	—	70	70	70	10	5	252
Trifolium repens	L.	2	1	—	74	74	74	11.5	5	134
Trifolium repens	L.	2	3	—	6	234	83	25	4	36
Trifolium repens	L.	2	2	—	80	100	90	10	5	149
Trifolium repens	L.	2	1	—	115	115	115	6.5	7	182
Trifolium repens	L.	2	1	—	156	156	156	6	5	28
Trifolium repens	L.	2	1	—	180	180	180	8	5	108
Trifolium repens	L.	2	11	—	38	1025	182	10	5	255
Trifolium repens	L.	2	1	—	625	625	625	10	4	88
Trifolium repens	L.	2	3	—	194	1281	646	30	5	45
Trifolium repens	L.	2	2	—	700	1410	1055	30	5	29
Trifolium repens	L.	3	1	>4	1300	1300	1300	20	5	69
Trifolium repens	L.	3	1	5	1313	1313	1313	17	5	162
Trifolium repens	L.	3	1	5	0	0	0	100	1	54
Trifolium repens	L.	3	1	>26	0	0	0	20	1	61
Trifolium repens	L.	3	1	>20	0	0	0	30	3	61
Trifolium repens	L.	3	1	>20	0	0	0	—	1	124
Trifolium repens	L.	3	1	>20	0	0	0	25	1	125
Trifolium repens	L.	3	2	>5	0	0	0	13	1	133
Trifolium repens	L.	3	3	>30	0	0	0	26	1	133
Trifolium repens	L.	3	1	21	0	0	0	39	2	133
Trifolium repens	L.	3	1	30	0	0	0	7.5	1	200
Trifolium repens	L.	3	1	>20	0	0	0	15	1	240
Trifolium repens	L.	3	1	>30	0	0	0	100	1	240
Trifolium repens	L.	3	1	>35	0	0	0	50	5	240
Trifolium repens	L.	3	1	>10	24	24	24	13	5	183
Trifolium repens	L.	3	2	>18	33	33	33	16	5	168
Trifolium repens	L.	3	1	>460	33	33	33	24	4	168
Trifolium repens	L.	3	2	—	49	49	49	10	5	241
Trifolium repens	L.	3	1	>26	56	56	56	32	5	169
Trifolium repens	L.	3	1	—	70	70	70	62	5	161
Trifolium repens	L.	3	1	—	50	100	75	6.5	5	68
Trifolium repens	L.	3	1	—	105	105	105	12	4	51
Trifolium repens	L.	3	1	—	72	204	138	13	5	68
Trifolium repens	L.	4	1	—	300	300	300	20	5	4
Trifolium repens	L.	4	6	—	375	375	375	10	4	88
Trifolium repens	L.	4	4	—	237	689	503	30	5	45
Trifolium repens	L.	4	3	—	86	1227	522	15	5	29
Trifolium repens	L.	4	2	—	5	7	6	25	4	62
Trifolium repens	L.	4	1	—	6	38	22	9	5	36
Trifolium repens	L.	4	4	—	32	32	32	12	5	84
Trifolium repens	L.	4	2	—	12	114	41	5	5	174
Trifolium repens	L.	4	1	—	41	51	46	11.5	5	222
Trifolium repens	L.	4	1	—	49	49	49	11.5	5	134
Trifolium repens	L.	4	1	—	51	51	51	5	5	244

184 / THE DATABASE

Species	Authority	Seed bank type	Number of records	Longevity (y)	Minimum density (seeds m⁻²)	Maximum density (seeds m⁻²)	Mean density (seeds m⁻²)	Depth (cm)	Method	Source code
Trifolium repens	L	4	1	—	54	54	54	3	5	59
Trifolium repens	L	4	2	—	50	67	59	3	5	96
Trifolium repens	L	4	1	—	63	63	63	10	5	255
Trifolium repens	L	4	1	—	80	80	80	10	5	149
Trifolium repens	L	4	3	—	17	196	88	7	5	58
Trifolium repens	L	4	2	—	113	204	159	5	4	53
Trifolium repens	L	4	4	—	146	234	197	10	5	150
Trifolium repens	L	4	1	—	208	208	208	6	5	99
Trifolium repens	L	4	1	—	212	212	212	20	4	111
Trifolium repens	L	4	1	—	291	291	291	8	4	42
Trifolium repens	L	4	3	—	255	485	368	5	4	41
Trifolium repens	L	4	1	—	395	395	395	15	5	207
Trifolium repens	L	4	1	—	621	621	621	15	5	267
Trifolium repens	L	4	1	—	700	700	700	1	5	57
Trifolium repens	L	4	7	—	96	2368	754	5	5	69
Trifolium repens	L	4	2	—	722	969	846	6	5	151
Trifolium repens	L	4	16	—	130	3200	871	5	4	6
Trifolium repens	L	4	2	—	956	1462	1209	17.8	5	155
Trifolium repens	L	4	1	—	2697	2697	2697	5	5	234
Trifolium repens	L	4	69	—	161	54035	3257	5	4	104
Trifolium striatum	L	1	1	—	0	0	0	3	5	96
Trifolium striatum	L	1	2	—	0	0	0	3	5	132
Trifolium striatum	L	3	1	>5	0	0	0	7.5	2	200
Trifolium subterraneum	L	1	1	—	0	0	0	3	5	132
Trifolium subterraneum	L	2	1	>3	0	0	0	0	3	229
Trifolium subterraneum	L	3	1	>4	0	0	0	2	3	229
Trifolium subterraneum	L	3	1	>4	0	0	0	6	3	229
Trifolium subterraneum	L	3	1	>4	0	0	0	10	3	229
Trifolium subterraneum	L	4	1	—	170	170	170	10	5	147
Trifolium suffocatum	L	4	20	—	269	10333	3703	5	4	104
Trifolium suffocatum	L	1	2	—	0	0	0	3	5	132
Trifolium suffocatum	L	4	1	—	975	975	975	3	5	132
Trigonella monspeliaca	L	1	1	—	0	0	0	3	5	132
Ulex europaeus	L	1	2	—	0	0	0	3	5	96
Ulex europaeus	L	2	1	—	60	60	60	15	5	257
Ulex europaeus	L	3	5	—	260	3000	1232	15	5	257
Ulex europaeus	L	4	1	—	150	150	150	7	5	58
Ulex europaeus	L	4	2	—	277	857	567	17.8	5	155
Ulex europaeus	L	4	1	—	29493	29493	29493	5	4	104
Ulex gallii	Planchon	3	1	—	273	273	273	10	5	93
Ulex gallii	Planchon	4	1	—	126	126	126	17.8	5	155
Ulex gallii	Planchon	4	1	—	707	707	707	17.8	5	156

Species	Authority									
Vicia cracca	L.	1	1	—	0	0	0	25	4	36
Vicia cracca	L.	1	1	—	0	0	0	3	5	59
Vicia cracca	L.	1	1	—	0	0	0	5	4	88
Vicia cracca	L.	1	4	—	0	0	0	10	4	88
Vicia cracca	L.	1	3	—	0	0	0	15	5	94
Vicia cracca	L.	1	6	—	0	0	0	3	5	96
Vicia cracca	L.	1	2	—	0	0	0	20	5	97
Vicia cracca	L.	1	2	—	0	0	0	10	5	110
Vicia cracca	L.	1	6	—	0	0	0	10	5	150
Vicia cracca	L.	1	10	—	0	0	0	10	5	152
Vicia cracca	L.	1	4	—	0	0	0	12	5	174
Vicia cracca	L.	1	1	—	0	0	0	6.5	7	182
Vicia cracca	L.	1	6	—	0	0	0	3	5	208
Vicia cracca	L.	1	2	—	0	0	0	20	4	225
Vicia cracca	L.	1	3	—	0	0	0	3	7	235
Vicia cracca	L.	1	10	—	0	0	0	5	5	251
Vicia cracca	L.	1	1	—	0	0	0	10	5	252
Vicia cracca	L.	1	28	—	0	0	0	10	5	255
Vicia cracca	L.	2	5	>1	0	0	0	—	1	275
Vicia cracca	L.	3	2	>5	0	0	0	7.5	2	200
Vicia cracca	L.	4	4	—	12	12	12	12	5	174
Vicia cracca	L.	4	1	—	1600	1600	1600	20	4	225
Vicia hirsuta	(L.) Gray	1	2	—	0	0	0	25	4	36
Vicia hirsuta	(L.) Gray	1	2	—	0	0	0	30	5	66
Vicia hirsuta	(L.) Gray	1	1	—	0	0	0	6	5	67
Vicia hirsuta	(L.) Gray	1	1	—	0	0	0	5	5	69
Vicia hirsuta	(L.) Gray	1	1	—	0	0	0	3	5	96
Vicia hirsuta	(L.) Gray	1	1	—	0	0	0	20	4	225
Vicia hirsuta	(L.) Gray	1	14	—	0	0	0	10	5	255
Vicia hirsuta	(L.) Gray	2	1	1	7	7	7	12	4	51
Vicia hirsuta	(L.) Gray	2	1	—	7	7	7	—	1	275
Vicia hirsuta	(L.) Gray	2	5	—	0	0	0	20	5	40
Vicia hirsuta	(L.) Gray	3	3	>25	3	3	3	25	4	36
Vicia hirsuta	(L.) Gray	3	1	>5	18	22	20	25	1	125
Vicia hirsuta	(L.) Gray	3	2	>5	0	0	0	7.5	2	200
Vicia hirsuta	(L.) Gray	3	1	>5	0	0	0	2.5	1	203
Vicia hirsuta	(L.) Gray	3	1	>5	0	0	0	2.5	2	203
Vicia hirsuta	(L.) Gray	3	1	>5	0	0	0	7.5	1	203
Vicia hirsuta	(L.) Gray	3	1	>5	0	0	0	7.5	2	203
Vicia hirsuta	(L.) Gray	3	1	>5	0	0	0	15	1	203
Vicia hirsuta	(L.) Gray	4	1	—	0	0	0	15	1	203
Vicia hirsuta	(L.) Gray	4	1	—	6	6	6	5	5	234
Vicia hirsuta	(L.) Gray	4	4	—	15	15	15	25	5	2
Vicia hirsuta	(L.) Gray	4	4	—	6	38	23	25	4	36
Vicia hirsuta	(L.) Gray	4	1	—	126	126	126	17.8	5	155
Vicia hirsuta	(L.) Gray	4	1	—	161	161	161	15	4	31
Vicia hirsuta	(L.) Gray	4	6	—	38	413	182	10	5	255

Species	Authority	Seed bank type	Number of records	Longevity (y)	Minimum density (seeds m⁻²)	Maximum density (seeds m⁻²)	Mean density (seeds m⁻²)	Depth (cm)	Method	Source code
Vicia lathyroides	L.	1	2	—	0	0	0	3	5	96
Vicia lathyroides	L.	1	2	—	0	0	0	3	5	132
Vicia lathyroides	L.	2	1	—	170	170	170	13	5	162
Vicia lutea	L.	1	1	—	0	0	0	6	5	67
Vicia sativa	L.	1	1	—	0	0	0	10	4	22
Vicia sativa	L.	1	4	—	0	0	0	25	4	36
Vicia sativa	L.	1	1	—	0	0	0	30	5	45
Vicia sativa	L.	1	4	—	0	0	0	5	5	69
Vicia sativa	L.	1	1	—	0	0	0	20	5	97
Vicia sativa	L.	1	2	—	0	0	0	10	5	110
Vicia sativa	L.	1	1	—	0	0	0	3	5	132
Vicia sativa	L.	1	1	—	0	0	0	6	5	181
Vicia sativa	L.	1	1	—	0	0	0	3	7	235
Vicia sativa	L.	1	5	—	0	0	0	10	5	255
Vicia sativa	L.	1	6	—	53	231	96	8	7	275
Vicia sativa	L.	2	5	>1	0	0	0	—	1	275
Vicia sativa	L.	2	1	—	38	38	38	3	5	132
Vicia sativa	L.	2	1	—	75	75	75	15	5	162
Vicia sativa	L.	2	1	—	140	140	140	6	5	162
Vicia sativa	L.	2	1	—	225	225	225	12	5	162
Vicia sativa	L.	2	1	—	439	439	439	22	5	162
Vicia sativa	L.	2	1	—	500	500	500	17	5	162
Vicia sativa	L.	3	1	—	83	83	83	5	5	234
Vicia sativa	L.	3	1	>35	1935	1935	1935	65	5	161
Vicia sativa	L.	4	1	—	14	14	14	20	5	111
Vicia sativa	L.	4	1	—	15	15	15	25	5	2
Vicia sativa	L.	4	1	—	100	100	100	10	5	140
Vicia sativa	L.	4	1	—	133	133	133	3	5	96
Vicia sepium	L.	1	2	—	0	0	0	6.5	5	68
Vicia sepium	L.	1	2	—	0	0	0	13	5	68
Vicia sepium	L.	1	2	—	0	0	0	10	5	110
Vicia sepium	L.	1	8	—	0	0	0	10	5	150
Vicia sepium	L.	1	1	—	0	0	0	12	5	171
Vicia sepium	L.	1	1	—	0	0	0	6	5	181
Vicia sepium	L.	1	7	—	0	0	0	10	5	255
Vicia sepium	L.	1	2	—	1771	1771	1771	20	5	97
Vicia sepium	L.	3	1	>20	24	24	24	13	5	183
Vicia tetrasperma	(L.) Schreber	1	1	—	0	0	0	25	4	36
Vicia tetrasperma	(L.) Schreber	1	1	—	0	0	0	20	5	67
Vicia tetrasperma	(L.) Schreber	1	1	—	0	0	0	15	5	94
Vicia tetrasperma	(L.) Schreber	1	1	—	0	0	0	20	4	225
Vicia tetrasperma	(L.) Schreber	2	2	—	10	14	12	6	5	49

Species	Author										
Vicia tetrasperma	(L.) Schreber	2	1	—	14	14	14	—	25	4	36
Vicia tetrasperma	(L.) Schreber	2	1	—	2313	2313	2313	—	10	5	255
Vicia tetrasperma	(L.) Schreber	4	2	—	460	1440	950	—	15	5	94
Lemnaceae	—	—	—	—	—	—	—	—	—	—	
Lemna minor	L.	1	1	—	0	0	0	—	10	5	87
Lemna minor	L.	1	3	—	0	0	0	—	5	5	177
Lemna trisulca	L.	1	6	—	0	0	0	—	10	5	87
Lemna trisulca	L.	1	3	—	0	0	0	—	5	5	177
Lentibulariaceae	—	—	—	—	—	—	—	—	—	—	
Pinguicula alpina	L.	1	1	—	0	0	0	—	5	5	55
Pinguicula vulgaris	L.	4	2	—	9	43	26	—	12	5	174
Utricularia minor	L.	1	1	—	0	0	0	—	5	5	248
Utricularia vulgaris	L.	1	1	—	0	0	0	—	10	5	87
Utricularia vulgaris	L.	1	1	—	0	0	0	—	5	5	177
Utricularia vulgaris	L.	2	3	—	37	44	42	—	5	5	177
Utricularia vulgaris	L.	4	1	—	2	2	2	—	5	5	166
Utricularia vulgaris	L.	4	3	—	30	59	42	—	5	5	177
Liliaceae	—	—	—	—	—	—	—	—	—	—	
Allium cepa	L.	1	1	0	0	0	0	—	15	1	240
Allium cepa	L.	1	1	0	0	0	0	—	50	1	240
Allium cepa	L.	1	1	0	0	0	0	—	100	1	240
Allium oleraceum	L.	1	1	—	0	0	0	—	30	5	66
Allium oleraceum	L.	1	1	—	0	0	0	—	12.5	5	126
Allium scorodoprasm	L.	1	1	—	0	0	0	—	50	5	66
Allium sphaerocephalon	L.	1	5	—	0	0	0	—	3	5	132
Allium ursinum	L.	1	1	—	0	0	0	—	3	5	96
Allium ursinum	L.	1	1	—	0	0	0	—	5	5	220
Allium ursinum	L.	1	1	—	0	0	0	—	3	5	235
Allium ursinum	L.	1	1	—	504	504	504	—	20	7	67
Allium vineale	L.	1	1	—	0	0	0	—	30	5	66
Allium vineale	L.	1	1	—	0	0	0	—	6	5	67
Allium vineale	L.	1	1	—	0	0	0	—	1	5	96
Allium vineale	L.	1	1	—	3	3	3	—	3	5	235
Allium vineale	L.	4	1	—	400	400	400	—	15	7	249
Anthericum ramosum	L.	1	1	—	0	0	0	—	6	5	67
Anthericum ramosum	L.	1	1	—	0	0	0	—	6	5	181
Asparagus officinalis	L.	1	1	0	0	0	0	—	15	1	240
Asparagus officinalis	L.	1	1	0	0	0	0	—	50	1	240
Asparagus officinalis	L.	1	1	0	0	0	0	—	100	1	240
Colchicum autumnale	L.	1	1	—	0	0	0	—	6.5	7	182
Colchicum autumnale	L.	1	7	—	0	0	0	—	10	5	255
Convallaria majalis	L.	1	4	—	0	0	0	—	6	5	49
Convallaria majalis	L.	1	1	—	0	0	0	—	20	5	67
Convallaria majalis	L.	1	1	—	0	0	0	—	12.5	5	126

Species	Authority	Seed bank type	Number of records	Longevity (y)	Minimum density (seeds m⁻²)	Maximum density (seeds m⁻²)	Mean density (seeds m⁻²)	Depth (cm)	Method	Source code
Convallaria majalis	L.	1	1	—	0	0	0	10	5	150
Convallaria majalis	L.	1	1	—	0	0	0	10	5	152
Convallaria majalis	L.	1	4	—	0	0	0	5	6	170
Convallaria majalis	L.	1	2	—	0	0	0	12	5	171
Convallaria majalis	L.	1	1	—	0	0	0	20	5	173
Convallaria majalis	L.	2	2	1	0	0	0	10	3	216
Convallaria majalis	L.	2	1	2	0	0	0	10	3	216
Fritillaria meleagris	L.	2	1	1	0	0	0	0.5	3	273
Gagea arvensis	(Pers.) Dumort.	1	1	—	0	0	0	50	5	66
Gagea lutea	(L.) Ker Gawler	1	1	—	0	0	0	10	5	126
Gagea lutea	(L.) Ker Gawler	1	1	—	0	0	0	12.5	5	126
Gagea lutea	(L.) Ker Gawler	1	1	—	0	0	0	10	5	172
Gagea spathacea	(Hayne) Salisb.	1	1	—	0	0	0	10	5	126
Gagea spathacea	(Hayne) Salisb.	1	2	—	0	0	0	5	5	220
Hyacinthoides non-scripta	(L.) Chouard ex Rothm.	1	1	—	0	0	0	15	5	32
Hyacinthoides non-scripta	(L.) Chouard ex Rothm.	1	1	—	0	0	0	7	5	58
Hyacinthoides non-scripta	(L.) Chouard ex Rothm.	1	1	—	0	0	0	3	5	96
Hyacinthoides non-scripta	(L.) Chouard ex Rothm.	1	5	—	0	0	0	15	5	257
Hyacinthoides non-scripta	(L.) Chouard ex Rothm.	1	2	—	0	3	2	3	7	235
Hyacinthoides non-scripta	(L.) Chouard ex Rothm.	1	1	—	7	7	7	12	4	51
Lilium martagon	L.	1	3	—	0	0	0	6	5	49
Maianthemum bifolium	(L.) F.W. Schmidt	1	1	—	0	0	0	2	5	82
Maianthemum bifolium	(L.) F.W. Schmidt	1	2	—	0	0	0	5	5	82
Maianthemum bifolium	(L.) F.W. Schmidt	1	2	—	0	0	0	12.5	5	126
Maianthemum bifolium	(L.) F.W. Schmidt	1	1	—	0	0	0	10	5	150
Maianthemum bifolium	(L.) F.W. Schmidt	1	4	—	0	0	0	5	6	170
Maianthemum bifolium	(L.) F.W. Schmidt	1	2	—	0	0	0	12	5	171
Maianthemum bifolium	(L.) F.W. Schmidt	1	1	—	0	0	0	20	5	173
Maianthemum bifolium	(L.) F.W. Schmidt	1	2	—	0	0	0	4	5	176
Maianthemum bifolium	(L.) F.W. Schmidt	1	1	—	0	0	0	5	5	220
Narthecium ossifragum	(L.) Hudson	1	1	—	0	0	0	7	5	232
Narthecium ossifragum	(L.) Hudson	2	1	—	133	133	133	3	5	96
Narthecium ossifragum	(L.) Hudson	4	2	—	76	126	101	17.8	5	156
Narthecium ossifragum	(L.) Hudson	4	3	—	151	227	184	17.8	5	155
Ornithogalum angustifolium	Boreau	1	1	—	0	0	0	50	5	66
Ornithogalum angustifolium	Boreau	1	2	—	0	0	0	3	5	132
Paris quadrifolia	L.	1	1	—	0	0	0	12.5	5	126
Paris quadrifolia	L.	1	2	—	0	0	0	5	6	170
Paris quadrifolia	L.	1	2	—	0	0	0	12	5	171
Polygonatum multiflorum	(L.) All.	1	1	—	0	0	0	30	5	173
Polygonatum multiflorum	(L.) All.	1	1	—	0	0	0	6	5	49
Polygonatum multiflorum	(L.) All.	1	1	—	0	0	0	12.5	5	126

Species	Author								
Polygonatum multiflorum	(L.) All.	1	2	—	0	0	0	4	5	176
Polygonatum odoratum	(Miller) Druce	1	3	—	0	0	0	6	5	49
Polygonatum odoratum	(Miller) Druce	1	1	—	0	0	0	12	5	171
Ruscus aculeatus	L.	1	1	—	0	0	0	15	5	257
Tofieldia calyculata	(L.) Wahlenb.	1	1	—	0	0	0	12	5	174
Tofieldia calyculata	(L.) Wahlenb.	4	1	—	31	31	31	12	5	174
Linaceae		—	—	—	—	—	—	—	—	
Linum catharticum	L.	1	4	—	0	0	0	3	5	96
Linum catharticum	L.	1	1	—	0	0	0	20	5	97
Linum catharticum	L.	1	1	—	0	0	0	10	5	110
Linum catharticum	L.	1	6	—	0	0	0	10	5	149
Linum catharticum	L.	1	2	—	0	0	0	12	5	174
Linum catharticum	L.	1	2	>3	0	3	2	3	7	235
Linum catharticum	L.	2	1	—	0	0	0	6	3	179
Linum catharticum	L.	2	1	—	160	160	160	6.5	5	128
Linum catharticum	L.	2	1	—	291	291	291	6	5	211
Linum catharticum	L.	2	1	—	659	659	659	6	5	48
Linum catharticum	L.	2	11	—	234	1638	681	10	5	152
Linum catharticum	L.	2	1	—	953	953	953	6.5	7	182
Linum catharticum	L.	3	1	>5	0	0	0	7.5	2	195
Linum catharticum	L.	3	1	>20	32	32	32	13	5	183
Linum catharticum	L.	3	1	—	33	33	33	16	5	168
Linum catharticum	L.	3	1	>100	33	33	33	32	5	169
Linum catharticum	L.	3	1	>30	144	144	144	10	5	181
Linum catharticum	L.	3	1	>30	167	167	167	16	5	168
Linum catharticum	L.	4	2	—	52	84	68	5	5	122
Linum catharticum	L.	4	2	—	29	110	70	7	5	58
Linum catharticum	L.	4	1	—	83	83	83	3	5	96
Linum catharticum	L.	4	3	—	38	163	105	10	5	255
Linum catharticum	L.	4	1	—	183	183	183	3	5	59
Linum catharticum	L.	4	3	—	60	480	227	6.5	5	128
Linum catharticum	L.	4	1	—	256	256	256	6	5	67
Linum catharticum	L.	4	2	—	44	584	314	1.5	5	121
Linum catharticum	L.	4	4	—	234	505	328	10	5	110
Linum catharticum	L.	4	3	—	496	1144	883	6	5	181
Linum catharticum	L.	4	3	—	936	1560	1222	10	5	152
Linum usitatissimum	L.	1	1	0	0	0	0	15	1	240
Linum usitatissimum	L.	1	1	0	0	0	0	50	1	240
Linum usitatissimum	L.	1	1	0	0	0	0	100	1	240
Lythraceae		—	—	—	—	—	—	—	—	
Lythrum portula	(L.) D. Webb	2	4	—	50	167	79	3	5	96
Lythrum portula	(L.) D. Webb	2	1	—	468	468	468	10	5	27
Lythrum portula	(L.) D. Webb	2	3	—	468	520	485	6	5	11
Lythrum portula	(L.) D. Webb	3	2	—	156	260	208	40	5	11
Lythrum portula	(L.) D. Webb	3	2	—	156	312	234	6	5	11

Species	Authority	Seed bank type	Number of records	Longevity (y)	Minimum density (seeds m⁻²)	Maximum density (seeds m⁻²)	Mean density (seeds m⁻²)	Depth (cm)	Method	Source code
Lythrum portula	(L.) D. Webb	3	2	—	364	1924	1144	10	5	27
Lythrum salicaria	L.	1	1	—	0	0	0	5	4	88
Lythrum salicaria	L.	1	2	—	0	0	0	6.5	5	128
Lythrum salicaria	L.	1	1	—	0	0	0	12	5	174
Lythrum salicaria	L.	1	6	—	0	0	0	3	5	208
Lythrum salicaria	L.	1	1	—	0	0	0	50	5	216
Lythrum salicaria	L.	1	3	—	0	0	0	20	4	225
Lythrum salicaria	L.	1	2	—	0	0	0	10	5	255
Lythrum salicaria	L.	1	3	—	26	26	26	2	7	99
Lythrum salicaria	L.	2	2	—	25	43	34	12	5	174
Lythrum salicaria	L.	2	1	—	80	80	80	6.5	5	128
Lythrum salicaria	L.	2	1	—	1061	1061	1061	16	5	216
Lythrum salicaria	L.	3	1	>5	0	0	0	7.5	2	195
Lythrum salicaria	L.	3	1	—	610	610	610	16	5	216
Lythrum salicaria	L.	4	3	—	80	120	100	6.5	5	128
Lythrum salicaria	L.	4	4	—	42	217	119	5	5	98
Lythrum salicaria	L.	4	1	—	150	150	150	10	5	255
Lythrum salicaria	L.	4	3	—	9	818	280	12	5	174
Lythrum salicaria	L.	4	1	—	410000	410000	410000	5	4	261
Malvaceae										
Lavatera cretica	L.	1	1	—	—	0	—	10	4	22
Malva moschata	L.	3	1	>5	0	0	0	7.5	2	198
Malva neglecta	Wallr.	1	2	—	0	0	0	10	4	22
Malva neglecta	Wallr.	2	1	—	125	125	125	10	5	162
Malva neglecta	Wallr.	2	1	—	226	226	226	20	5	162
Malva neglecta	Wallr.	2	1	—	350	350	350	15	5	162
Malva neglecta	Wallr.	2	1	—	875	875	875	40	5	162
Malva neglecta	Wallr.	2	1	—	1311	1311	1311	17	5	162
Malva neglecta	Wallr.	3	3	>5	0	0	0	7.5	2	198
Malva pusilla	Smith	3	1	>100	0	0	0	—	3	124
Malva sylvestris	L.	2	1	—	188	188	188	17	5	162
Malva sylvestris	L.	2	1	—	301	301	301	45	5	162
Malva sylvestris	L.	2	1	—	351	351	351	22	5	162
Malva sylvestris	L.	2	1	—	598	598	598	8	7	275
Malva sylvestris	L.	2	1	—	1489	1489	1489	13	5	162
Malva sylvestris	L.	3	3	>5	0	0	0	7.5	2	198
Malva sylvestris	L.	3	1	>20	150	150	150	15	5	162
Malva sylvestris	L.	3	1	>90	289	289	289	55	5	161
Menyanthaceae										
Menyanthes trifoliata	L.	1	2	—	0	0	0	10	5	149

Species	Authority									
Menyanthes trifoliata	L.	1	1	—	0	0	0	16	5	216
Nymphoides peltata	Kuntze	2	1	>1	0	0	0	0	3	218
Nymphoides peltata	Kuntze	2	1	>1	0	0	0	3	3	218
Nymphoides peltata	Kuntze	4	2	—	637	796	717	15	4	218
Monotropaceae										
Monotropa hypopitys	L.	1	1	—	0	0	0	24	5	168
Myricaceae										
Myrica gale	L.	1	1	—	0	0	0	5	5	117
Myrica gale	L.	4	1	—	31	31	31	16	5	216
Najadaceae										
Najas flexilis	(Willd.) Rostkov & W. Schmidt	1	3	—	0	0	0	35	5	247
Najas flexilis	(Willd.) Rostkov & W. Schmidt	3	2	—	184	645	415	35	5	247
Najas marina	L.	4	1	—	545	545	545	5	5	116
Nymphaeaceae										
Nuphar lutea	(L.) Smith	1	3	—	0	0	0	15	4	218
Nuphar lutea	(L.) Smith	1	1	0	0	0	0	0	3	218
Nuphar lutea	(L.) Smith	1	1	0	0	0	0	3	3	218
Nymphaea alba	L.	1	2	—	0	0	0	15	4	218
Nymphaea alba	L.	1	1	0	0	0	0	0	3	218
Nymphaea alba	L.	1	1	0	0	0	0	3	3	218
Nymphaea alba	L.	1	1	—	0	0	0	5	5	248
Oleaceae										
Fraxinus excelsior	L.	1	2	—	0	0	0	20	5	67
Fraxinus excelsior	L.	1	1	—	0	0	0	3	5	96
Fraxinus excelsior	L.	1	2	—	0	0	0	4	5	176
Fraxinus excelsior	L.	1	6	—	0	0	0	3	5	208
Fraxinus excelsior	L.	1	1	—	0	0	0	10	5	255
Fraxinus excelsior	L.	1	3	—	0	0	0	15	5	257
Fraxinus excelsior	L.	1	1	—	3	3	3	3	7	235
Fraxinus excelsior	L.	1	1	—	24	24	24	13	5	183
Fraxinus excelsior	L.	2	1	2	0	0	0	0	1	77
Ligustrum vulgare	L.	1	2	—	0	0	0	6	5	181
Ligustrum vulgare	L.	1	1	—	0	0	0	10	5	181
Ligustrum vulgare	L.	1	2	—	0	0	0	10	5	255
Ligustrum vulgare	L.	1	2	—	0	0	0	15	5	257
Ligustrum vulgare	L.	1	1	—	1289	1289	1289	6.5	7	182
Onagraceae										
Chamerion angustifolium	(L.) Holub	1	1	—	0	0	0	2	5	82
Chamerion angustifolium	(L.) Holub	1	1	—	0	0	0	3	5	82
Chamerion angustifolium	(L.) Holub	1	1	0	0	0	0	2	3	83
Chamerion angustifolium	(L.) Holub	1	1	—	0	0	0	3	5	96

Species	Authority	Seed bank type	Number of records	Longevity (y)	Minimum density (seeds m⁻²)	Maximum density (seeds m⁻²)	Mean density (seeds m⁻²)	Depth (cm)	Method	Source code
Chamerion angustifolium	(L.) Holub	1	1	—	0	0	0	10	5	150
Chamerion angustifolium	(L.) Holub	1	2	—	0	0	0	12	5	171
Chamerion angustifolium	(L.) Holub	1	5	—	0	0	0	10	5	255
Chamerion angustifolium	(L.) Holub	1	1	—	15	15	15	3	7	235
Chamerion angustifolium	(L.) Holub	1	4	—	26	33	28	2	7	99
Chamerion angustifolium	(L.) Holub	1	1	—	79	79	79	10	5	93
Chamerion angustifolium	(L.) Holub	2	1	—	5	5	5	15	5	62
Chamerion angustifolium	(L.) Holub	2	1	—	7	7	7	0	5	157
Chamerion angustifolium	(L.) Holub	2	1	—	10	10	10	5	5	244
Chamerion angustifolium	(L.) Holub	2	1	—	58	58	58	10	5	126
Chamerion angustifolium	(L.) Holub	2	1	—	60	60	60	15	5	257
Chamerion angustifolium	(L.) Holub	2	1	—	80	80	80	5	5	69
Chamerion angustifolium	(L.) Holub	2	2	—	34	132	83	10	5	252
Chamerion angustifolium	(L.) Holub	2	7	—	50	200	93	3	5	96
Chamerion angustifolium	(L.) Holub	2	4	—	12	351	114	3	7	235
Chamerion angustifolium	(L.) Holub	2	4	—	138	525	241	10	5	255
Chamerion angustifolium	(L.) Holub	2	1	—	419	419	419	6	5	48
Chamerion angustifolium	(L.) Holub	4	1	—	3	3	3	10	5	33
Chamerion angustifolium	(L.) Holub	4	2	—	17	17	17	7	5	58
Chamerion angustifolium	(L.) Holub	4	1	—	17	17	17	0	5	157
Chamerion angustifolium	(L.) Holub	4	3	—	41	169	84	10	5	110
Chamerion angustifolium	(L.) Holub	4	1	—	128	128	128	10	5	9
Chamerion angustifolium	(L.) Holub	4	7	—	67	417	198	3	5	96
Chamerion angustifolium	(L.) Holub	4	8	—	50	1850	777	10	5	255
Circaea alpina	L.	1	2	—	0	0	0	4	5	176
Circaea lutetiana	L.	1	1	—	0	0	0	3	5	96
Circaea lutetiana	L.	1	2	—	0	0	0	12.5	5	126
Circaea lutetiana	L.	1	2	—	0	0	0	15	5	257
Epilobium ciliatum	Raf.	2	1	—	17	17	17	15	5	32
Epilobium ciliatum	Raf.	2	1	—	44	44	44	5	5	177
Epilobium ciliatum	Raf.	2	22	—	50	1217	267	3	5	96
Epilobium ciliatum	Raf.	2	1	—	521	521	521	20	5	97
Epilobium ciliatum	Raf.	3	3	—	29	36	31	12.5	5	126
Epilobium ciliatum	Raf.	3	1	—	88	88	88	8	5	108
Epilobium ciliatum	Raf.	3	2	—	100	350	225	3	5	96
Epilobium ciliatum	Raf.	3	5	—	417	625	479	20	5	97
Epilobium ciliatum	Raf.	4	1	—	6	6	6	0.538	5	269
Epilobium ciliatum	Raf.	4	16	—	20	166	63	30	4	13
Epilobium ciliatum	Raf.	4	2	—	100	267	184	3	5	96
Epilobium collinum	C.C. Gmelin	1	8	—	4938	4938	4938	50	5	216
Epilobium glandulosum	Lehm.	1	8	—	0	0	0	10	5	255
		4	1	—	0	0	0	5	5	166

Taxon	Authority									
Epilobium glandulosum	Lehm.	4	1	—	27	27	27	10	5	33
Epilobium hirsutum	L.	1	2	—	0	0	0	6.5	5	128
Epilobium hirsutum	L.	1	2	—	0	0	0	15	5	142
Epilobium hirsutum	L.	1	3	—	0	0	0	20	4	225
Epilobium hirsutum	L.	1	2	—	26	52	39	2	7	99
Epilobium hirsutum	L.	1	3	—	53	105	70	8	7	275
Epilobium hirsutum	L.	2	3	—	16	28	21	10	5	110
Epilobium hirsutum	L.	2	7	—	150	483	312	3	5	96
Epilobium hirsutum	L.	2	1	—	467	467	467	1	5	96
Epilobium hirsutum	L.	2	2	—	10	1109	559	3	7	235
Epilobium hirsutum	L.	2	3	—	100	2100	860	6.5	5	128
Epilobium hirsutum	L.	3	1	—	6400	6400	6400	20	4	225
Epilobium hirsutum	L.	3	2	—	11	12	12	10	5	252
Epilobium hirsutum	L.	3	1	—	29	29	29	12.5	5	126
Epilobium hirsutum	L.	4	2	—	180	180	180	6	5	48
Epilobium hirsutum	L.	4	2	—	124	124	124	12	4	230
Epilobium hirsutum	L.	4	1	—	220	380	300	6.5	5	128
Epilobium hirsutum	L.	4	1	—	1200	1200	1200	20	4	225
Epilobium hirsutum	L.	4	1	—	3917	3917	3917	3	5	96
Epilobium montanum	L.	1	1	—	0	0	0	3	5	96
Epilobium montanum	L.	1	2	—	0	0	0	12.5	5	126
Epilobium montanum	L.	1	1	—	48	48	48	10	5	255
Epilobium montanum	L.	1	1	—	54	54	54	25	5	219
Epilobium montanum	L.	2	1	'	0	0	0	30	6	45
Epilobium montanum	L.	2	10	—	0	463	153	5	5	170
Epilobium montanum	L.	2	1	—	355	355	355	10	5	255
Epilobium montanum	L.	2	4	—	50	1383	388	30	5	45
Epilobium montanum	L.	3	1	—	11	11	11	3	5	96
Epilobium montanum	L.	3	1	>40	22	22	22	10	5	172
Epilobium montanum	L.	3	1	—	178	178	178	12.5	5	126
Epilobium montanum	L.	3	1	—	222	222	222	24	5	168
Epilobium montanum	L.	4	1	—	3	3	3	16	7	168
Epilobium montanum	L.	4	1	—	4	4	4	3	4	235
Epilobium montanum	L.	4	8	—	38	1713	614	20	5	111
Epilobium montanum	L.	4	1	—				10	7	255
Epilobium obscurum	Schreber	1	1	—	3	3	3	3	5	235
Epilobium obscurum	Schreber	2	1	—	6	6	6	3	7	235
Epilobium obscurum	Schreber	4	1	—	14	14	14	5	5	213
Epilobium obscurum	Schreber	4	1	—	2167	2167	2167	3	5	96
Epilobium palustre	L.	1	1	—	0	0	0	3	5	208
Epilobium palustre	L.	1	2	—	0	0	0	6	6	208
Epilobium palustre	L.	1	1	—	0	0	0	10	5	255
Epilobium palustre	L.	1	1	—	12	12	12	3	7	235
Epilobium palustre	L.	1	1	—	43	43	43	30	5	45
Epilobium palustre	L.	1	4	—	34	166	73	2	7	99
Epilobium palustre	L.	2	1	—	6	6	6	5	5	213
Epilobium palustre	L.	2	1	—	14	14	14	3	7	235

Species	Authority	Seed bank type	Number of records	Longevity (y)	Minimum density (seeds m⁻²)	Maximum density (seeds m⁻²)	Mean density (seeds m⁻²)	Depth (cm)	Method	Source code
Epilobium palustre	L.	2	2	—	50	88	69	10	5	255
Epilobium palustre	L.	3	1	—	142	142	142	6	5	208
Epilobium palustre	L.	4	1	—	85	85	85	3	5	208
Epilobium palustre	L.	4	3	—	42	150	92	5	5	98
Epilobium palustre	L.	4	1	—	225	225	225	10	5	255
Epilobium parviflorum	Schreber	1	2	—	0	0	0	6.5	5	128
Epilobium parviflorum	Schreber	1	3	—	0	0	0	6	5	208
Epilobium parviflorum	Schreber	1	1	—	0	0	0	10	5	255
Epilobium parviflorum	Schreber	1	1	—	12	12	12	25	5	219
Epilobium parviflorum	Schreber	1	1	—	78	78	78	2	7	99
Epilobium parviflorum	Schreber	2	1	—	333	333	333	3	5	96
Epilobium roseum	Schreber	1	2	—	53	53	53	8	7	275
Epilobium tetragonum	L.	1	2	—	0	0	0	10	5	27
Epilobium tetragonum	L.	1	5	—	26	130	68	2	7	99
Epilobium tetragonum	L.	2	1	—	156	156	156	6	5	99
Epilobium tetragonum	L.	2	17	—	38	1613	500	10	5	255
Epilobium tetragonum	L.	2	1	—	1200	1200	1200	30	5	66
Epilobium tetragonum	L.	4	1	—	50	50	50	10	5	255
Epilobium tetragonum	L.	4	4	—	260	832	494	6	5	99
Ludwigia palustris	(L.) Elliott	1	1	—	135	135	135	21	5	160
Ludwigia palustris	(L.) Elliott	4	1	—	22	22	22	5	5	117
Oenothera biennis	L.	1	1	—	0	0	0	11.5	5	134
Oenothera biennis	L.	2	1	—	37	37	37	11.5	5	134
Oenothera biennis	L.	3	1	>5	0	0	0	23	1	64
Oenothera biennis	L.	3	1	80	0	0	0	—	3	124
Oenothera biennis	L.	3	1	>30	0	0	0	15	1	240
Oenothera biennis	L.	3	1	>39	0	0	0	50	1	240
Oenothera biennis	L.	3	1	>39	0	0	0	100	1	240
Oenothera biennis	L.	3	1	—	224	224	224	16	5	23
Oenothera biennis	L.	4	1	—	11	11	11	10	5	33
Oenothera cambrica	Rostanski	2	5	>1	0	0	0	—	1	275
Oenothera cambrica	Rostanski	2	5	—	1400	4200	2680	10	4	20
Oxalidaceae	—						—			
Oxalis acetosella	L.	1	1	—	0	0	0	15	5	32
Oxalis acetosella	L.	1	1	—	0	0	0	20	5	67
Oxalis acetosella	L.	1	1	—	0	0	0	7.5	5	84
Oxalis acetosella	L.	1	1	—	0	0	0	5	5	92
Oxalis acetosella	L.	1	1	—	0	0	0	12.5	5	126
Oxalis acetosella	L.	1	2	—	0	0	0	17.5	5	126.
Oxalis acetosella	L.	1	1	—	0	0	0	5	6	170
Oxalis acetosella	L.	1	1	—	0	0	0	5	5	220

Species	Authority									
Oxalis acetosella	L.	1	1	—	0	0	0	3	7	235
Oxalis acetosella	L.	1	5	—	0	0	0	15	5	257
Oxalis acetosella	L.	1	2	—	1028	2139	1584	12	5	171
Oxalis acetosella	L.	2	2	—	2	5	4	5	6	170
Oxalis acetosella	L.	2	2	—	97	143	120	5	5	92
Oxalis acetosella	L.	4	2	—	2	27	15	4	5	176
Oxalis acetosella	L.	4	1	—	22	22	22	5	5	126
Oxalis acetosella	L.	4	2	—	43	22	22	12.5	5	126
Oxalis acetosella	L.	4	2	—	71	66	55	10	5	92
Oxalis acetosella	L.	4	1	—	101	87	79	5	5	155
Oxalis acetosella	L.	4	4	—	58	101	101	17.8	5	220
Oxalis corniculata	L.	2	1	—	4140	4140	4140	5	5	137
Oxalis corniculata	L.	4	1	—	8	8	8	8	5	33
Oxalis corniculata	L.	4	7	—	27	81	50	10	5	163
Oxalis stricta	L.	1	2	—	0	0	0	13.5	5	19
Oxalis stricta	L.	1	2	—	32	32	32	5	5	175
Oxalis stricta	L.	2	1	—	556	556	556	15	5	260
Oxalis stricta	L.	2	3	—	14	14	14	50	5	189
Oxalis stricta	L.	2	1	—	13	28	21	5	5	1
Oxalis stricta	L.	2	1	—	35	35	35	5	5	272
Oxalis stricta	L.	2	1	—	129	129	129	15	5	175
Oxalis stricta	L.	3	2	—	480	480	480	15	5	94
Oxalis stricta	L.	4	3	—	16	32	24	15	5	175
Oxalis stricta	L.	4	3	—	9	30	18	15	5	19
Oxalis stricta	L.	4	3	—	19	19	19	5	5	33
Oxalis stricta	L.	4	8	—	36	135	72	10	5	163
Oxalis stricta	L.	4	2	—	43	108	76	13.5	5	59
Oxalis stricta	L.	4	1	—	366	366	366	3	4	13
Oxalis stricta	L.	4	1	—	635	635	635	30	5	272
Oxalis stricta	L.	4	3	—	110	2160	873	15	5	94
Papaveraceae	—									
Chelidonium majus	L.	2	1	—	875	875	875	10	5	162
Chelidonium majus	L.	2	1	—	1750	1750	1750	15	5	162
Chelidonium majus	L.	2	1	—	2188	2188	2188	17	5	162
Chelidonium majus	L.	3	1	—	17	17	17	6	5	211
Papaver argemone	L.	1	1	—	0	0	0	25	5	2
Papaver argemone	L.	2	1	—	64	64	64	5	5	69
Papaver argemone	L.	3	1	—	432	432	432	50	5	260
Papaver argemone	L.	4	1	>5	0	0	0	7.5	2	197
Papaver argemone	L.	1	1	—	263	263	263	25	5	260
Papaver dubium	L.	1	1	—	0	0	0	3	5	96
Papaver dubium	L.	1	1	—	0	0	0	3	7	132
Papaver dubium	L.	1	2	—	53	79	66	8	5	275
Papaver dubium	L.	2	2	—	270	556	413	50	5	260
Papaver dubium	L.	2	2	—	88	88	88	10	5	255
Papaver dubium	L.	2	2	—	64	224	144	5	5	69

Species	Authority	Seed bank type	Number of records	Longevity (y)	Minimum density (seeds m^{-2})	Maximum density (seeds m^{-2})	Mean density (seeds m^{-2})	Depth (cm)	Method	Source code
Papaver dubium	L.	3	1	>5	0	0	0	7.5	2	197
Papaver dubium	L.	3	1	>10	0	0	0	25	3	209
Papaver dubium subsp. lecoqii	(Lamotte) Syme	3	1	>5	0	0	0	7.5	2	197
Papaver hybridum	L.	1	1	—	0	0	0	3	5	132
Papaver radicatum	Rottb.	4	2	—	15	48	32	2	5	73
Papaver rhoeas	L.	1	2	—	0	0	0	10	4	22
Papaver rhoeas	L.	1	4	—	0	0	0	25	4	36
Papaver rhoeas	L.	1	1	—	0	0	0	3	5	96
Papaver rhoeas	L.	1	4	—	53	123	77	8	7	275
Papaver rhoeas	L.	2	1	>2	0	0	0	0	1	75
Papaver rhoeas	L.	2	1	>2	0	0	0	0	2	75
Papaver rhoeas	L.	2	1	>2	0	0	0	5	1	75
Papaver rhoeas	L.	2	1	>2	0	0	0	5	2	75
Papaver rhoeas	L.	2	1	1	0	0	0	20	1	219
Papaver rhoeas	L.	2	1	>3	0	0	0	20	3	268
Papaver rhoeas	L.	2	1	—	35	35	35	15	5	62
Papaver rhoeas	L.	2	1	—	52	52	52	25	4	36
Papaver rhoeas	L.	2	1	—	75	75	75	10	5	255
Papaver rhoeas	L.	2	1	—	125	125	125	10	4	22
Papaver rhoeas	L.	2	1	—	350	350	350	25	4	3
Papaver rhoeas	L.	2	1	—	912	912	912	5	5	69
Papaver rhoeas	L.	2	1	—	1203	1203	1203	8	7	275
Papaver rhoeas	L.	2	1	—	8000	8000	8000	20	5	69
Papaver rhoeas	L.	3	1	>5	0	0	0	30	3	14
Papaver rhoeas	L.	3	1	>26	0	0	0	25	1	125
Papaver rhoeas	L.	3	1	>5	0	0	0	7.5	2	197
Papaver rhoeas	L.	3	1	>5	0	0	0	2.5	1	203
Papaver rhoeas	L.	3	1	>5	0	0	0	2.5	2	203
Papaver rhoeas	L.	3	1	>5	0	0	0	7.5	1	203
Papaver rhoeas	L.	3	1	>5	0	0	0	7.5	2	203
Papaver rhoeas	L.	3	1	>5	0	0	0	15	1	203
Papaver rhoeas	L.	3	1	>5	0	0	0	15	2	203
Papaver rhoeas	L.	3	1	>8	10	10	10	15	5	62
Papaver rhoeas	L.	3	1	>30	29	29	29	6	5	211
Papaver rhoeas	L.	3	1	>20	1011	1011	1011	20	5	40
Papaver rhoeas	L.	4	2	—	12	52	32	25	4	36
Papaver rhoeas	L.	4	1	—	48	48	48	5	5	69
Papaver rhoeas	L.	4	5	—	42	546	217	10	5	15
Papaver rhoeas	L.	4	19	—	20	2564	335	30	4	13
Papaver rhoeas	L.	4	2	—	182	1286	734	25	5	260
Papaver rhoeas	L.	4	2	—	134	1881	1008	10	4	22
Papaver somniferum	L.	1	3	—	0	0	0	3	5	132

Parnassiaceae

Species	Authority									
Parnassiaceae	—	—	—	—	—	—	—	—	—	—
Parnassia palustris	L.	1	2	—	0	0	—	10	5	149
Parnassia palustris	L.	1	1	—	0	0	0	12	5	174
Parnassia palustris	L.	1	1	—	0	0	0	16	5	216
Parnassia palustris	L.	1	1	—	0	0	0	3	7	235
Parnassia palustris	L.	1	1	—	0	0	0	10	5	255
Parnassia palustris	L.	2	1	—	19	19	19	6.5	7	182
Parnassia palustris	L.	4	1	—	56	56	56	12	5	174

Pinaceae

Species	Authority									
Pinaceae	—	—	—	—	—	—	—	—	—	—
Abies grandis	(Douglas ex D. Don) Lindley	1	1	—	0	0	0	10	4	127
Larix decidua	Miller	1	2	—	0	0	0	24	5	168
Picea abies	(L.) Karsten	1	2	—	0	0	0	4	5	176
Picea abies	(L.) Karsten	1	1	—	0	0	0	13	5	183
Picea abies	(L.) Karsten	1	2	—	0	0	0	16	5	216
Picea abies	(L.) Karsten	1	2	—	0	0	0	10	5	255
Picea abies	(L.) Karsten	1	1	0	0	0	0	0	3	271
Pinus contorta	Douglas ex Loudon	1	1	—	0	0	0	10	4	127
Pinus mugo	Turra	1	2	—	0	0	0	30	5	180
Pinus nigra	Arnold	1	1	—	0	0	0	16	5	168
Pinus nigra	Arnold	1	2	—	0	0	0	24	5	168
Pinus strobus	L.	1	2	—	0	0	0	15	5	72
Pinus strobus	L.	1	1	—	0	0	0	0	5	157
Pinus sylvestris	L.	1	1	0	0	0	0	2	3	83
Pinus sylvestris	L.	1	2	—	0	0	0	10	5	255
Pinus sylvestris	L.	1	1	0	0	0	0	0	3	271
Pseudotsuga menziesii	(Mirbel) Franco	1	1	—	0	0	0	10	4	127
Pseudotsuga menziesii	(Mirbel) Franco	1	1	—	9	9	9	10	5	93
Pseudotsuga menziesii	(Mirbel) Franco	2	3	1	0	0	0	0	3	106
Pseudotsuga menziesii	(Mirbel) Franco	2	3	1	0	0	0	2.5	3	106
Pseudotsuga menziesii	(Mirbel) Franco	2	3	1	0	0	0	5	3	106

Plantaginaceae

Species	Authority									
Plantaginaceae	—	—	—	—	—	—	—	—	—	—
Littorella uniflora	(L.) Asch.	1	1	—	0	0	0	3	5	96
Plantago coronopus	L.	1	1	—	0	0	0	1	5	57
Plantago coronopus	L.	1	1	—	0	0	0	3	5	132
Plantago coronopus	L.	2	1	—	600	600	600	1	5	57
Plantago coronopus	L.	3	1	—	500	500	500	20	5	210
Plantago coronopus	L.	4	1	—	303	303	303	25	5	260
Plantago coronopus	L.	4	2	—	317	767	542	3	5	96
Plantago coronopus	L.	4	3	—	300	1900	833	1	5	57
Plantago lanceolata	L.	1	4	—	0	0	0	10	5	27
Plantago lanceolata	L.	1	4	—	0	0	0	6	5	28
Plantago lanceolata	L.	1	2	—	0	0	0	25	4	36
Plantago lanceolata	L.	1	1	—	0	0	0	1	5	57
Plantago lanceolata	L.	1	1	—	0	0	0	3	5	59

Species	Authority	Seed bank type	Number of records	Longevity (y)	Minimum density (seeds m⁻²)	Maximum density (seeds m⁻²)	Mean density (seeds m⁻²)	Depth (cm)	Method	Source code
Plantago lanceolata	L	1	5	—	0	0	0	5	5	69
Plantago lanceolata	L	1	1	—	0	0	0	9	5	84
Plantago lanceolata	L	1	3	—	0	0	0	10	4	88
Plantago lanceolata	L	1	7	—	0	0	0	3	5	96
Plantago lanceolata	L	1	6	—	0	0	0	20	5	97
Plantago lanceolata	L	1	2	—	0	0	0	2	7	99
Plantago lanceolata	L	1	7	—	0	0	0	6	5	99
Plantago lanceolata	L	1	3	—	0	0	0	10	5	110
Plantago lanceolata	L	1	1	—	0	0	0	5	5	123
Plantago lanceolata	L	1	1	0	0	0	0	26	1	133
Plantago lanceolata	L	1	1	—	0	0	0	10	5	150
Plantago lanceolata	L	1	5	—	0	0	0	10	5	152
Plantago lanceolata	L	1	1	—	0	0	0	2	5	153
Plantago lanceolata	L	1	5	—	0	0	0	12	5	174
Plantago lanceolata	L	1	1	—	0	0	0	6	5	181
Plantago lanceolata	L	1	3	—	0	0	0	20	4	225
Plantago lanceolata	L	1	2	—	0	0	0	5	5	250
Plantago lanceolata	L	1	7	—	0	0	0	5	5	251
Plantago lanceolata	L	1	12	—	0	0	0	10	5	255
Plantago lanceolata	L	1	1	—	9	9	9	3	7	235
Plantago lanceolata	L	1	2	—	0	54	27	30	5	45
Plantago lanceolata	L	1	3	—	53	105	70	8	7	275
Plantago lanceolata	L	1	1	—	156	156	156	13	5	68
Plantago lanceolata	L	1	4	—	208	208	208	6	5	11
Plantago lanceolata	L	2	1	4	0	0	0	20	1	61
Plantago lanceolata	L	2	1	4	0	0	0	13	1	133
Plantago lanceolata	L	2	1	4	0	0	0	26	1	133
Plantago lanceolata	L	2	1	4	0	0	0	39	1	133
Plantago lanceolata	L	2	1	>3	0	0	0	6	3	179
Plantago lanceolata	L	2	1	4	0	0	0	7.5	2	198
Plantago lanceolata	L	2	5	>1	0	0	0	—	1	275
Plantago lanceolata	L	2	1	—	5	5	5	5	5	234
Plantago lanceolata	L	2	1	—	6	6	6	16	5	23
Plantago lanceolata	L	2	1	—	7	7	7	15	5	62
Plantago lanceolata	L	2	3	—	32	48	41	3	7	235
Plantago lanceolata	L	2	1	—	50	50	50	10	5	255
Plantago lanceolata	L	2	1	—	60	60	60	10	5	149
Plantago lanceolata	L	2	1	—	67	67	67	3	5	96
Plantago lanceolata	L	2	1	—	79	79	79	10	5	252
Plantago lanceolata	L	2	1	—	80	80	80	6	5	162
Plantago lanceolata	L	2	5	—	43	409	200	30	5	45
Plantago lanceolata	L	2	2	—	60	500	280	6.5	7	182

Species									
Plantago lanceolata	2	1	—	289	289	289	8	7	275
Plantago lanceolata	2	2	—	229	510	370	2	7	99
Plantago lanceolata	2	1	—	391	391	391	12	5	81
Plantago lanceolata	2	3	—	368	708	500	3	5	208
Plantago lanceolata	2	1	—	575	575	575	6.5	5	68
Plantago lanceolata	2	2	—	800	1400	1100	1	5	57
Plantago lanceolata	2	1	—	2548	2548	2548	6	5	11
Plantago lanceolata	2	1	—	4970	4970	4970	6	5	48
Plantago lanceolata	3	1	>9	0	0	0	30	1	60
Plantago lanceolata	3	1	11	0	0	0	30	1	61
Plantago lanceolata	3	1	5	0	0	0	8	1	61
Plantago lanceolata	3	1	15	0	0	0	25	1	125
Plantago lanceolata	3	2	>5	0	0	0	7.5	2	198
Plantago lanceolata	3	1	10	0	0	0	15	1	240
Plantago lanceolata	3	1	10	0	0	0	50	1	240
Plantago lanceolata	3	1	16	0	0	0	100	1	240
Plantago lanceolata	3	1	—	20	20	20	30	5	219
Plantago lanceolata	4	1	—	7	7	7	20	4	111
Plantago lanceolata	4	1	—	9	9	9	12	5	174
Plantago lanceolata	4	2	—	9	14	12	5	5	122
Plantago lanceolata	4	2	—	16	16	16	10	5	110
Plantago lanceolata	4	1	—	25	25	25	15	5	267
Plantago lanceolata	4	1	—	26	26	26	5	5	123
Plantago lanceolata	4	3	—	17	52	40	7	5	58
Plantago lanceolata	4	2	—	48	64	56	6	5	181
Plantago lanceolata	4	3	—	38	113	75	10	5	255
Plantago lanceolata	4	1	—	113	113	113	30	4	13
Plantago lanceolata	4	1	—	185	185	185	2	7	99
Plantago lanceolata	4	5	—	50	467	197	3	5	96
Plantago lanceolata	4	2	—	140	256	198	6	5	67
Plantago lanceolata	4	1	—	250	250	250	5	4	88
Plantago lanceolata	4	3	—	283	396	321	3	5	208
Plantago lanceolata	4	2	—	48	864	456	5	5	69
Plantago lanceolata	4	4	—	120	1960	975	6.5	5	128
Plantago lanceolata	4	1	—	1020	1020	1020	8.5	4	254
Plantago lanceolata	4	7	—	468	2964	1166	6	5	99
Plantago lanceolata	4	3	—	1400	1400	1400	1	5	57
Plantago lanceolata	4	3	—	126	3977	1670	17.8	5	155
Plantago major	1	2	—	0	0	0	25	4	36
Plantago major	1	3	0	0	0	0	5	5	69
Plantago major	1	1	0	0	0	0	0	1	75
Plantago major	1	1	—	0	0	0	0	2	75
Plantago major	1	3	—	0	0	0	20	5	97
Plantago major	1	1	—	0	0	0	10	5	150
Plantago major	1	1	—	0	0	0	10	5	152
Plantago major	1	2	—	0	0	0	10	5	255

Species	Authority	Seed bank type	Number of records	Longevity (y)	Minimum density (seeds m^{-2})	Maximum density (seeds m^{-2})	Mean density (seeds m^{-2})	Depth (cm)	Method	Source code
Plantago major	L.	1	1	—	8	8	8	10	5	252
Plantago major	L.	1	1	—	26	26	26	2	7	99
Plantago major	L.	1	2	—	104	104	104	6	5	28
Plantago major	L.	1	6	—	53	316	126	8	7	275
Plantago major	L.	1	1	—	333	333	333	50	5	260
Plantago major	L.	1	1	—	400	400	400	5	5	251
Plantago major	L.	2	3	>3	0	0	0	7	2	44
Plantago major	L.	2	1	>2	0	0	0	5	1	75
Plantago major	L.	2	1	>2	0	0	0	5	2	75
Plantago major	L.	2	1	—	8	8	8	25	4	36
Plantago major	L.	2	3	—	8	22	13	3	7	235
Plantago major	L.	2	1	—	51	51	51	5	5	222
Plantago major	L.	2	1	—	60	60	60	6.5	5	128
Plantago major	L.	2	2	—	62	74	68	15	5	62
Plantago major	L.	2	1	—	87	87	87	15	5	267
Plantago major	L.	2	1	—	100	100	100	16	5	168
Plantago major	L.	2	1	—	102	102	102	15	5	142
Plantago major	L.	2	1	—	135	135	135	11	5	162
Plantago major	L.	2	4	>1	78	390	208	6	5	28
Plantago major	L.	2	10	—	50	500	228	3	5	96
Plantago major	L.	2	11	—	38	1075	260	10	5	255
Plantago major	L.	2	1	—	268	268	268	5	5	244
Plantago major	L.	2	1	—	286	286	286	5	4	53
Plantago major	L.	2	2	—	300	300	300	5	5	251
Plantago major	L.	2	1	—	312	312	312	10	5	152
Plantago major	L.	2	4	—	215	480	329	15	5	272
Plantago major	L.	2	1	—	439	439	439	52	5	162
Plantago major	L.	2	1	—	451	451	451	65	5	162
Plantago major	L.	2	1	—	563	563	563	17	5	162
Plantago major	L.	2	2	—	140	1023	582	30	5	45
Plantago major	L.	2	1	—	608	608	608	6	5	151
Plantago major	L.	2	3	—	195	875	648	10	5	162
Plantago major	L.	2	2	—	112	1311	712	12	5	162
Plantago major	L.	2	1	—	1311	1311	1311	27	5	162
Plantago major	L.	2	1	—	1750	1750	1750	30	5	162
Plantago major	L.	3	3	>5	0	0	0	7	2	44
Plantago major	L.	3	1	40	0	0	0	—	3	124
Plantago major	L.	3	1	21	0	0	0	25	1	125
Plantago major	L.	3	3	>5	0	0	0	7.5	2	198
Plantago major	L.	3	1	21	0	0	0	15	1	240
Plantago major	L.	3	1	21	0	0	0	100	1	240
Plantago major	L.	3	1	30	0	0	0	50	1	240

Species										
Plantago major	L.	3	1	—	32	32	32	15	5	175
Plantago major	L.	3	1	—	55	55	55	8	5	108
Plantago major	L.	3	1	>18	67	67	67	32	5	169
Plantago major	L.	3	1	>14	156	156	156	6	5	11
Plantago major	L.	3	3	—	32	301	161	30	5	29
Plantago major	L.	3	1	>40	189	189	189	24	5	168
Plantago major	L.	3	1	—	260	260	260	10	5	27
Plantago major	L.	3	1	>1	260	260	260	6	5	28
Plantago major	L.	3	1	—	270	270	270	50	5	260
Plantago major	L.	3	1	—	313	313	313	20	5	97
Plantago major	L.	3	1	—	500	500	500	90	5	162
Plantago major	L.	3	1	>20	700	700	700	24	5	168
Plantago major	L.	3	6	—	312	2185	757	30	5	45
Plantago major	L.	3	1	—	1100	1100	1100	20	5	69
Plantago major	L.	4	1	—	11	11	11	10	5	33
Plantago major	L.	4	1	—	19	19	19	10	5	110
Plantago major	L.	4	1	—	22	22	22	10	5	8
Plantago major	L.	4	3	—	8	60	26	25	4	36
Plantago major	L.	4	1	—	37	37	37	7.5	5	10
Plantago major	L.	4	1	—	188	188	188	10	5	255
Plantago major	L.	4	2	—	126	252	189	17.8	5	155
Plantago major	L.	4	1	—	268	268	268	5	5	244
Plantago major	L.	4	2	—	151	431	291	3	5	59
Plantago major	L.	4	2	—	49	583	316	15	5	62
Plantago major	L.	4	3	—	117	467	339	3	5	96
Plantago major	L.	4	4	—	117	1374	687	10	5	150
Plantago major	L.	4	1	—	760	760	760	6	5	151
Plantago major	L.	4	9	—	200	2267	885	25	5	260
Plantago major	L.	4	6	—	80	3824	923	5	5	69
Plantago major	L.	4	4	—	404	2367	1392	15	5	272
Plantago major	L.	4	1	—	2030	2030	2030	20	4	111
Plantago major	L.	4	6	—	833	8167	2945	20	5	74
Plantago major	L.	4	2	—	3088	3088	3088	15	5	207
Plantago major	L.	4	22	—	20	49670	3837	30	4	13
Plantago major	L.	4	4	—	1700	5700	4188	15	5	94
Plantago maritima	L.	1	6	—	0	0	0	1	5	57
Plantago maritima	L.	1	2	—	0	0	0	10	4	103
Plantago maritima	L.	1	4	—	0	0	0	7	5	112
Plantago media	L.	2	1	—	39	39	39	5	5	123
Plantago media	L.	1	1	—	0	0	0	6	5	67
Plantago media	L.	1	2	—	0	0	0	5	5	69
Plantago media	L.	1	3	—	0	0	0	20	5	97
Plantago media	L.	1	2	—	0	0	0	10	5	150
Plantago media	L.	1	4	—	0	0	0	10	5	152
Plantago media	L.	1	1	—	0	0	0	10	5	255
Plantago media	L.	2	1	>3	0	0	0	6	3	179
Plantago media	L.	2	1	4	0	0	0	7.5	2	195

Species	Authority	Seed bank type	Number of records	Longevity (y)	Minimum density (seeds m⁻²)	Maximum density (seeds m⁻²)	Mean density (seeds m⁻²)	Depth (cm)	Method	Source code
Plantago media	L.	2	1	—	85	85	85	3	5	208
Plantago media	L.	2	2	—	72	355	214	6.5	7	182
Plantago media	L.	3	1	—	313	313	313	20	5	97
Plantago media	L.	4	1	—	81	81	81	7	5	58
Plantago media	L.	4	2	—	38	125	82	10	5	255
Plantago media	L.	4	2	—	80	128	104	5	5	69
Plantago media	L.	4	3	—	96	400	267	6	5	181
Plumbaginaceae	—		—	—	—	—	—	—	—	
Armeria maritima	(Miller) Willd.	1	8	—	0	0	0	1	5	57
Armeria maritima	(Miller) Willd.	1	2	—	0	0	0	10	4	103
Armeria maritima	(Miller) Willd.	4	2	—	76	101	89	17.8	5	156
Armeria maritima	(Miller) Willd.	4	2	—	400	600	500	1	5	57
Limonium auriculae-ursifolium	(Pourret) Druce	1	2	—	0	0	0	5	5	251
Limonium vulgare	Miller	1	3	—	0	0	0	1	5	57
Limonium vulgare	Miller	1	2	—	0	0	0	10	4	103
Limonium vulgare	Miller	4	1	—	300	300	300	1	5	57
Polygalaceae	—		—	—	—	—	—	—	—	
Polygala amarella	Crantz	1	1	—	0	0	0	5	5	55
Polygala amarella	Crantz	1	4	—	0	0	0	12	5	174
Polygala amarella	Crantz	1	1	—	0	0	0	6	5	181
Polygala amarella	Crantz	1	1	—	0	0	0	10	5	255
Polygala amarella	Crantz	1	2	—	12	14	13	6.5	7	182
Polygala amarella	Crantz	4	1	—	24	24	24	6	5	181
Polygala amarella	Crantz	4	1	—	49	49	49	12	5	174
Polygala chamaebuxus	L.	1	2	—	0	0	0	5	4	91
Polygala comosa	Schkuhr	1	1	—	0	0	0	8.5	4	254
Polygala comosa	Schkuhr	1	3	—	0	0	0	10	5	255
Polygala comosa	Schkuhr	1	2	—	26	45	36	6.5	7	182
Polygala serpyllifolia	Hose	1	1	—	0	0	0	2	5	153
Polygala serpyllifolia	Hose	2	1	—	41	41	41	5	5	92
Polygala vulgaris	L.	1	1	—	0	0	0	6	5	48
Polygala vulgaris	L.	1	2	—	0	0	0	3	5	96
Polygala vulgaris	L.	1	2	—	0	0	0	10	5	110
Polygala vulgaris	L.	1	2	—	0	0	0	17.8	5	156
Polygala vulgaris	L.	1	1	—	0	0	0	10	5	255
Polygala vulgaris	L.	1	1	—	5	5	5	3	7	235
Polygala vulgaris	L.	1	1	—	43	43	43	30	5	45
Polygala vulgaris	L.	2	1	—	125	125	125	10	5	255
Polygala vulgaris	L.	4	2	—	19	25	22	10	5	110

Polygonaceae	—									
Fagopyrum esculentum	Moench	1	1	0	0	0	0	15	1	240
Fagopyrum esculentum	Moench	1	1	0	0	0	0	50	1	240
Fagopyrum esculentum	Moench	1	1	0	0	0	0	100	1	240
Fallopia convolvulus	(L.) A. Löve	1	2	—	0	0	0	25	4	36
Fallopia convolvulus	(L.) A. Löve	1	2	—	0	0	0	25	4	185
Fallopia convolvulus	(L.) A. Löve	1	3	—	0	0	0	10	5	255
Fallopia convolvulus	(L.) A. Löve	1	2	—	53	210	132	8	7	275
Fallopia convolvulus	(L.) A. Löve	1	1	—	222	222	222	50	5	260
Fallopia convolvulus	(L.) A. Löve	2	1	1	0	0	0	7	2	44
Fallopia convolvulus	(L.) A. Löve	2	4	2	0	0	0	7	2	44
Fallopia convolvulus	(L.) A. Löve	2	3	3	0	0	0	7	2	44
Fallopia convolvulus	(L.) A. Löve	2	1	>2	0	0	0	2	1	47
Fallopia convolvulus	(L.) A. Löve	2	1	>2	0	0	0	15	1	47
Fallopia convolvulus	(L.) A. Löve	2	1	4	0	0	0	2.5	2	203
Fallopia convolvulus	(L.) A. Löve	2	1	1	0	0	0	2.5	1	256
Fallopia convolvulus	(L.) A. Löve	2	1	1	0	0	0	5	1	256
Fallopia convolvulus	(L.) A. Löve	2	1	1	0	0	0	7.5	1	256
Fallopia convolvulus	(L.) A. Löve	2	1	1	0	0	0	12.5	1	256
Fallopia convolvulus	(L.) A. Löve	2	1	1	0	0	0	17.5	1	256
Fallopia convolvulus	(L.) A. Löve	2	1	—	0	0	0	25	5	32
Fallopia convolvulus	(L.) A. Löve	2	1	—	9	9	9	15	5	10
Fallopia convolvulus	(L.) A. Löve	2	3	—	37	37	37	7.5	4	36
Fallopia convolvulus	(L.) A. Löve	3	8	—	12	98	44	25	5	255
Fallopia convolvulus	(L.) A. Löve	3	2	—	50	425	189	10	5	142
Fallopia convolvulus	(L.) A. Löve	3	1	>5	153	357	255	15	5	260
Fallopia convolvulus	(L.) A. Löve	3	1	>5	324	324	324	50	3	14
Fallopia convolvulus	(L.) A. Löve	3	1	>4	0	0	0	30	2	44
Fallopia convolvulus	(L.) A. Löve	3	1	>5	0	0	0	7	1	54
Fallopia convolvulus	(L.) A. Löve	3	1	>5	0	0	0	200	1	203
Fallopia convolvulus	(L.) A. Löve	3	1	>5	0	0	0	2.5	2	203
Fallopia convolvulus	(L.) A. Löve	3	1	>5	0	0	0	7.5	2	203
Fallopia convolvulus	(L.) A. Löve	3	1	>5	0	0	0	7.5	1	203
Fallopia convolvulus	(L.) A. Löve	3	1	>12	0	0	0	15	2	203
Fallopia convolvulus	(L.) A. Löve	3	1	>6	12	12	12	15	5	219
Fallopia convolvulus	(L.) A. Löve	3	1	>20	25	25	25	50	5	62
Fallopia convolvulus	(L.) A. Löve	3	1	>40	42	42	42	15	5	40
Fallopia convolvulus	(L.) A. Löve	3	1	>24	43	43	43	20	5	45
Fallopia convolvulus	(L.) A. Löve	3	1	>6	65	65	65	30	5	45
Fallopia convolvulus	(L.) A. Löve	4	1	—	97	97	97	30	4	45
Fallopia convolvulus	(L.) A. Löve	4	1	—	422	422	422	30	5	12
Fallopia convolvulus	(L.) A. Löve	4	1	—	556	556	556	50	5	260
Fallopia convolvulus	(L.) A. Löve	4	1	—	16	16	16	25	4	36
Fallopia convolvulus	(L.) A. Löve	4	1	—	22	22	22	10	5	8
Fallopia convolvulus	(L.) A. Löve	4	2	—	12	37	25	15	4	31

Species	Authority	Seed bank type	Number of records	Longevity (y)	Minimum density (seeds m^{-2})	Maximum density (seeds m^{-2})	Mean density (seeds m^{-2})	Depth (cm)	Method	Source code
Fallopia convolvulus	(L.) A. Love	4	4	—	9	140	47	15	4	30
Fallopia convolvulus	(L.) A. Love	4	2	—	63	88	76	10	5	255
Fallopia convolvulus	(L.) A. Love	4	5	—	42	189	97	10	5	15
Fallopia convolvulus	(L.) A. Love	4	13	—	20	779	114	30	4	13
Fallopia convolvulus	(L.) A. Love	4	6	—	37	268	126	7.5	5	10
Fallopia convolvulus	(L.) A. Love	4	1	—	377	377	377	20	4	111
Fallopia convolvulus	(L.) A. Love	4	1	—	387	387	387	25	5	2
Fallopia convolvulus	(L.) A. Love	4	5	—	200	571	388	25	5	260
Fallopia convolvulus	(L.) A. Love	4	1	—	445	445	445	15	5	207
Fallopia convolvulus	(L.) A. Love	4	2	—	460	760	610	25	4	3
Fallopia convolvulus	(L.) A. Love	4	1	—	900	900	900	7	4	50
Fallopia dumetorum	(L.) Holub	1	1	—	0	0	0	6	5	49
Koenigia islandica	L.	1	1	—	0	0	0	10	5	255
Oxyria digyna	(L.) Hill	1	1	—	0	0	0	5	5	55
Persicaria amphibia	(L.) Gray	1	1	—	0	0	0	10	4	88
Persicaria amphibia	(L.) Gray	1	1	—	0	0	0	3	5	96
Persicaria amphibia	(L.) Gray	1	6	—	0	0	0	6	5	208
Persicaria amphibia	(L.) Gray	1	2	—	0	0	0	5	5	251
Persicaria bistorta	(L.) Samp.	1	2	—	0	0	0	20	5	191
Persicaria bistorta	(L.) Samp.	1	16	—	0	0	0	10	5	255
Persicaria bistorta	(L.) Samp.	3	1	>8	248	248	248	30	5	45
Persicaria hydropiper	(L.) Spach	1	3	—	0	0	0	10	5	27
Persicaria hydropiper	(L.) Spach	1	1	—	0	0	0	15	5	94
Persicaria hydropiper	(L.) Spach	1	5	—	0	0	0	2	7	99
Persicaria hydropiper	(L.) Spach	1	6	—	0	0	0	6	5	99
Persicaria hydropiper	(L.) Spach	2	1	—	48	48	48	5	5	69
Persicaria hydropiper	(L.) Spach	2	1	—	50	50	50	3	5	96
Persicaria hydropiper	(L.) Spach	2	2	—	28	114	71	25	4	36
Persicaria hydropiper	(L.) Spach	3	1	50	0	0	0	—	3	124
Persicaria hydropiper	(L.) Spach	3	1	—	312	312	312	6	5	99
Persicaria hydropiper	(L.) Spach	3	2	—	208	468	338	10	5	27
Persicaria hydropiper	(L.) Spach	4	1	—	28	28	28	20	4	111
Persicaria hydropiper	(L.) Spach	4	1	—	550	550	550	15	5	94
Persicaria hydropiper	(L.) Spach	4	1	—	1600	1600	1600	25	4	3
Persicaria hydropiper	(L.) Spach	4	2	—	2079	2339	2209	50	5	216
Persicaria lapathifolia	(L.) Gray	1	1	—	0	0	0	10	5	252
Persicaria lapathifolia	(L.) Gray	2	1	—	1	1	1	15	5	32
Persicaria lapathifolia	(L.) Gray	2	1	—	6	6	6	25	4	36
Persicaria lapathifolia	(L.) Gray	2	1	—	38	38	38	10	5	255
Persicaria lapathifolia	(L.) Gray	2	1	16	750	750	750	35	5	247
Persicaria lapathifolia	(L.) Gray	3	1	—	0	0	0	25	1	125

Species	Author									
Persicaria lapathifolia	(L.) Gray	3	3	>5	0	0	0	7.5	2	204
Persicaria lapathifolia	(L.) Gray	3	1	>4	0	0	0	20	1	228
Persicaria lapathifolia	(L.) Gray	3	1	>4	0	0	0	15	1	258
Persicaria lapathifolia	(L.) Gray	3	1	—	132	132	132	10	5	252
Persicaria lapathifolia	(L.) Gray	3	3	—	156	156	156	40	5	11
Persicaria lapathifolia	(L.) Gray	4	1	—	694	7396	4051	35	5	247
Persicaria lapathifolia	(L.) Gray	4	1	—	160	160	160	20	4	111
Persicaria lapathifolia	(L.) Gray	4	16	—	20	2550	488	30	4	13
Persicaria laxiflora	(Weihe) Opiz	1	1	—	0	0	0	10	5	27
Persicaria laxiflora	(Weihe) Opiz	1	5	—	0	26	10	2	7	99
Persicaria laxiflora	(Weihe) Opiz	2	1	—	16	16	16	25	4	36
Persicaria laxiflora	(Weihe) Opiz	3	1	—	728	728	728	10	5	27
Persicaria maculosa	Gray	1	4	—	0	0	0	25	4	36
Persicaria maculosa	Gray	1	1	—	0	0	0	3	5	96
Persicaria maculosa	Gray	1	1	—	556	556	556	10	5	255
Persicaria maculosa	Gray	2	1	—	10	10	10	50	5	260
Persicaria maculosa	Gray	2	1	—	10	10	10	8	5	108
Persicaria maculosa	Gray	2	1	—	14	14	14	5	5	117
Persicaria maculosa	Gray	2	4	—	9	39	18	25	4	36
Persicaria maculosa	Gray	2	1	—	29	29	29	12	5	174
Persicaria maculosa	Gray	2	1	—	40	40	40	5	5	220
Persicaria maculosa	Gray	2	6	—	38	375	127	40	5	219
Persicaria maculosa	Gray	2	3	>5	126	6753	2495	17.8	5	255
Persicaria maculosa	Gray	3	1	>4	0	0	0	30	3	14
Persicaria maculosa	Gray	3	1	>20	0	0	0	200	1	54
Persicaria maculosa	Gray	3	2	>20	0	0	0	13	1	133
Persicaria maculosa	Gray	3	1	>20	0	0	0	26	1	133
Persicaria maculosa	Gray	3	3	>5	0	0	0	39	1	133
Persicaria maculosa	Gray	3	1	>30	0	0	0	7.5	2	204
Persicaria maculosa	Gray	3	3	30	7	7	7	15	1	240
Persicaria maculosa	Gray	3	1	30	20	20	20	50	1	240
Persicaria maculosa	Gray	3	1	>6	48	48	48	100	1	240
Persicaria maculosa	Gray	3	1	>12	78	78	78	15	5	62
Persicaria maculosa	Gray	3	1	—	86	86	86	50	5	219
Persicaria maculosa	Gray	3	1	—	86	86	86	10	5	252
Persicaria maculosa	Gray	3	1	—	140	140	140	6	5	28
Persicaria maculosa	Gray	3	1	>50	151	151	151	30	5	29
Persicaria maculosa	Gray	3	1	>8	156	156	156	30	5	45
Persicaria maculosa	Gray	3	1	>6	215	215	215	30	5	45
Persicaria maculosa	Gray	3	1	>22	608	608	608	30	5	45
Persicaria maculosa	Gray	3	1	—	756	756	756	70	5	219
Persicaria maculosa	Gray	3	1	—	3	3	3	17.8	5	155
Persicaria maculosa	Gray	4	1	—	3	3	3	10	5	33
Persicaria maculosa	Gray	4	3	—	4	8	7	25	4	36

Species	Authority	Seed bank type	Number of records	Longevity (y)	Minimum density (seeds m^{-2})	Maximum density (seeds m^{-2})	Mean density (seeds m^{-2})	Depth (cm)	Method	Source code
Persicaria maculosa	Gray	4	1	—	35	35	35	10	5	110
Persicaria maculosa	Gray	4	9	—	33	520	180	10	5	15
Persicaria maculosa	Gray	4	1	—	600	600	600	20	4	111
Persicaria maculosa	Gray	4	4	—	333	1273	693	25	5	260
Persicaria maculosa	Gray	4	1	—	1951	1951	1951	15	5	207
Persicaria vivipara	(L.) Ronse Decraene	1	2	—	0	0	0	5	4	91
Persicaria vivipara	(L.) Ronse Decraene	1	4	—	0	0	0	10	5	150
Persicaria vivipara	(L.) Ronse Decraene	2	1	—	123	123	123	20	5	191
Persicaria vivipara	(L.) Ronse Decraene	4	2	—	63	241	152	5	5	55
Persicaria vivipara	(L.) Ronse Decraene	4	2	—	92	256	174	20	5	191
Polygonum aviculare	L.	1	2	—	0	0	0	25	4	36
Polygonum aviculare	L.	1	1	—	0	0	0	7	4	50
Polygonum aviculare	L.	1	1	—	0	0	0	50	5	66
Polygonum aviculare	L.	1	1	0	0	0	0	0	1	75
Polygonum aviculare	L.	1	3	—	0	0	0	3	5	96
Polygonum aviculare	L.	1	5	—	0	0	0	2	7	99
Polygonum aviculare	L.	1	1	—	0	0	0	10	5	255
Polygonum aviculare	L.	1	6	—	53	316	137	8	7	275
Polygonum aviculare	L.	1	2	—	556	757	657	50	5	260
Polygonum aviculare	L.	2	1	>2	0	0	0	2	1	47
Polygonum aviculare	L.	2	1	>2	0	0	0	15	1	47
Polygonum aviculare	L.	2	1	>2	0	0	0	0	2	75
Polygonum aviculare	L.	2	1	>2	0	0	0	5	1	75
Polygonum aviculare	L.	2	1	>2	0	0	0	5	2	75
Polygonum aviculare	L.	2	1	4	0	0	0	2.5	2	203
Polygonum aviculare	L.	2	3	—	38	50	46	10	5	255
Polygonum aviculare	L.	2	1	—	83	83	83	3	5	96
Polygonum aviculare	L.	2	1	—	123	123	123	8	7	275
Polygonum aviculare	L.	2	1	—	125	125	125	10	5	162
Polygonum aviculare	L.	2	1	—	136	136	136	25	4	36
Polygonum aviculare	L.	2	1	—	200	200	200	30	5	162
Polygonum aviculare	L.	2	1	—	1000	1000	1000	8	5	137
Polygonum aviculare	L.	2	3	—	900	7900	3333	7	4	50
Polygonum aviculare	L.	3	1	>4	0	0	0	100	1	54
Polygonum aviculare	L.	3	1	>4	0	0	0	200	1	54
Polygonum aviculare	L.	3	1	>5	0	0	0	2.5	1	203
Polygonum aviculare	L.	3	1	>5	0	0	0	7.5	1	203
Polygonum aviculare	L.	3	1	>5	0	0	0	7.5	2	203
Polygonum aviculare	L.	3	1	>5	0	0	0	15	1	203
Polygonum aviculare	L.	3	1	>5	0	0	0	15	2	203
Polygonum aviculare	L.	3	1	>10	2	2	2	10	4	241
Polygonum aviculare	L.	3	1	—	12	12	12	15	5	62

Species										
Polygonum aviculare	L.	3	1	—	28	28	28	25	5	219
Polygonum aviculare	L.	3	1	>20	30	30	30	20	5	40
Polygonum aviculare	L.	3	1	>58	32	32	32	30	5	29
Polygonum aviculare	L.	3	1	—	54	54	54	30	5	45
Polygonum aviculare	L.	3	1	>6	57	57	57	15	5	62
Polygonum aviculare	L.	3	1	>8	65	65	65	30	5	45
Polygonum aviculare	L.	3	1	>460	88	88	88	62	5	161
Polygonum aviculare	L.	3	1	>20	100	100	100	15	5	162
Polygonum aviculare	L.	3	1	—	208	208	208	10	5	27
Polygonum aviculare	L.	3	1	>6	237	237	237	30	5	45
Polygonum aviculare	L.	3	1	—	308	308	308	30	4	12
Polygonum aviculare	L.	3	1	—	325	325	325	12	5	81
Polygonum aviculare	L.	3	1	—	495	495	495	30	5	29
Polygonum aviculare	L.	3	1	>20	560	560	560	30	5	29
Polygonum aviculare	L.	3	1	—	667	667	667	50	5	260
Polygonum aviculare	L.	3	1	>10	710	710	710	30	5	29
Polygonum aviculare	L.	4	1	—	7	7	7	15	4	21
Polygonum aviculare	L.	4	1	—	13	13	13	3	7	235
Polygonum aviculare	L.	4	6	—	6	90	28	25	4	36
Polygonum aviculare	L.	4	1	—	29	29	29	7	5	58
Polygonum aviculare	L.	4	2	—	31	61	46	7.5	5	10
Polygonum aviculare	L.	4	2	—	50	75	63	10	5	255
Polygonum aviculare	L.	4	1	—	96	96	96	5	5	69
Polygonum aviculare	L.	4	9	—	33	250	122	10	5	15
Polygonum aviculare	L.	4	2	—	196	365	281	5	5	250
Polygonum aviculare	L.	4	7	—	133	883	520	15	4	30
Polygonum aviculare	L.	4	1	—	610	610	610	10	5	147
Polygonum aviculare	L.	4	1	—	686	686	686	23	5	202
Polygonum aviculare	L.	4	1	—	826	826	826	20	4	111
Polygonum aviculare	L.	4	1	—	833	833	833	3	5	96
Polygonum aviculare	L.	4	1	—	836	836	836	25	5	2
Polygonum aviculare	L.	4	18	—	20	10130	1021	30	4	13
Polygonum aviculare	L.	4	3	—	333	2170	1057	25	5	260
Polygonum aviculare	L.	4	1	—	1100	1100	1100	15	5	205
Polygonum aviculare	L.	4	2	—	1288	1628	1458	15	4	31
Polygonum aviculare	L.	4	12	—	500	4833	1931	20	5	74
Polygonum aviculare	L.	4	1	—	2100	2100	2100	15	5	207
Rumex acetosa	L.	1	1	—	0	0	0	6	5	49
Rumex acetosa	L.	1	2	—	0	0	0	6.5	5	68
Rumex acetosa	L.	1	5	—	0	0	0	13	5	68
Rumex acetosa	L.	1	9	—	0	0	0	5	5	69
Rumex acetosa	L.	1	2	—	0	0	0	3	5	96
Rumex acetosa	L.	1	4	—	0	0	0	20	5	97
Rumex acetosa	L.	1	2	—	0	0	0	6	5	99
Rumex acetosa	L.	1	7	—	0	0	0	6.5	5	128
Rumex acetosa	L.	1	3	—	0	0	0	10	5	150
Rumex acetosa	L.	1		—	0	0	0	10	5	152

Species	Authority	Seed bank type	Number of records	Longevity (y)	Minimum density (seeds m^{-2})	Maximum density (seeds m^{-2})	Mean density (seeds m^{-2})	Depth (cm)	Method	Source code
Rumex acetosa	L	1	2	—	0	0	0	12	5	174
Rumex acetosa	L	1	2	—	0	0	0	5	5	222
Rumex acetosa	L	1	10	—	0	0	0	20	4	225
Rumex acetosa	L	1	1	—	0	0	0	7	5	232
Rumex acetosa	L	1	18	—	0	0	0	10	5	255
Rumex acetosa	L	1	5	—	0	4	1	3	7	235
Rumex acetosa	L	1	1	—	52	52	52	2	7	99
Rumex acetosa	L	1	3	—	54	54	54	10	5	252
Rumex acetosa	L	1	7	—	78	104	95	6	5	28
Rumex acetosa	L	1	1	—	142	198	170	3	5	208
Rumex acetosa	L	1	2	—	172	172	172	22.5	5	45
Rumex acetosa	L	1	4	—	208	208	208	40	5	11
Rumex acetosa	L	1	2	—	113	368	241	6	5	208
Rumex acetosa	L	1	5	—	312	312	312	10	5	27
Rumex acetosa	L	1	14	—	156	832	520	6	5	11
Rumex acetosa	L	1	4	—	400	900	600	5	5	251
Rumex acetosa	L	1	1	—	625	625	625	10	4	88
Rumex acetosa	L	2	1	3	0	0	0	7.5	2	199
Rumex acetosa	L	2	1	4	0	0	0	7.5	2	199
Rumex acetosa	L	2	5	>1	0	0	0	—	1	275
Rumex acetosa	L	2	1	—	13	13	13	8	5	108
Rumex acetosa	L	2	1	—	19	19	19	10	5	252
Rumex acetosa	L	2	1	—	104	104	104	6	5	28
Rumex acetosa	L	2	7	—	140	140	140	30	5	29
Rumex acetosa	L	2	2	—	38	300	147	10	5	255
Rumex acetosa	L	2	2	—	60	240	150	10	5	149
Rumex acetosa	L	2	1	—	255	255	255	3	5	208
Rumex acetosa	L	2	2	—	312	572	442	10	5	27
Rumex acetosa	L	2	4	—	73	1266	545	2	7	99
Rumex acetosa	L	2	3	—	75	2217	893	30	5	45
Rumex acetosa	L	2	3	—	780	988	919	6	5	11
Rumex acetosa	L	2	2	—	142	2038	1090	6	5	208
Rumex acetosa	L	3	1	>5	4600	4600	4600	30	5	66
Rumex acetosa	L	3	1	—	0	0	0	7.5	2	199
Rumex acetosa	L	3	1	—	120	120	120	20	5	67
Rumex acetosa	L	3	1	—	150	150	150	20	5	4
Rumex acetosa	L	3	1	—	208	208	208	6	5	99
Rumex acetosa	L	3	1	>12	256	256	256	50	5	219
Rumex acetosa	L	3	2	—	208	468	338	10	5	27
Rumex acetosa	L	3	3	—	60	700	460	15	5	257
Rumex acetosa	L	3	1	—	947	947	947	30	5	45
Rumex acetosa	L	3	1	—	1100	1100	1100	20	5	69

Species											
Rumex acetosa	L		4	1	—	7	7	7	15	5	62
Rumex acetosa	L		4	1	—	17	17	17	7	5	58
Rumex acetosa	L		4	4	—	9	97	41	12	5	174
Rumex acetosa	L		4	1	—	48	48	48	5	5	69
Rumex acetosa	L		4	1	—	50	50	50	15	5	267
Rumex acetosa	L		4	1	—	67	67	67	5	5	98
Rumex acetosa	L		4	5	—	38	150	70	10	5	255
Rumex acetosa	L		4	4	—	85	170	121	3	5	208
Rumex acetosa	L		4	2	—	83	183	133	3	5	96
Rumex acetosa	L		4	3	—	101	731	471	17.8	5	155
Rumex acetosa	L		4	1	—	813	813	813	5	4	88
Rumex acetosa	L		4	13	—	156	8268	1916	6	5	99
Rumex acetosella	L		1	4	—	0	0	0	6	5	28
Rumex acetosella	L		1	2	—	0	0	0	3	5	59
Rumex acetosella	L		1	2	—	0	0	0	3	5	96
Rumex acetosella	L		1	1	—	0	0	0	20	5	97
Rumex acetosella	L		1	1	—	0	0	0	10	5	152
Rumex acetosella	L		1	1	—	0	0	0	5	5	244
Rumex acetosella	L		1	4	—	0	0	0	10	5	255
Rumex acetosella	L		1	1	—	512	512	512	38	5	186
Rumex acetosella	L		2	1	—	6	6	6	25	4	36
Rumex acetosella	L		2	2	—	30	30	30	5	5	177
Rumex acetosella	L		2	1	—	43	43	43	16	5	23
Rumex acetosella	L		2	1	—	43	43	43	9.5	5	84
Rumex acetosella	L		2	1	—	51	51	51	2	5	82
Rumex acetosella	L		2	1	—	297	297	297	11.5	5	134
Rumex acetosella	L		2	1	—	336	336	336	5	5	69
Rumex acetosella	L		2	3	—	50	967	356	3	5	96
Rumex acetosella	L		2	4	—	63	2300	638	10	5	255
Rumex acetosella	L		2	2	—	140	1880	1010	15	5	257
Rumex acetosella	L		2	1	—	3350	3350	3350	25	4	3
Rumex acetosella	L		3	1	—	4160	4160	4160	6	5	11
Rumex acetosella	L		3	1	>5	0	0	0	2	3	83
Rumex acetosella	L		3	1	>26	0	0	0	25	1	125
Rumex acetosella	L		3	1	>5	0	0	0	7.5	2	199
Rumex acetosella	L		3	1	>10	6	6	6	10	4	241
Rumex acetosella	L		3	2	>18	33	111	72	32	5	169
Rumex acetosella	L		3	1	—	80	80	80	15	5	257
Rumex acetosella	L		3	1	—	86	86	86	30	5	45
Rumex acetosella	L		3	1	—	112	112	112	12	4	51
Rumex acetosella	L		3	1	—	333	333	333	50	5	260
Rumex acetosella	L		3	6	—	87	767	386	11.5	5	134
Rumex acetosella	L		3	1	—	1400	1400	1400	20	5	69
Rumex acetosella	L		4	1	—	3	3	3	10	5	33
Rumex acetosella	L		4	2	—	53	69	61	15	4	31
Rumex acetosella	L		4	1	—	96	96	96	9	5	84

Species	Authority	Seed bank type	Number of records	Longevity (y)	Minimum density (seeds m^{-2})	Maximum density (seeds m^{-2})	Mean density (seeds m^{-2})	Depth (cm)	Method	Source code
Rumex acetosella	L.	4	4	—	50	233	104	3	5	96
Rumex acetosella	L.	4	1	—	135	135	135	20	5	111
Rumex acetosella	L.	4	5	—	62	198	139	11.5	5	134
Rumex acetosella	L.	4	1	—	172	172	172	3	5	59
Rumex acetosella	L.	4	1	—	272	272	272	5	5	69
Rumex acetosella	L.	4	2	—	200	438	319	10	5	255
Rumex acetosella	L.	4	1	—	500	500	500	15	5	249
Rumex acetosella	L.	4	2	—	184	1604	894	5	5	123
Rumex acetosella	L.	4	6	—	504	2117	1193	17.8	5	155
Rumex acetosella	L.	4	1	—	1370	1370	1370	5	5	92
Rumex acetosella	L.	4	1	—	2133	2133	2133	0	5	157
Rumex acetosella	L.	4	1	—	10102	10102	10102	8.5	4	254
Rumex conglomeratus	Murray	1	1	—	0	0	0	20	4	225
Rumex conglomeratus	Murray	2	1	3	0	0	0	7.5	2	199
Rumex crispus	L.	1	1	—	0	0	0	6.5	5	68
Rumex crispus	L.	1	1	—	0	0	0	13	5	68
Rumex crispus	L.	1	1	—	0	0	0	5	5	69
Rumex crispus	L.	1	1	—	0	0	0	5	4	88
Rumex crispus	L.	1	1	—	0	0	0	6	5	151
Rumex crispus	L.	1	2	—	0	0	0	10	5	255
Rumex crispus	L.	1	3	—	53	175	111	8	7	275
Rumex crispus	L.	1	6	—	85	198	142	6	5	208
Rumex crispus	L.	2	1	4	0	0	0	39	1	133
Rumex crispus	L.	2	1	—	7	7	7	15	5	62
Rumex crispus	L.	2	2	—	18	24	21	4	5	217
Rumex crispus	L.	2	1	—	21	21	21	10	5	252
Rumex crispus	L.	2	1	—	75	75	75	10	5	162
Rumex crispus	L.	2	1	—	96	96	96	5	5	69
Rumex crispus	L.	2	4	—	113	263	188	10	5	255
Rumex crispus	L.	2	1	—	301	301	301	5	4	53
Rumex crispus	L.	2	1	—	967	967	967	3	5	96
Rumex crispus	L.	2	1	—	2188	2188	2188	17	5	162
Rumex crispus	L.	3	1	>4	0	0	0	8	1	60
Rumex crispus	L.	3	1	>4	0	0	0	20	1	60
Rumex crispus	L.	3	1	>4	0	0	0	30	1	60
Rumex crispus	L.	3	1	>6	0	0	0	20	1	61
Rumex crispus	L.	3	1	5	0	0	0	8	1	61
Rumex crispus	L.	3	1	5	0	0	0	30	1	61
Rumex crispus	L.	3	1	80	0	0	0	—	3	124
Rumex crispus	L.	3	1	>25	0	0	0	25	1	125
Rumex crispus	L.	3	1	>20	0	0	0	13	1	133
Rumex crispus	L.	3	2	>20	0	0	0	26	1	133

Species	Auth.									
Rumex crispus	L.	3	4	>5	0	0	0	7.5	2	204
Rumex crispus	L.	3	1	>39	0	0	0	50	1	240
Rumex crispus	L.	3	1	>39	0	0	0	100	1	240
Rumex crispus	L.	3	1	21	0	0	0	15	1	240
Rumex crispus	L.	3	1	—	14	14	14	10	5	252
Rumex crispus	L.	4	1	—	32	32	32	30	5	45
Rumex crispus	L.	4	2	—	8	8	8	10	5	33
Rumex crispus	L.	4	2	—	7	10	9	15	5	62
Rumex crispus	L.	4	3	—	17	17	17	10	5	147
Rumex crispus	L.	4	1	—	20	66	35	30	4	13
Rumex crispus	L.	4	1	—	80	80	80	5	5	69
Rumex crispus	L.	4	1	—	104	104	104	7	5	58
Rumex crispus	L.	4	2	—	114	114	114	6	5	151
Rumex crispus	L.	4	3	—	200	200	200	25	5	260
Rumex crispus	L.	4	1	—	94	404	225	4	5	217
Rumex crispus	L.	4	8	>5	247	247	247	23	5	202
Rumex crispus	L.	4	1	—	38	988	346	10	5	255
Rumex maritimus	L.	2	1	—	491	491	491	50	5	216
Rumex maritimus	L.	2	8	—	66	2881	605	5	5	177
Rumex maritimus	L.	2	3	—	1885	16073	6865	35	5	247
Rumex maritimus	L.	3	1	>5	0	0	0	7.5	2	199
Rumex maritimus	L.	3	3	—	2438	10564	6003	35	5	247
Rumex maritimus	L.	4	1	—	37	37	37	5	5	166
Rumex maritimus	L.	4	1	—	804	804	804	5	5	177
Rumex maritimus	L.	4	2	—	192	4480	2336	12	4	230
Rumex obtusifolius	L.	1	4	—	0	0	0	10	5	27
Rumex obtusifolius	L.	1	1	—	0	0	0	25	4	36
Rumex obtusifolius	L.	1	3	—	0	0	0	3	5	96
Rumex obtusifolius	L.	1	6	—	0	0	0	20	5	225
Rumex obtusifolius	L.	1	5	—	0	0	0	10	5	255
Rumex obtusifolius	L.	1	4	—	60	60	12	2	2	99
Rumex obtusifolius	L.	2	5	—	156	156	156	6	7	11
Rumex obtusifolius	L.	2	1	>1	105	484	276	8	7	275
Rumex obtusifolius	L.	2	1	—	0	0	0	—	1	275
Rumex obtusifolius	L.	2	1	—	6	6	6	3	7	235
Rumex obtusifolius	L.	2	1	—	7	7	7	15	5	62
Rumex obtusifolius	L.	2	1	—	11	11	11	15	5	32
Rumex obtusifolius	L.	2	2	—	12	12	12	10	5	252
Rumex obtusifolius	L.	2	1	—	75	75	75	10	5	162
Rumex obtusifolius	L.	2	1	—	102	102	102	5	5	222
Rumex obtusifolius	L.	2	4	—	63	288	141	10	5	255
Rumex obtusifolius	L.	2	1	—	170	170	170	3	5	208
Rumex obtusifolius	L.	2	1	—	188	188	188	17	5	162
Rumex obtusifolius	L.	2	8	—	50	1250	335	3	7	96
Rumex obtusifolius	L.	2	2	—	954	1091	1023	8	7	275
Rumex obtusifolius	L.	2	1	—	1311	1311	1311	12	5	162
Rumex obtusifolius	L.	2	1	—	1311	1311	1311	27	5	162

Species	Authority	Seed bank type	Number of records	Longevity (y)	Minimum density (seeds m⁻²)	Maximum density (seeds m⁻²)	Mean density (seeds m⁻²)	Depth (cm)	Method	Source code
Rumex obtusifolius	L.	2	1	—	3070	3070	3070	52	5	162
Rumex obtusifolius	L.	2	3	—	260	11232	4420	6	5	11
Rumex obtusifolius	L.	3	3	>5	0	0	0	7.5	2	204
Rumex obtusifolius	L.	3	1	>30	0	0	0	15	1	240
Rumex obtusifolius	L.	3	1	>39	0	0	0	50	1	240
Rumex obtusifolius	L.	3	1	>39	0	0	0	100	1	240
Rumex obtusifolius	L.	3	1	>4	0	0	0	15	1	258
Rumex obtusifolius	L.	3	1	—	6	6	6	5	5	234
Rumex obtusifolius	L.	3	1	—	10	10	10	8	5	108
Rumex obtusifolius	L.	3	2	—	60	60	60	15	5	257
Rumex obtusifolius	L.	3	1	—	156	156	156	6	5	99
Rumex obtusifolius	L.	3	1	—	198	198	198	6	5	208
Rumex obtusifolius	L.	3	2	—	156	260	208	10	5	27
Rumex obtusifolius	L.	3	2	—	43	1259	651	30	5	45
Rumex obtusifolius	L.	3	1	>14	988	988	988	6	5	11
Rumex obtusifolius	L.	3	1	>10	1359	1359	1359	10	4	241
Rumex obtusifolius	L.	4	2	—	7	10	9	15	5	62
Rumex obtusifolius	L.	4	1	—	382	382	382	15	5	142
Rumex obtusifolius	L.	4	6	—	38	1025	411	10	5	255
Rumex obtusifolius	L.	4	2	—	33	819	426	30	4	13
Rumex obtusifolius	L.	4	1	—	1083	1083	1083	3	5	96
Rumex obtusifolius	L.	4	2	—	956	1235	1096	17.8	5	155
Rumex patientia	L.	1	1	—	48	48	48	30	5	219
Rumex sanguineus	L.	1	1	—	0	0	0	12.5	5	126
Rumex sanguineus	L.	2	1	—	24	24	24	6	5	49
Rumex sanguineus	L.	2	2	—	49	49	49	3	7	235
Rumex sanguineus	L.	3	3	>5	0	0	0	7.5	2	204
Rumex sanguineus	L.	4	1	—	102	102	102	15	5	142
Portulaceae	—				—		—		—	
Claytonia parviflora	(L.) Pohl	2	1	—	42	42	42	10	4	127
Claytonia parviflora	(L.) Pohl	4	1	—	450	450	450	10	5	140
Montia fontana	L.	1	4	—	0	0	0	2	7	99
Montia fontana	L.	1	1	—	260	260	260	10	5	27
Montia fontana	L.	2	1	—	156	156	156	6	5	99
Montia fontana	L.	2	1	—	200	200	200	3	5	132
Montia fontana	L.	2	1	—	364	364	364	10	5	27
Montia fontana	L.	3	1	—	143	143	143	8	5	108
Montia fontana	L.	3	1	—	468	468	468	10	5	27
Montia fontana	L.	4	4	—	126	301	220	17.8	5	155
Portulaca oleracea	L.	1	2	—	0	0	0	5	5	189
Portulaca oleracea	L.	2	3	>3	0	0	0	7	2	44

Portulaca oleracea	L.	2	1	—	272	272	272	11.5	5	134
Portulaca oleracea	L.	2	4	—	2071	21610	7041	15	5	272
Portulaca oleracea	L.	3	3	>5	0	0	0	7	2	44
Portulaca oleracea	L.	3	3	>6	0	0	0	7	2	44
Portulaca oleracea	L.	3	1	>5	0	0	0	23	1	64
Portulaca oleracea	L.	3	1	40	0	0	0	—	—	124
Portulaca oleracea	L.	3	1	>4	0	0	0	20	3	228
Portulaca oleracea	L.	3	1	>30	0	0	0	15	1	240
Portulaca oleracea	L.	3	1	16	0	0	0	50	1	240
Portulaca oleracea	L.	3	1	21	0	0	0	100	1	240
Portulaca oleracea	L.	3	3	—	70	215	124	15	5	175
Portulaca oleracea	L.	4	4	—	71	104	83	15	5	272
Portulaca oleracea	L.	4	1	—	235	235	235	11.5	5	134
Potamogetonaceae										
Potamogeton pectinatus	L.	1	4	—	0	0	0	10	5	87
Potamogeton pectinatus	L.	1	7	—	0	0	0	5	5	116
Potamogeton pectinatus	L.	1	2	—	0	0	0	5	5	177
Potamogeton pectinatus	L.	2	1	—	522	522	522	10	5	87
Potamogeton pectinatus	L.	4	1	—	132	132	132	5	5	166
Potamogeton pectinatus	L.	4	4	—	76	2088	697	10	5	87
Potamogeton perfoliatus	L.	1	5	—	0	0	0	5	5	116
Potamogeton pusillus	L.	1	1	—	0	0	0	10	5	87
Potamogeton pusillus	L.	2	1	—	102	102	102	10	5	87
Potamogeton pusillus	L.	4	7	—	38	751	275	10	5	87
Potamogeton vaginatus	Turcz.	1	5	—	0	0	0	10	5	87
Primulaceae										
Anagallis arvensis	L.	1	2	—	0	0	0	10	4	22
Anagallis arvensis	L.	1	4	—	0	0	0	25	4	36
Anagallis arvensis	L.	1	1	—	0	0	0	2	5	95
Anagallis arvensis	L.	1	1	>2	0	0	0	3	5	132
Anagallis arvensis	L.	2	1	—	0	0	0	20	1	219
Anagallis arvensis	L.	2	2	—	14	20	17	25	4	36
Anagallis arvensis	L.	2	1	—	28	28	28	70	5	219
Anagallis arvensis	L.	2	5	—	38	113	73	10	5	255
Anagallis arvensis	L.	2	1	—	77	77	77	17.8	5	155
Anagallis arvensis	L.	2	1	—	160	160	160	5	5	69
Anagallis arvensis	L.	2	1	—	272	272	272	15	5	62
Anagallis arvensis	L.	2	1	—	300	300	300	7	4	50
Anagallis arvensis	L.	2	1	—	393	393	393	30	4	12
Anagallis arvensis	L.	2	4	—	438	438	438	17	5	162
Anagallis arvensis	L.	2	1	—	177	899	474	15	5	272
Anagallis arvensis	L.	2	1	—	823	823	823	50	5	260
Anagallis arvensis	L.	2	1	—	1200	1200	1200	8	5	137
Anagallis arvensis	L.	3	1	>11	0	0	0	25	3	209
Anagallis arvensis	L.	3	1	>8	5	5	5	15	5	62

Species	Authority	Seed bank type	Number of records	Longevity (y)	Minimum density (seeds m⁻²)	Maximum density (seeds m⁻²)	Mean density (seeds m⁻²)	Depth (cm)	Method	Source code
Anagallis arvensis	L.	3	1	>30	28	28	28	6	5	211
Anagallis arvensis	L.	3	1	—	29	29	29	5	5	234
Anagallis arvensis	L.	3	1	>20	32	32	32	30	5	29
Anagallis arvensis	L.	3	1	—	32	32	32	30	5	45
Anagallis arvensis	L.	3	1	>58	43	43	43	30	5	29
Anagallis arvensis	L.	3	2	>18	33	56	45	32	5	169
Anagallis arvensis	L.	3	1	>26	63	63	63	12	4	51
Anagallis arvensis	L.	3	1	>35	67	67	67	24	5	168
Anagallis arvensis	L.	3	1	>15	89	89	89	24	5	168
Anagallis arvensis	L.	3	1	>40	89	89	89	24	5	168
Anagallis arvensis	L.	3	1	>50	161	161	161	30	5	45
Anagallis arvensis	L.	3	1	>24	194	194	194	30	5	45
Anagallis arvensis	L.	3	1	—	240	240	240	40	5	219
Anagallis arvensis	L.	3	1	>20	300	300	300	24	5	168
Anagallis arvensis	L.	3	1	—	521	521	521	20	5	97
Anagallis arvensis	L.	3	1	>40	549	549	549	30	5	45
Anagallis arvensis	L.	3	1	>6	568	568	568	15	5	62
Anagallis arvensis	L.	3	1	>68	592	592	592	30	5	45
Anagallis arvensis	L.	3	1	>60	624	624	624	30	5	45
Anagallis arvensis	L.	3	1	>6	700	700	700	30	5	45
Anagallis arvensis	L.	3	1	>8	1195	1195	1195	30	5	45
Anagallis arvensis	L.	3	1	—	4800	4800	4800	20	5	69
Anagallis arvensis	L.	3	1	—	5660	5660	5660	12	5	81
Anagallis arvensis	L.	4	4	—	6	30	14	25	4	36
Anagallis arvensis	L.	4	2	—	23	87	55	7	5	58
Anagallis arvensis	L.	4	2	—	9	312	161	15	4	30
Anagallis arvensis	L.	4	1	—	196	196	196	20	4	111
Anagallis arvensis	L.	4	12	—	33	539	226	10	5	15
Anagallis arvensis	L.	4	4	—	210	420	323	2	5	95
Anagallis arvensis	L.	4	1	—	400	400	400	7	4	50
Anagallis arvensis	L.	4	42	—	33	2291	449	30	4	13
Anagallis arvensis	L.	4	3	—	176	1212	521	25	5	260
Anagallis arvensis	L.	4	1	—	600	600	600	6.5	5	128
Anagallis arvensis	L.	4	1	—	940	940	940	10	5	86
Anagallis arvensis	L.	4	1	—	1161	1161	1161	15	5	207
Anagallis arvensis	L.	4	1	—	1212	1212	1212	25	5	2
Anagallis arvensis	L.	4	6	—	500	2333	1556	20	5	74
Anagallis arvensis	L.	4	4	—	628	4868	2076	15	5	272
Anagallis arvensis subsp. *caerulea*	Hartman	2	2	—	15	366	191	10	4	22
Anagallis arvensis subsp. *caerulea*	Hartman	4	2	—	30	329	180	10	4	22
Anagallis tenella	(L.) L.	3	2	—	721	1066	894	30	5	45
Androsace septentrionalis	L.	2	9	—	17	466	168	2.5	5	114

Androsace septentrionalis	L.	4	3	—	42	376	163	10	5	8
Androsace septentrionalis	L.	4	6	—	97	823	357	6	5	39
Androsace septentrionalis	L.	4	1	—	528	528	528	2.5	5	114
Glaux maritima	L.	1	6	—	0	0	0	10	4	20
Glaux maritima	L.	1	4	—	0	0	0	1	5	57
Glaux maritima	L.	2	1	—	2500	2500	2500	20	5	210
Glaux maritima	L.	2	8	—	294	8706	3750	7	5	112
Glaux maritima	L.	3	2	—	500	5000	2750	20	5	210
Glaux maritima	L.	3	2	—	2647	5941	4294	7	5	112
Glaux maritima	L.	4	2	—	11	80	46	10	5	8
Glaux maritima	L.	4	2	—	379	808	594	17.8	5	156
Lysimachia nemorum	L.	1	3	—	0	0	0	3	5	208
Lysimachia nemorum	L.	2	1	—	3	3	3	3	7	235
Lysimachia nemorum	L.	2	1	—	23	23	23	15	5	32
Lysimachia nemorum	L.	1	2	—	250	250	250	3	5	96
Lysimachia nemorum	L.	3	1	—	260	260	260	15	5	257
Lysimachia nummularia	L.	1	2	—	0	0	0	6	5	49
Lysimachia nummularia	L.	1	4	—	0	0	0	6.5	5	68
Lysimachia nummularia	L.	1	7	—	0	0	0	6.5	5	128
Lysimachia nummularia	L.	1	5	—	0	0	0	10	5	149
Lysimachia nummularia	L.	1	1	—	0	0	0	3	5	208
Lysimachia nummularia	L.	1	1	—	0	0	0	20	4	225
Lysimachia nummularia	L.	1	2	—	0	0	0	3	7	235
Lysimachia nummularia	L.	1	1	—	0	0	0	5	5	250
Lysimachia nummularia	L.	2	1	—	0	0	0	10	5	252
Lysimachia nummularia	L.	2	2	—	0	0	0	6	5	208
Lysimachia nummularia	L.	2	1	—	142	142	142	3	5	208
Lysimachia nummularia	L.	3	1	—	113	113	113	6	5	208
Lysimachia nummularia	L.	4	2	—	142	142	142	13	5	68
Lysimachia nummularia	L.	4	1	—	168	168	168	13	5	68
Lysimachia nummularia	L.	1	1	—	48	48	48	3	5	208
Lysimachia nummularia	L.	4	1	—	113	113	113	6.5	5	128
Lysimachia nummularia	L.	1	3	—	60	240	120	16	5	216
Lysimachia thyrsiflora	L.	1	1	—	0	0	0	16	5	216
Lysimachia thyrsiflora	L.	4	1	—	155	155	155	5	4	88
Lysimachia vulgaris	L.	1	1	—	0	0	0	15	5	148
Lysimachia vulgaris	L.	1	1	—	0	0	0	10	5	149
Lysimachia vulgaris	L.	1	2	—	0	0	0	12	5	171
Lysimachia vulgaris	L.	1	1	—	0	0	0	6	5	208
Lysimachia vulgaris	L.	1	3	—	0	0	0	20	4	225
Lysimachia vulgaris	L.	1	5	—	0	0	0	5	5	248
Lysimachia vulgaris	L.	1	1	—	0	0	0	10	5	255
Lysimachia vulgaris	L.	1	3	—	0	0	0	5	6	170
Lysimachia vulgaris	L.	2	5	>1	0	0	0	—	1	275
Lysimachia vulgaris	L.	3	2	>4	62	72	67	10	3	216
Lysimachia vulgaris	L.	4	2	—	0	0	0	16	5	216
Primula auricula	L.	1	1	—	0	0	0	5	5	55

Species	Authority	Seed bank type	Number of records	Longevity (y)	Minimum density (seeds m⁻²)	Maximum density (seeds m⁻²)	Mean density (seeds m⁻²)	Depth (cm)	Method	Source code
Primula elatior	(L.) Hill	1	1	—	0	0	0	20	5	67
Primula elatior	(L.) Hill	1	1	—	0	0	0	12.5	5	126
Primula elatior	(L.) Hill	1	4	—	0	0	0	6.5	5	128
Primula elatior	(L.) Hill	1	3	—	0	0	0	6	5	208
Primula elatior	(L.) Hill	1	1	—	0	0	0	5	5	213
Primula elatior	(L.) Hill	1	4	—	0	0	0	20	4	225
Primula farinosa	L.	1	1	—	0	0	0	10	5	255
Primula farinosa	L.	1	1	—	0	0	0	5	5	55
Primula farinosa	L.	4	1	—	188	188	188	12	5	174
Primula veris	L.	1	1	—	0	0	0	12	5	174
Primula veris	L.	1	1	—	0	0	0	6	5	49
Primula veris	L.	1	3	—	0	0	0	3	5	96
Primula veris	L.	1	9	—	0	0	0	10	5	150
Primula veris	L.	1	1	—	0	0	0	10	5	152
Primula veris	L.	1	1	—	0	0	0	10	5	181
Primula veris	L.	1	1	—	0	0	0	3	7	235
Primula veris	L.	1	1	—	0	0	0	10	5	252
Primula veris	L.	1	4	—	0	0	0	10	5	255
Primula veris	L.	2	1	—	8	8	8	6.5	7	182
Primula veris	L.	2	1	—	105	105	105	6.5	7	182
Primula veris	L.	4	3	—	263	263	263	10	5	150
Primula veris	L.	4	3	—	32	144	69	6	5	181
Primula vulgaris	Hudson	4	3	—	312	546	468	10	5	152
Primula vulgaris	Hudson	2	1	—	100	100	100	15	5	257
Primula vulgaris	Hudson	3	3	—	60	140	100	15	5	257
Samolus valerandi	L.	4	1	—	1	1	1	15	5	32
Soldanella alpina	L.	4	1	—	202	202	202	17.8	5	156
Trientalis europaea	L.	1	1	—	0	0	0	5	5	55
Trientalis europaea	L.	1	2	—	0	0	0	5	5	82
Trientalis europaea	L.	1	6	—	0	0	0	5	5	92
Trientalis europaea	L.	1	1	—	0	0	0	2	5	153
Trientalis europaea	L.	1	3	—	0	0	0	5	6	170
Trientalis europaea	L.	1	2	—	0	0	0	12	5	171
Trientalis europaea	L.	1	2	—	0	0	0	16	5	216
Trientalis europaea	L.	3	1	>5	0	0	0	2	3	83
Trientalis europaea	L.	3	1	>11	0	0	0	25	3	209

Pyrolaceae

Species	Authority	Seed bank type	Number of records	Longevity (y)	Minimum density (seeds m⁻²)	Maximum density (seeds m⁻²)	Mean density (seeds m⁻²)	Depth (cm)	Method	Source code
Orthilia secunda	—	1	—	—	—	—	—	—	—	—
Orthilia secunda	(L.) House	1	3	—	0	0	0	5	6	170
Orthilia secunda	(L.) House	1	2	—	0	0	0	12	5	171
Orthilia secunda	(L.) House	1	1	—	0	0	0	5	5	264
Pyrola minor	L.	1	1	—	0	0	0	5	6	170

Taxon										
Pyrola rotundifolia	⌐	1	1	—	0	0	0	5	6	170
Pyrola rotundifolia	⌐	1	1	—	0	0	0	12	5	171
Ranunculaceae	⌐	—	—	—	—	—	—	—	—	—
Actaea spicata	⌐	1	1	—	0	0	0	10	4	22
Adonis annua	⌐	1	1	—	0	0	0	3	5	96
Anemone apennina	⌐	4	1	—	3	3	3	5	5	71
Anemone narcissifolia	⌐	1	1	—	0	0	0	15	5	32
Anemone nemorosa	⌐	1	1	—	0	0	0	7	5	58
Anemone nemorosa	⌐	1	3	—	0	0	0	20	5	67
Anemone nemorosa	⌐	1	1	—	0	0	0	9	5	84
Anemone nemorosa	⌐	1	1	—	0	0	0	3	5	96
Anemone nemorosa	⌐	1	1	—	0	0	0	12.5	5	126
Anemone nemorosa	⌐	1	3	—	0	0	0	10	5	149
Anemone nemorosa	⌐	1	3	—	0	0	0	10	5	150
Anemone nemorosa	⌐	1	1	—	0	0	0	4	5	176
Anemone nemorosa	⌐	1	6	—	0	0	0	3	5	208
Anemone nemorosa	⌐	1	6	—	0	0	0	5	5	220
Anemone nemorosa	⌐	1	1	—	0	0	0	3	7	235
Anemone nemorosa	⌐	1	10	—	0	0	0	10	5	255
Anemone nemorosa	⌐	1	2	—	0	0	0	15	5	257
Anemone nemorosa	⌐	2	1	—	975	975	975	10	5	255
Anemone nemorosa	⌐	4	2	—	41	87	64	12.5	5	126
Anemone nemorosa	⌐	4	2	—	22	108	65	10	5	126
Anemone ranunculoides	⌐	1	1	—	0	0	0	30	5	173
Anemone ranunculoides	⌐	2	2	—	0	0	0	5	5	220
Anemone ranunculoides	⌐	1	1	—	26	26	26	10	5	172
Caltha palustris	⌐	1	1	—	0	0	0	10	5	27
Caltha palustris	⌐	1	2	—	0	0	0	6	5	99
Caltha palustris	⌐	1	4	—	0	0	0	6.5	5	128
Caltha palustris	⌐	1	2	—	0	0	0	10	5	149
Caltha palustris	⌐	1	11	—	0	0	0	3	5	208
Caltha palustris	⌐	1	2	0	0	0	0	16	5	216
Caltha palustris	⌐	1	1	—	0	0	0	10	3	216
Caltha palustris	⌐	2	2	—	0	0	0	20	4	225
Caltha palustris	⌐	1	1	—	0	0	0	10	5	252
Caltha palustris	⌐	1	8	—	0	0	0	10	5	255
Caltha palustris	⌐	1	5	—	170	104	31	2	7	99
Caltha palustris	⌐	3	3	—	651	651	411	6	5	208
Caltha palustris	⌐	2	2	1	0	0	0	10	3	216
Caltha palustris	⌐	2	1	—	260	260	260	10	5	27
Caltha palustris	⌐	3	1	—	156	156	156	10	5	27
Caltha palustris	⌐	3	1	—	416	416	416	6	5	99
Caltha palustris	⌐	4	1	—	255	255	255	3	5	208
Caltha palustris	⌐	4	1	—	1600	1600	1600	20	4	225
Clematis vitalba	⌐	2	1	—	368	368	368	10	5	181

Species	Authority	Seed bank type	Number of records	Longevity (y)	Minimum density (seeds m^{-2})	Maximum density (seeds m^{-2})	Mean density (seeds m^{-2})	Depth (cm)	Method	Source code
Clematis vitalba	L.	4	2	—	56	88	72	6	5	181
Clematis vitalba	L.	4	1	—	175	175	175	6	5	49
Consolida regalis	S.F. Gray	3	1	>11	0	0	0	25	3	209
Consolida regalis	S.F. Gray	4	1	—	1300	1300	1300	6.5	5	128
Helleborus foetidus	L.	1	2	—	0	0	0	6	5	181
Helleborus viridis	L.	1	9	—	0	0	0	10	4	20
Hepatica nobilis	Schreber	1	1	—	0	0	0	12.5	5	126
Hepatica nobilis	Schreber	1	1	—	0	0	0	10	5	152
Hepatica nobilis	Schreber	1	2	—	0	0	0	4	5	176
Myosurus minimus	L.	4	1	—	14	14	14	20	5	111
Myosurus minimus	L.	4	1	—	73	73	73	30	4	13
Myosurus minimus	L.	4	2	—	84	252	168	10	5	15
Nigella damascena	L.	1	1	—	0	0	0	10	4	22
Pulsatilla vulgaris	Miller	1	1	—	0	0	0	6	5	67
Pulsatilla vulgaris	Miller	1	1	—	0	0	0	10	5	255
Ranunculus aconitifolius	L.	4	1	—	14	14	14	5	5	213
Ranunculus acris	L.	1	3	—	0	0	0	25	4	36
Ranunculus acris	L.	1	3	—	0	0	0	3	5	59
Ranunculus acris	L.	1	1	—	0	0	0	20	5	67
Ranunculus acris	L.	1	6	—	0	0	0	5	5	69
Ranunculus acris	L.	1	4	—	0	0	0	10	4	88
Ranunculus acris	L.	1	7	—	0	0	0	20	5	97
Ranunculus acris	L.	1	4	—	0	0	0	6	5	99
Ranunculus acris	L.	1	1	—	0	0	0	8	5	108
Ranunculus acris	L.	1	3	—	0	0	0	6.5	5	128
Ranunculus acris	L.	1	4	—	0	0	0	10	5	150
Ranunculus acris	L.	1	2	—	0	0	0	12	5	174
Ranunculus acris	L.	1	5	—	0	0	0	3	5	208
Ranunculus acris	L.	1	2	—	0	0	0	5	5	222
Ranunculus acris	L.	1	3	—	0	0	0	20	4	225
Ranunculus acris	L.	1	1	—	0	0	0	3	7	235
Ranunculus acris	L.	1	1	—	0	0	0	5	5	244
Ranunculus acris	L.	1	12	—	0	0	0	10	5	255
Ranunculus acris	L.	1	3	—	0	143	67	2	7	99
Ranunculus acris	L.	1	8	—	104	104	104	6	5	28
Ranunculus acris	L.	1	6	—	312	312	312	10	5	27
Ranunculus acris	L.	2	2	3	0	0	0	7.5	2	199
Ranunculus acris	L.	2	1	4	0	0	0	7.5	2	199
Ranunculus acris	L.	2	5	>1	0	0	0	—	1	275
Ranunculus acris	L.	2	1	—	43	43	43	30	5	29
Ranunculus acris	L.	2	6	—	38	188	98	10	5	255
Ranunculus acris	L.	2	1	—	156	156	156	6	5	99

Ranunculus acris	L.	2	1	—	168	168	168	5	4	53
Ranunculus acris	L.	2	2	—	248	372	310	13	5	68
Ranunculus acris	L.	2	2	—	255	1040	648	2	7	99
Ranunculus acris	L.	2	1	—	1425	1425	1425	6.5	5	68
Ranunculus acris	L.	3	1	—	50	50	50	6.5	5	68
Ranunculus acris	L.	3	1	—	198	198	198	3	5	208
Ranunculus acris	L.	3	1	—	2808	2808	2808	6	5	99
Ranunculus acris	L.	4	5	—	9	54	25	12	5	174
Ranunculus acris	L.	4	1	—	67	67	67	5	5	244
Ranunculus acris	L.	4	1	—	99	99	99	15	5	267
Ranunculus acris	L.	4	2	—	146	263	205	10	5	150
Ranunculus acris	L.	4	4	—	38	600	272	10	5	255
Ranunculus acris	L.	4	2	—	306	526	416	5	4	53
Ranunculus acris	L.	4	1	—	813	813	813	5	4	88
Ranunculus acris	L.	4	7	—	156	3120	1434	6	5	99
Ranunculus arvensis	L.	1	1	—	0	0	0	10	4	22
Ranunculus arvensis	L.	2	1	—	0	0	0	25	4	36
Ranunculus arvensis	L.	4	1	—	6	6	6	25	4	36
Ranunculus arvensis	L.	4	2	—	44	44	44	25	4	36
Ranunculus arvensis	L.	4	2	—	40	366	203	30	4	13
Ranunculus auricomus	L.	1	2	—	0	0	0	6	5	49
Ranunculus auricomus	L.	1	1	—	0	0	0	10	5	126
Ranunculus auricomus	L.	1	2	—	0	0	0	12.5	5	126
Ranunculus auricomus	L.	1	1	—	0	0	0	10	5	150
Ranunculus auricomus	L.	1	1	—	0	0	0	6	5	151
Ranunculus auricomus	L.	1	2	—	0	0	0	10	5	152
Ranunculus auricomus	L.	1	6	—	0	0	0	3	5	208
Ranunculus auricomus	L.	1	3	—	0	0	0	6	5	208
Ranunculus auricomus	L.	1	5	—	312	312	312	10	5	255
Ranunculus auricomus	L.	1	1	—	48	48	48	6	5	11
Ranunculus auricomus	L.	3	1	—	50	50	50	13	5	68
Ranunculus auricomus	L.	3	1	—	10	10	10	6.5	5	68
Ranunculus auricomus	L.	4	1	—	88	88	88	6	5	49
Ranunculus auricomus	L.	4	1	—	88	88	88	10	5	150
Ranunculus auricomus	L.	4	1	—	0	0	0	10	5	255
Ranunculus baudotii	Godron	1	2	—	0	0	0	5	5	116
Ranunculus bulbosus	L.	1	3	—	0	0	0	5	4	53
Ranunculus bulbosus	L.	1	1	—	0	0	0	6	5	67
Ranunculus bulbosus	L.	1	7	—	0	0	0	5	5	69
Ranunculus bulbosus	L.	1	1	—	0	0	0	3	5	96
Ranunculus bulbosus	L.	1	1	—	0	0	0	20	5	97
Ranunculus bulbosus	L.	1	1	—	0	0	0	5	5	122
Ranunculus bulbosus	L.	1	2	—	0	0	0	10	5	255
Ranunculus bulbosus	L.	1	1	—	8	8	8	8	5	108
Ranunculus bulbosus	L.	2	1	—	24	24	24	6	5	181
Ranunculus bulbosus	L.	2	1	—	60	60	60	6	5	162

Species	Authority	Seed bank type	Number of records	Longevity (y)	Minimum density (seeds m⁻²)	Maximum density (seeds m⁻²)	Mean density (seeds m⁻²)	Depth (cm)	Method	Source code
Ranunculus bulbosus	L.	2	2	—	15	138	77	6.5	7	182
Ranunculus bulbosus	L.	2	1	—	100	100	100	10	5	255
Ranunculus bulbosus	L.	3	1	>5	0	0	0	7.5	2	199
Ranunculus bulbosus	L.	3	1	>30	32	32	32	10	5	181
Ranunculus bulbosus	L.	3	1	>20	48	48	48	13	5	183
Ranunculus bulbosus	L.	3	2	—	43	495	269	30	5	29
Ranunculus bulbosus	L.	4	2	—	24	32	28	6	5	181
Ranunculus bulbosus	L.	4	1	—	37	37	37	15	5	267
Ranunculus bulbosus	L.	4	1	—	47	47	47	5	5	122
Ranunculus bulbosus	L.	4	1	—	88	88	88	6	5	67
Ranunculus bulbosus	L.	4	1	—	100	100	100	10	5	255
Ranunculus circinatus	Sibth.	1	2	—	0	0	0	10	5	87
Ranunculus circinatus	Sibth.	1	1	—	0	0	0	5	5	116
Ranunculus circinatus	Sibth.	4	2	—	38	76	57	10	5	87
Ranunculus ficaria	L.	1	1	—	0	0	0	15	5	32
Ranunculus ficaria	L.	1	1	—	0	0	0	50	5	66
Ranunculus ficaria	L.	1	1	—	0	0	0	20	5	67
Ranunculus ficaria	L.	1	5	—	0	0	0	5	5	69
Ranunculus ficaria	L.	1	1	—	0	0	0	3	5	96
Ranunculus ficaria	L.	1	2	—	0	0	0	10	5	126
Ranunculus ficaria	L.	1	2	—	0	0	0	12.5	5	126
Ranunculus ficaria	L.	1	2	—	0	0	0	4	5	176
Ranunculus ficaria	L.	1	6	—	0	0	0	3	5	208
Ranunculus ficaria	L.	1	3	—	0	0	0	6	5	208
Ranunculus ficaria	L.	1	1	—	0	0	0	5	5	220
Ranunculus ficaria	L.	1	1	—	0	0	0	3	7	235
Ranunculus ficaria	L.	1	3	—	0	0	0	10	5	255
Ranunculus ficaria	L.	1	2	—	0	0	0	15	5	257
Ranunculus flammula	L.	1	3	—	0	0	0	6	5	99
Ranunculus flammula	L.	1	1	—	0	0	0	6.5	5	128
Ranunculus flammula	L.	1	1	—	0	0	0	12	5	174
Ranunculus flammula	L.	1	6	—	0	0	0	3	5	208
Ranunculus flammula	L.	1	1	—	0	0	0	20	4	225
Ranunculus flammula	L.	1	5	—	0	130	33	2	7	99
Ranunculus flammula	L.	2	1	3	0	0	0	7.5	2	199
Ranunculus flammula	L.	2	1	—	2	2	2	15	5	32
Ranunculus flammula	L.	2	1	—	15	15	15	12	5	174
Ranunculus flammula	L.	2	1	—	31	31	31	5	5	244
Ranunculus flammula	L.	2	1	—	80	80	80	6.5	5	128
Ranunculus flammula	L.	2	1	—	260	260	260	40	5	11
Ranunculus flammula	L.	2	1	—	260	260	260	6	5	99
Ranunculus flammula	L.	2	1	—	450	450	450	10	5	255

Species	Authority									
Ranunculus flammula	L.	2	1	—	1200	1200	1200	20	4	225
Ranunculus flammula	L.	2	1	—	2094	2094	2094	6	5	208
Ranunculus flammula	L.	2	2	—	958	4919	2939	30	5	45
Ranunculus flammula	L.	3	1	—	32	32	32	30	5	45
Ranunculus flammula	L.	3	1	—	285	285	285	50	5	107
Ranunculus flammula	L.	3	1	—	364	364	364	6	5	11
Ranunculus flammula	L.	3	4	—	312	572	390	10	5	27
Ranunculus flammula	L.	3	1	—	416	416	416	6	5	99
Ranunculus flammula	L.	3	2	—	823	858	841	16	5	216
Ranunculus flammula	L.	3	1	—	1254	1254	1254	20	5	107
Ranunculus flammula	L.	3	5	—	877	1868	1341	6	5	208
Ranunculus flammula	L.	3	1	—	6042	6042	6042	30	5	107
Ranunculus flammula	L.	4	2	—	60	260	160	6.5	5	128
Ranunculus flammula	L.	4	2	—	126	906	516	17.8	5	155
Ranunculus flammula	L.	4	1	—	767	767	767	3	5	96
Ranunculus flammula	L.	4	3	—	623	1075	849	3	5	208
Ranunculus flammula	L.	4	1	—	1454	1454	1454	12	5	174
Ranunculus glacialis	L.	1	1	—	0	0	0	5	5	55
Ranunculus glacialis	L.	4	1	—	24	24	24	5	5	55
Ranunculus lanuginosus	L.	1	2	—	0	0	0	4	5	176
Ranunculus lingua	L.	1	1	—	0	0	0	50	5	216
Ranunculus montanus	Willd.	1	2	—	0	0	0	5	5	55
Ranunculus nivalis	L.	1	1	—	0	0	0	5	5	55
Ranunculus parviflorus	L.	1	2	—	213	213	213	3	5	132
Ranunculus parviflorus	L.	4	1	—	67	133	100	3	5	132
Ranunculus peltatus	Schrank	2	2	—	50	233	142	3	5	96
Ranunculus penicillatus	(Dumort.) Bab.	4	2	—	0	0	0	3	5	96
Ranunculus repens	L.	1	3	—	0	0	0	10	5	27
Ranunculus repens	L.	1	1	—	0	0	0	25	4	36
Ranunculus repens	L.	1	1	—	0	0	0	20	5	67
Ranunculus repens	L.	1	3	—	0	0	0	6.5	5	68
Ranunculus repens	L.	1	1	—	0	0	0	5	5	69
Ranunculus repens	L.	1	3	—	0	0	0	10	4	88
Ranunculus repens	L.	1	1	—	0	0	0	3	5	96
Ranunculus repens	L.	1	1	—	0	0	0	6	5	99
Ranunculus repens	L.	1	3	—	0	0	0	10	5	126
Ranunculus repens	L.	1	2	—	0	0	0	6.5	5	128
Ranunculus repens	L.	1	3	—	0	0	0	10	5	150
Ranunculus repens	L.	1	11	—	0	0	0	3	5	208
Ranunculus repens	L.	1	1	—	0	0	0	20	4	225
Ranunculus repens	L.	1	6	—	0	0	0	5	5	244
Ranunculus repens	L.	1	1	—	0	0	0	10	5	255
Ranunculus repens	L.	1	2	—	32	32	32	30	5	45
Ranunculus repens	L.	1	1	—	72	72	72	13	5	68
Ranunculus repens	L.	1	1	—	80	80	80	25	5	219
Ranunculus repens	L.	1	4	—	104	104	104	6	5	28
Ranunculus repens	L.	1	7	—	53	333	127	8	7	275

Species	Authority	Seed bank type	Number of records	Longevity (y)	Minimum density (seeds m⁻²)	Maximum density (seeds m⁻²)	Mean density (seeds m⁻²)	Depth (cm)	Method	Source code
Ranunculus repens	L	1	1	—	417	417	417	20	5	97
Ranunculus repens	L	2	1	—	0	0	0	5	6	170
Ranunculus repens	L	2	1	—	3	3	3	15	5	32
Ranunculus repens	L	2	2	—	3	8	6	3	7	235
Ranunculus repens	L	2	1	—	10	10	10	25	4	36
Ranunculus repens	L	2	1	—	67	67	67	3	5	96
Ranunculus repens	L	2	1	—	82	82	82	5	5	222
Ranunculus repens	L	2	1	—	150	150	150	15	5	162
Ranunculus repens	L	2	1	—	180	180	180	11	5	162
Ranunculus repens	L	2	1	—	187	187	187	12	5	162
Ranunculus repens	L	2	1	—	188	188	188	17	5	162
Ranunculus repens	L	2	3	>2	78	442	217	6	5	28
Ranunculus repens	L	2	10	—	38	675	232	10	5	255
Ranunculus repens	L	2	1	—	337	337	337	27	5	162
Ranunculus repens	L	2	5	—	225	766	425	8	7	275
Ranunculus repens	L	2	1	—	433	433	433	8	5	108
Ranunculus repens	L	2	1	—	481	481	481	3	5	208
Ranunculus repens	L	2	1	—	617	617	617	10	5	162
Ranunculus repens	L	2	1	—	700	700	700	20	5	69
Ranunculus repens	L	2	5	—	36	2044	730	2	7	99
Ranunculus repens	L	2	2	—	260	2704	1482	40	5	11
Ranunculus repens	L	2	1	—	1824	1824	1824	30	5	107
Ranunculus repens	L	2	2	—	2150	2400	2275	20	5	4
Ranunculus repens	L	2	7	—	1456	9256	4762	6	5	11
Ranunculus repens	L	2	5	—	215	24919	6288	30	5	45
Ranunculus repens	L	2	1	—	10100	10100	10100	12	5	81
Ranunculus repens	L	3	1	>20	0	0	0	13	1	133
Ranunculus repens	L	3	2	>20	0	0	0	26	1	133
Ranunculus repens	L	3	1	>20	0	0	0	39	1	133
Ranunculus repens	L	3	3	>5	0	0	0	7.5	2	199
Ranunculus repens	L	3	1	>10	2	2	2	10	4	241
Ranunculus repens	L	3	1	—	32	32	32	30	5	29
Ranunculus repens	L	3	1	>45	44	44	44	16	5	168
Ranunculus repens	L	3	2	>40	44	78	61	24	5	168
Ranunculus repens	L	3	1	—	76	76	76	70	5	219
Ranunculus repens	L	3	1	>100	78	78	78	32	5	169
Ranunculus repens	L	3	1	—	80	80	80	15	5	257
Ranunculus repens	L	3	1	—	100	100	100	6.5	5	68
Ranunculus repens	L	3	1	>26	156	156	156	12	4	51
Ranunculus repens	L	3	1	>35	167	167	167	24	5	168
Ranunculus repens	L	3	1	>45	311	311	311	16	5	168
Ranunculus repens	L	3	1	>20	433	433	433	24	5	168

Species										
Ranunculus repens	L	3	3	—	200	550	433	20	5	4
Ranunculus repens	L	3	2	—	453	566	510	3	5	208
Ranunculus repens	L	3	1	—	513	513	513	20	5	107
Ranunculus repens	L	3	1	—	513	513	513	50	5	107
Ranunculus repens	L	3	5	>15	156	1872	655	6	5	99
Ranunculus repens	L	3	1	—	700	700	700	24	5	168
Ranunculus repens	L	3	1	>14	912	912	912	30	5	107
Ranunculus repens	L	3	6	—	1092	1092	1092	6	5	11
Ranunculus repens	L	3	1	—	417	3646	1250	20	5	97
Ranunculus repens	L	3	9	—	1600	1600	1600	30	5	66
Ranunculus repens	L	3	7	—	198	5151	1644	6	5	208
Ranunculus repens	L	3	4	—	260	10920	5170	10	5	27
Ranunculus repens	L	3	2	—	484	18245	5527	30	5	45
Ranunculus repens	L	3	2	—	832	13780	7306	6	5	11
Ranunculus repens	L	4	1	—	4	7	5	3	7	235
Ranunculus repens	L	4	2	—	10	10	10	25	4	36
Ranunculus repens	L	4	1	—	25	28	27	10	5	110
Ranunculus repens	L	4	1	—	28	28	28	20	5	111
Ranunculus repens	L	4	1	—	42	42	42	5	5	98
Ranunculus repens	L	4	1	—	63	63	63	5	5	244
Ranunculus repens	L	4	1	—	75	75	75	7	5	58
Ranunculus repens	L	4	1	—	80	80	80	30	4	13
Ranunculus repens	L	4	1	—	83	83	83	3	5	96
Ranunculus repens	L	4	3	—	140	260	193	6.5	5	128
Ranunculus repens	L	4	1	—	249	249	249	15	5	267
Ranunculus repens	L	4	9	—	71	1857	526	5	5	222
Ranunculus repens	L	4	1	—	571	571	571	25	5	260
Ranunculus repens	L	4	6	—	160	1008	600	5	5	69
Ranunculus repens	L	4	1	—	938	938	938	5	4	88
Ranunculus repens	L	4	13	—	263	2800	1061	10	5	255
Ranunculus repens	L	4	6	—	396	2038	1146	3	5	208
Ranunculus repens	L	4	2	—	1200	1600	1400	20	4	225
Ranunculus repens	L	4	2	—	2599	3119	2859	50	5	216
Ranunculus repens	L	4	13	—	312	8840	3108	6	5	99
Ranunculus repens	L	4	4	—	77	13881	5505	17.8	5	155
Ranunculus reptans	L	2	4	—	162	162	162	0.538	5	269
Ranunculus sceleratus	L	2	1	—	53	70	62	8	7	275
Ranunculus sceleratus	L	2	2	—	1	1	1	2	5	113
Ranunculus sceleratus	L	2	1	—	18	18	18	3	7	235
Ranunculus sceleratus	L	2	1	—	50	50	50	5	5	244
Ranunculus sceleratus	L	2	8	—	37	3892	686	5	5	177
Ranunculus sceleratus	L	3	1	>50	0	0	0	50	3	115
Ranunculus sceleratus	L	3	2	>5	0	0	0	7.5	2	199
Ranunculus sceleratus	L	3	1	>5	0	0	0	5	1	246
Ranunculus sceleratus	L	3	1	—	110	110	110	32	5	131
Ranunculus sceleratus	L	3	1	—	360	360	360	10	5	131
Ranunculus sceleratus	L	4	1	—	61	61	61	5	5	166

Species	Authority	Seed bank type	Number of records	Longevity (y)	Minimum density (seeds m^{-2})	Maximum density (seeds m^{-2})	Mean density (seeds m^{-2})	Depth (cm)	Method	Source code
Ranunculus sceleratus	L.	4	1	—	173	173	173	5	5	177
Ranunculus sceleratus	L.	4	2	—	712	1040	876	12	4	230
Ranunculus serpens	Schrank	1	2	—	0	0	0	5	5	69
Ranunculus serpens	Schrank	1	6	—	0	0	0	10	5	255
Ranunculus serpens	Schrank	4	1	—	38	38	38	10	5	255
Thalictrum alpinum	L.	4	1	—	30	30	30	20	5	191
Thalictrum aquilegiifolium	L.	1	1	—	0	0	0	5	6	170
Thalictrum flavum	L.	3	1	—	23	23	23	10	5	252
Trollius europaeus	L.	1	4	—	0	0	0	10	5	255
Resedaceae										
Reseda alba	—	2	1	—	30	30	30	10	4	22
Reseda lutea	L.	2	2	—	233	617	425	3	5	96
Reseda lutea	L.	2	1	—	456	456	456	12	5	81
Reseda lutea	L.	4	1	—	340	340	340	10	5	86
Reseda luteola	L.	2	1	—	313	313	313	10	5	110
Reseda luteola	L.	3	1	>5	0	0	0	7.5	2	195
Reseda luteola	L.	4	2	—	56	166	111	10	5	110
Rhamnaceae										
Frangula alnus	Miller	1	2	—	0	0	0	16	5	216
Frangula alnus	Miller	1	1	—	0	0	0	5	5	248
Frangula alnus	Miller	1	1	—	0	0	0	10	5	255
Rosaceae										
Agrimonia eupatoria	L.	1	1	—	0	0	0	6	5	67
Agrimonia eupatoria	L.	1	1	—	0	0	0	20	5	97
Agrimonia eupatoria	L.	1	2	—	0	0	0	6	5	181
Agrimonia eupatoria	L.	1	1	—	0	0	0	10	5	181
Agrimonia eupatoria	L.	1	1	—	0	0	0	6.5	7	182
Agrimonia eupatoria	L.	1	1	—	0	0	0	13	5	183
Agrimonia eupatoria	L.	1	2	—	0	0	0	10	5	255
Agrimonia eupatoria	L.	2	2	2	0	0	0	7.5	2	195
Alchemilla alpina	L.	1	3	—	0	0	0	20	5	97
Alchemilla vulgaris	0	1	1	—	0	0	0	6.5	5	68
Alchemilla vulgaris	0	1	26	—	0	0	0	6.5	5	128
Alchemilla vulgaris	0	1	1	—	0	0	0	10	5	255
Alchemilla vulgaris	0	3	1	—	60	60	60	13	5	68
Amelanchier lamarckii	F.-G. Schroeder	1	5	—	0	0	0	5	5	222
Amelanchier ovalis	Medicus	4	1	—	113	113	113	10	5	255
Aphanes arvensis	L.	1	3	—	0	0	0	25	4	36
Aphanes arvensis	L.	1	1	—	0	0	0	1	5	96

Aphanes arvensis	L.	1	1	—	0	0	0	3	5	96
Aphanes arvensis	L.	2	1	—	7	7	7	3	7	235
Aphanes arvensis	L.	2	1	—	24	24	24	25	4	36
Aphanes arvensis	L.	2	1	—	50	50	50	3	5	96
Aphanes arvensis	L.	2	1	—	151	151	151	17.8	5	155
Aphanes arvensis	L.	2	1	—	256	256	256	5	5	69
Aphanes arvensis	L.	2	2	—	154	469	312	10	5	110
Aphanes arvensis	L.	2	2	—	150	975	400	10	5	255
Aphanes arvensis	L.	2	4	—	486	486	486	50	5	260
Aphanes arvensis	L.	3	1	—	7	7	7	20	5	40
Aphanes arvensis	L.	3	1	>58	43	43	43	30	5	29
Aphanes arvensis	L.	3	2	—	32	54	43	30	5	45
Aphanes arvensis	L.	3	2	>20	33	100	67	24	5	168
Aphanes arvensis	L.	3	1	>40	86	86	86	30	5	45
Aphanes arvensis	L.	3	1	>8	484	484	484	30	5	45
Aphanes arvensis	L.	3	1	>6	1109	1109	1109	30	5	29
Aphanes arvensis	L.	4	1	—	2756	2756	2756	30	4	36
Aphanes arvensis	L.	4	4	—	12	32	18	25	5	15
Aphanes arvensis	L.	4	2	—	42	63	53	10	5	110
Aphanes arvensis	L.	4	1	—	91	91	91	10	5	260
Aphanes arvensis	L.	4	1	—	242	242	242	25	4	111
Aphanes arvensis	L.	4	1	—	289	289	289	20	5	255
Aphanes arvensis	L.	4	1	—	400	400	400	10	4	13
Aphanes arvensis	L.	4	15	—	80	5847	1376	30	5	207
Aphanes arvensis	L.	4	1	—	1951	1951	1951	15	4	31
Aphanes arvensis	L.	4	2	—	54	5592	2823	15	4	30
Aphanes arvensis	L.	4	7	—	67	6409	2857	15	5	132
Aphanes inexspectata	Lippert	4	4	—	413	1200	744	3	5	181
Crataegus laevigata	(Poiret) DC.	1	3	—	0	0	0	6	5	181
Crataegus laevigata	(Poiret) DC.	1	1	—	0	0	0	10	5	255
Crataegus laevigata	(Poiret) DC.	1	4	—	0	0	0	10	5	58
Crataegus monogyna	Jacq.	1	1	—	0	0	0	7	5	96
Crataegus monogyna	Jacq.	1	1	—	0	0	0	1	5	96
Crataegus monogyna	Jacq.	1	4	—	0	0	0	3	5	110
Crataegus monogyna	Jacq.	1	2	—	0	0	0	10	5	225
Crataegus monogyna	Jacq.	1	1	—	0	0	0	20	4	235
Crataegus monogyna	Jacq.	1	1	—	0	0	0	3	7	255
Crataegus monogyna	Jacq.	1	4	—	0	0	0	10	5	257
Dryas octopetala	L.	1	2	—	0	0	0	15	5	55
Dryas octopetala	L.	1	1	—	0	0	0	5	5	191
Dryas octopetala	L.	4	2	—	10	10	10	20	5	71
Filipendula ulmaria	(L.) Maxim.	1	2	—	0	0	0	6.5	5	68
Filipendula ulmaria	(L.) Maxim.	1	2	—	0	0	0	13	5	68
Filipendula ulmaria	(L.) Maxim.	1	1	—	0	0	0	5	4	88
Filipendula ulmaria	(L.) Maxim.	1	1	—	0	0	0	3	5	96
Filipendula ulmaria	(L.) Maxim.	1	4	—	0	0	0	6	5	99

Species	Authority	Seed bank type	Number of records	Longevity (y)	Minimum density (seeds m⁻²)	Maximum density (seeds m⁻²)	Mean density (seeds m⁻²)	Depth (cm)	Method	Source code
Filipendula ulmaria	(L.) Maxim.	1	5	—	0	0	0	6.5	5	128
Filipendula ulmaria	(L.) Maxim.	1	8	—	0	0	0	10	5	149
Filipendula ulmaria	(L.) Maxim.	1	5	—	0	0	0	10	5	150
Filipendula ulmaria	(L.) Maxim.	1	3	—	0	0	0	10	5	152
Filipendula ulmaria	(L.) Maxim.	1	4	0	0	0	0	12	5	174
Filipendula ulmaria	(L.) Maxim.	1	1	—	0	0	0	10	3	216
Filipendula ulmaria	(L.) Maxim.	1	8	—	0	0	0	20	4	225
Filipendula ulmaria	(L.) Maxim.	1	1	—	0	0	0	10	5	252
Filipendula ulmaria	(L.) Maxim.	1	10	—	0	0	0	10	5	255
Filipendula ulmaria	(L.) Maxim.	1	1	—	0	0	0	15	5	257
Filipendula ulmaria	(L.) Maxim.	1	2	—	3	3	3	3	7	235
Filipendula ulmaria	(L.) Maxim.	1	5	—	0	91	18	2	7	99
Filipendula ulmaria	(L.) Maxim.	1	5	—	142	142	142	6	5	208
Filipendula ulmaria	(L.) Maxim.	1	7	—	198	226	212	3	5	208
Filipendula ulmaria	(L.) Maxim.	1	1	1	250	250	250	8	7	275
Filipendula ulmaria	(L.) Maxim.	2	1	>1	0	0	0	10	3	216
Filipendula ulmaria	(L.) Maxim.	2	2	1	0	0	0	—	1	275
Filipendula ulmaria	(L.) Maxim.	2	3	—	0	0	0	—	1	275
Filipendula ulmaria	(L.) Maxim.	2	1	—	226	226	226	6	5	208
Filipendula ulmaria	(L.) Maxim.	2	1	—	260	260	260	6	5	99
Filipendula ulmaria	(L.) Maxim.	2	1	—	509	509	509	3	5	208
Filipendula ulmaria	(L.) Maxim.	4	2	—	19	37	28	12	5	174
Filipendula ulmaria	(L.) Maxim.	4	3	—	50	125	92	10	5	255
Filipendula ulmaria	(L.) Maxim.	4	1	—	103	103	103	16	5	216
Filipendula ulmaria	(L.) Maxim.	4	1	—	146	146	146	10	5	150
Filipendula ulmaria	(L.) Maxim.	4	4	—	85	679	248	3	5	208
Filipendula ulmaria	(L.) Maxim.	4	1	—	780	780	780	50	5	216
Filipendula ulmaria	(L.) Maxim.	4	2	—	50	1750	900	3	5	96
Filipendula ulmaria	(L.) Maxim.	4	1	—	1404	1404	1404	10	5	152
Filipendula vulgaris	Moench	1	9	—	0	0	0	10	5	150
Filipendula vulgaris	Moench	1	19	—	0	0	0	10	5	152
Filipendula vulgaris	Moench	1	1	—	0	0	0	3	7	235
Filipendula vulgaris	Moench	1	1	—	0	0	0	8.5	4	254
Fragaria vesca	L.	1	2	—	0	0	0	5	5	19
Fragaria vesca	L.	1	2	—	0	0	0	3	5	96
Fragaria vesca	L.	1	1	—	0	0	0	12.5	5	126
Fragaria vesca	L.	1	1	—	0	0	0	10	5	150
Fragaria vesca	L.	1	6	—	0	0	0	10	5	152
Fragaria vesca	L.	1	2	—	0	0	0	5	6	170
Fragaria vesca	L.	1	1	—	0	0	0	12	5	171
Fragaria vesca	L.	1	2	—	0	0	0	10	5	184
Fragaria vesca	L.	1	1	—	0	0	0	16	5	216

Species										
Fragaria vesca	L.	1	1	—	24	24	24	10	5	181
Fragaria vesca	L.	2	2	—	0	0	0	5	6	170
Fragaria vesca	L.	2	1	—	1	1	1	15	5	32
Fragaria vesca	L.	2	1	—	8	8	8	10	4	127
Fragaria vesca	L.	2	1	—	50	50	50	3	5	96
Fragaria vesca	L.	2	3	—	33	133	74	16	5	168
Fragaria vesca	L.	3	1	—	111	111	111	24	5	168
Fragaria vesca	L.	3	1	—	388	388	388	20	5	67
Fragaria vesca	L.	4	1	—	17	17	17	7	5	58
Fragaria vesca	L.	4	2	—	28	127	78	5	5	122
Fragaria vesca	L.	4	5	—	150	433	293	3	5	96
Fragaria viridis	Duchesne	1	4	—	0	0	0	10	5	255
Fragaria viridis	Duchesne	4	1	—	4082	4082	4082	8.5	4	254
Geum montanum	L.	1	1	—	0	0	0	5	5	55
Geum montanum	L.	1	2	—	0	0	0	5	4	91
Geum rivale	L.	1	1	—	0	0	0	6.5	5	68
Geum rivale	L.	1	1	—	0	0	0	13	5	68
Geum rivale	L.	1	4	—	0	0	0	6.5	5	128
Geum rivale	L.	1	7	—	0	0	0	10	5	150
Geum rivale	L.	1	5	—	0	0	0	10	5	152
Geum rivale	L.	1	1	—	0	0	0	30	5	173
Geum rivale	L.	1	4	—	0	0	0	10	5	255
Geum urbanum	L.	1	3	—	0	0	0	6	5	49
Geum urbanum	L.	1	1	—	0	0	0	7	5	58
Geum urbanum	L.	1	2	—	0	0	0	20	5	67
Geum urbanum	L.	1	2	—	0	0	0	12.5	5	126
Geum urbanum	L.	1	2	—	0	0	0	4	5	176
Geum urbanum	L.	1	1	—	0	0	0	20	4	225
Geum urbanum	L.	1	1	—	0	0	0	10	5	255
Geum urbanum	L.	2	1	3	0	0	0	15	5	257
Geum urbanum	L.	2	1	—	0	0	0	7.5	2	195
Geum urbanum	L.	4	1	—	60	60	60	15	5	257
Geum urbanum	L.	4	1	—	4	4	4	20	4	111
Geum urbanum	L.	4	1	—	50	50	50	3	5	96
Potentilla anserina	L.	1	1	—	0	0	0	3	5	111
Potentilla anserina	L.	1	1	—	0	0	0	20	4	111
Potentilla anserina	L.	1	1	—	0	0	0	20	5	112
Potentilla anserina	L.	1	2	—	0	0	0	7	5	128
Potentilla anserina	L.	1	4	—	0	0	0	6.5	5	151
Potentilla anserina	L.	1	1	—	0	0	0	6	5	208
Potentilla anserina	L.	1	3	—	0	0	0	3	5	208
Potentilla anserina	L.	1	4	—	260	260	260	6	5	11
Potentilla anserina	L.	1	1	—	37	37	37	6	5	177
Potentilla anserina	L.	2	1	—	142	283	213	5	5	208
Potentilla anserina	L.	2	2	—	513	513	513	20	5	107
Potentilla anserina	L.	2	1	—	1456	1456	1456	6	5	11

Species	Authority	Seed bank type	Number of records	Longevity (y)	Minimum density (seeds m^{-2})	Maximum density (seeds m^{-2})	Mean density (seeds m^{-2})	Depth (cm)	Method	Source code
Potentilla anserina	L.	3	1	—	285	285	285	30	5	107
Potentilla anserina	L.	4	1	—	133	133	133	6	5	151
Potentilla argentea	L.	1	1	—	0	0	0	11.5	5	134
Potentilla argentea	L.	3	1	—	99	99	99	11.5	5	134
Potentilla argentea	L.	3	1	>40	1260	1260	1260	20	5	173
Potentilla argentea	L.	4	1	—	28	28	28	20	4	111
Potentilla argentea	L.	4	1	—	2400	2400	2400	15	5	249
Potentilla argentea	L.	4	1	—	9286	9286	9286	8.5	4	254
Potentilla aurea	L.	1	1	—	0	0	0	5	5	55
Potentilla aurea	L.	1	2	—	0	0	0	5	4	91
Potentilla crantzii	(Crantz) G. Beck ex Fritsch	1	1	—	0	0	0	5	5	55
Potentilla erecta	(L.) Rausch.	1	1	—	0	0	0	6	5	11
Potentilla erecta	(L.) Rausch.	1	1	—	0	0	0	40	5	11
Potentilla erecta	(L.) Rausch.	1	2	—	0	0	0	20	5	67
Potentilla erecta	(L.) Rausch.	1	1	—	0	0	0	5	5	69
Potentilla erecta	(L.) Rausch.	1	2	—	0	0	0	9	5	84
Potentilla erecta	(L.) Rausch.	1	2	—	0	0	0	5	5	92
Potentilla erecta	(L.) Rausch.	1	2	—	0	0	0	10	5	150
Potentilla erecta	(L.) Rausch.	1	3	—	0	0	0	10	5	152
Potentilla erecta	(L.) Rausch.	1	1	—	0	0	0	5	6	170
Potentilla erecta	(L.) Rausch.	1	1	—	0	0	0	12	5	171
Potentilla erecta	(L.) Rausch.	1	1	—	0	0	0	12	5	174
Potentilla erecta	(L.) Rausch.	1	10	—	0	0	0	10	5	255
Potentilla erecta	(L.) Rausch.	1	1	—	0	0	0	15	5	257
Potentilla erecta	(L.) Rausch.	1	1	—	10	10	10	3	7	235
Potentilla erecta	(L.) Rausch.	2	2	—	1	2	2	5	6	170
Potentilla erecta	(L.) Rausch.	2	1	—	13	13	13	7	5	232
Potentilla erecta	(L.) Rausch.	2	1	—	33	33	33	5	5	123
Potentilla erecta	(L.) Rausch.	2	2	—	18	51	35	3	7	235
Potentilla erecta	(L.) Rausch.	2	1	—	88	88	88	10	5	150
Potentilla erecta	(L.) Rausch.	2	1	—	108	108	108	30	5	45
Potentilla erecta	(L.) Rausch.	2	1	—	150	150	150	3	5	96
Potentilla erecta	(L.) Rausch.	2	2	—	234	312	273	10	5	152
Potentilla erecta	(L.) Rausch.	2	2	—	183	377	280	22.5	5	45
Potentilla erecta	(L.) Rausch.	2	1	—	291	291	291	15	5	45
Potentilla erecta	(L.) Rausch.	3	1	>22	33	33	33	24	5	168
Potentilla erecta	(L.) Rausch.	3	1	>32	33	33	33	24	5	168
Potentilla erecta	(L.) Rausch.	3	2	>20	61	61	61	5	5	222
Potentilla erecta	(L.) Rausch.	3	1	>32	100	100	100	16	5	168
Potentilla erecta	(L.) Rausch.	3	1	—	129	129	129	30	5	45
Potentilla erecta	(L.) Rausch.	3	1	—	220	220	220	15	5	257
Potentilla erecta	(L.) Rausch.	3	1	>100	822	822	822	32	5	169

Potentilla erecta	(L.) Rausch.	4	1	—	3	3	3	3	7	235
Potentilla erecta	(L.) Rausch.	4	1	—	41	41	41	5	5	233
Potentilla erecta	(L.) Rausch.	4	1	—	82	82	82	7	5	232
Potentilla erecta	(L.) Rausch.	4	5	—	61	133	90	5	5	222
Potentilla erecta	(L.) Rausch.	4	10	—	38	463	165	10	5	255
Potentilla erecta	(L.) Rausch.	4	4	—	97	295	182	5	5	92
Potentilla erecta	(L.) Rausch.	4	2	—	234	234	234	10	5	150
Potentilla erecta	(L.) Rausch.	4	1	—	234	234	234	10	5	152
Potentilla erecta	(L.) Rausch.	4	5	—	25	941	263	12	5	174
Potentilla erecta	(L.) Rausch.	4	1	—	316	316	316	5	5	123
Potentilla erecta	(L.) Rausch.	4	5	—	177	758	354	17.8	5	156
Potentilla erecta	(L.) Rausch.	4	2	—	167	883	525	3	5	96
Potentilla erecta	(L.) Rausch.	4	7	—	553	1363	835	17.8	5	155
Potentilla neumanniana	Reichb.	1	1	—	0	0	0	5	5	69
Potentilla neumanniana	Reichb.	1	1	—	0	0	0	1	5	96
Potentilla neumanniana	Reichb.	1	1	—	22	22	22	6.5	7	182
Potentilla neumanniana	Reichb.	1	1	—	120	120	120	13	5	183
Potentilla neumanniana	Reichb.	2	1	—	216	216	216	6.5	7	182
Potentilla neumanniana	Reichb.	3	1	>30	96	96	96	10	5	181
Potentilla neumanniana	Reichb.	4	3	—	24	160	101	6	5	181
Potentilla neumanniana	Reichb.	4	1	—	108	108	108	6	5	67
Potentilla neumanniana	Reichb.	4	3	—	38	563	292	10	5	255
Potentilla norvegica	L.	2	1	>2	0	0	0	2	1	47
Potentilla norvegica	L.	2	1	>2	0	0	0	15	1	47
Potentilla norvegica	L.	2	1	—	59	59	59	5	5	177
Potentilla norvegica	L.	2	3	>30	37	322	136	11.5	5	134
Potentilla norvegica	L.	3	1	>39	0	0	0	15	1	240
Potentilla norvegica	L.	3	1	>39	0	0	0	50	1	240
Potentilla norvegica	L.	3	1	—	0	0	0	100	1	240
Potentilla norvegica	L.	3	2	>40	37	62	50	11.5	5	134
Potentilla norvegica	L.	3	1	—	1800	1800	1800	20	5	173
Potentilla norvegica	L.	4	3	—	17	102	52	10	5	8
Potentilla norvegica	L.	4	4	—	65	205	151	3	5	59
Potentilla norvegica	L.	4	1	—	854	854	854	11.5	5	134
Potentilla palustris	(L.) Scop.	1	3	—	0	0	0	3	5	208
Potentilla palustris	(L.) Scop.	1	3	—	0	0	0	6	5	208
Potentilla palustris	(L.) Scop.	3	1	>4	0	0	0	50	5	216
Potentilla palustris	(L.) Scop.	4	1	—	0	0	0	10	3	216
Potentilla palustris	(L.) Scop.	4	1	—	50	50	50	10	5	255
Potentilla palustris	(L.) Scop.	4	2	—	72	351	212	16	5	216
Potentilla recta	L.	2	1	>2	0	0	0	7	1	18
Potentilla reptans	L.	1	2	—	0	0	0	5	4	53
Potentilla reptans	L.	1	1	—	0	0	0	5	5	69
Potentilla reptans	L.	1	2	—	0	0	0	10	4	88
Potentilla reptans	L.	1	3	—	0	0	0	3	5	96
Potentilla reptans	L.	1	3	—	0	0	0	6.5	5	128

Species	Authority	Seed bank type	Number of records	Longevity (y)	Minimum density (seeds m^{-2})	Maximum density (seeds m^{-2})	Mean density (seeds m^{-2})	Depth (cm)	Method	Source code
Potentilla reptans	L.	1	1	—	0	0	0	10	5	150
Potentilla reptans	L.	1	1	—	0	0	0	5	5	244
Potentilla reptans	L.	1	1	—	0	0	0	5	5	250
Potentilla reptans	L.	1	5	—	0	0	0	5	5	251
Potentilla reptans	L.	1	2	—	0	0	0	10	5	252
Potentilla reptans	L.	2	1	—	7	7	7	15	5	62
Potentilla reptans	L.	2	1	—	65	65	65	30	5	45
Potentilla reptans	L.	2	1	—	130	130	130	10	5	162
Potentilla reptans	L.	2	2	—	180	2096	1138	18	5	162
Potentilla reptans	L.	2	1	—	1311	1311	1311	12	5	162
Potentilla reptans	L.	2	1	—	1750	1750	1750	15	5	162
Potentilla reptans	L.	3	2	—	32	75	54	30	5	45
Potentilla reptans	L.	3	1	>35	323	323	323	65	5	161
Potentilla reptans	L.	4	1	—	5	5	5	15	5	62
Potentilla reptans	L.	4	1	—	7	7	7	20	4	111
Potentilla sterilis	(L.) Garcke	4	1	—	35	35	35	7	5	58
Potentilla sterilis	(L.) Garcke	1	3	—	3	3	3	3	7	235
Potentilla sterilis	(L.) Garcke	1	1	—	40	40	40	20	5	219
Potentilla sterilis	(L.) Garcke	2	1	—	3	3	3	15	5	32
Potentilla sterilis	(L.) Garcke	2	1	—	40	40	40	25	5	219
Potentilla sterilis	(L.) Garcke	3	1	—	43	43	43	30	5	29
Potentilla sterilis	(L.) Garcke	1	1	—	100	100	100	15	5	257
Prunus avium	(L.) L.	1	1	—	0	0	0	6	5	67
Prunus avium	(L.) L.	1	1	—	0	0	0	20	5	67
Prunus avium	(L.) L.	1	1	—	0	0	0	13	5	183
Prunus domestica	L.	1	2	—	0	0	0	10	5	255
Prunus fruticosa	Pallas	4	3	—	0	0	0	10	5	255
Prunus padus	L.	2	1	1	200	200	200	25	5	260
Prunus serotina	Ehrh.	1	5	—	0	0	0	2	3	83
Prunus serotina	Ehrh.	1	1	—	0	0	0	5	5	19
Prunus serotina	Ehrh.	2	1	>3	0	0	0	10	5	33
Prunus serotina	Ehrh.	4	1	>3	0	0	0	10	5	141
Prunus serotina	Ehrh.	1	1	—	0	0	0	0	3	263
Prunus serotina	Ehrh.	1	2	—	74	74	74	10	5	141
Prunus spinosa	L.	1	1	—	0	0	0	6	5	49
Prunus spinosa	L.	1	2	—	0	0	0	6	5	67
Prunus spinosa	L.	1	3	—	0	0	0	6	5	181
Prunus spinosa	L.	1	1	—	0	0	0	10	5	181
Prunus spinosa	L.	1	1	—	0	0	0	13	5	183
Prunus spinosa	L.	1	1	—	0	0	0	3	7	235
Prunus spinosa	L.	1	6	—	0	0	0	10	5	255
Prunus spinosa	L.	1	2	—	0	0	0	15	5	257

Species	Authority									
Rosa canina	L.	1	1	—	0	0	0	6	5	67
Rosa canina	L.	1	1	—	0	0	0	6	5	181
Rosa canina	L.	1	1	—	0	0	0	6.5	7	182
Rosa canina	L.	1	5	—	0	0	0	10	5	255
Rosa canina	L.	1	2	—	0	0	0	15	5	257
Rosa rubiginosa	L.	1	3	—	0	0	0	6	5	181
Rosa rubiginosa	L.	1	1	—	0	0	0	10	5	181
Rosa tomentosa	Smith	1	1	—	0	0	0	10	5	255
Rubus caesius	L.	1	1	—	0	0	0	12.5	5	126
Rubus chamaemorus	L.	1	1	—	0	0	0	25	5	144
Rubus fruticosus	—	1	3	—	0	0	0	7	5	58
Rubus fruticosus	—	1	1	—	0	0	0	3	5	96
Rubus fruticosus	—	1	3	—	0	0	0	10	5	110
Rubus fruticosus	—	1	8	—	0	0	0	20	4	225
Rubus fruticosus	—	1	2	—	0	0	0	10	5	255
Rubus fruticosus	—	1	1	—	0	4	2	3	7	235
Rubus fruticosus	—	1	2	—	48	48	48	10	5	181
Rubus fruticosus	—	1	1	—	53	53	53	8	7	275
Rubus fruticosus	—	1	1	—	100	100	100	15	5	257
Rubus fruticosus	—	2	2	—	106	106	106	12	4	51
Rubus fruticosus	—	2	2	—	6	6	6	5	5	122
Rubus fruticosus	—	2	2	—	155	192	174	10	4	51
Rubus fruticosus	—	3	10	—	140	1440	684	15	5	257
Rubus fruticosus	—	3	2	—	812	812	812	16	4	51
Rubus fruticosus	—	3	1	—	6	6	6	5	5	234
Rubus fruticosus	—	3	1	—	137	137	137	10	5	93
Rubus fruticosus	—	4	2	—	29	245	137	12.5	5	126
Rubus fruticosus	—	4	9	—	120	980	356	15	5	257
Rubus fruticosus	—	4	1	—	14	14	14	5	5	122
Rubus fruticosus	—	4	3	—	17	40	29	7	5	58
Rubus fruticosus	—	4	1	—	66	66	66	10	5	126
Rubus fruticosus	—	4	4	—	10	333	118	6	5	49
Rubus fruticosus	—	1	1	—	153	153	153	15	5	142
Rubus fruticosus	—	1	1	—	184	184	184	15	5	32
Rubus idaeus	L.	1	2	—	0	0	0	7.5	5	84
Rubus idaeus	L.	1	2	—	0	0	0	5	4	91
Rubus idaeus	L.	1	1	—	0	0	0	3	5	96
Rubus idaeus	L.	1	1	—	0	0	0	10	5	150
Rubus idaeus	L.	1	1	—	0	0	0	5	6	170
Rubus idaeus	L.	1	1	—	0	0	0	5	5	213
Rubus idaeus	L.	1	1	—	0	0	0	16	5	216
Rubus idaeus	L.	1	1	—	0	0	0	20	4	225
Rubus idaeus	L.	1	8	—	43	43	43	10	5	255
Rubus idaeus	L.	1	1	—	4	4	4	10	5	84
Rubus idaeus	L.	2	1	—	0	0	0	15	5	32
Rubus idaeus	L.	2	4	—	0	51	17	5	6	170
Rubus idaeus	L.	2	1	—	20	20	20	2	5	82

Species	Authority	Seed bank type	Number of records	Longevity (y)	Minimum density (seeds m^{-2})	Maximum density (seeds m^{-2})	Mean density (seeds m^{-2})	Depth (cm)	Method	Source code
Rubus idaeus	L.	2	1	—	29	29	29	5	5	220
Rubus idaeus	L.	2	1	—	33	33	33	32	5	169
Rubus idaeus	L.	2	1	—	44	44	44	24	5	169
Rubus idaeus	L.	2	1	—	54	54	54	30	5	29
Rubus idaeus	L.	2	1	—	56	56	56	16	5	169
Rubus idaeus	L.	2	11	—	38	438	140	10	5	255
Rubus idaeus	L.	2	1	—	312	312	312	10	5	152
Rubus idaeus	L.	2	1	—	375	375	375	12	5	171
Rubus idaeus	L.	3	1	>5	0	0	0	2	3	83
Rubus idaeus	L.	3	3	—	29	113	62	0	5	157
Rubus idaeus	L.	3	1	>87	68	68	68	5	5	80
Rubus idaeus	L.	3	2	—	33	244	139	16	5	168
Rubus idaeus	L.	3	2	—	44	278	161	32	5	169
Rubus idaeus	L.	3	1	>30	286	286	286	5	5	80
Rubus idaeus	L.	3	1	—	444	444	444	20	5	67
Rubus idaeus	L.	3	1	—	794	794	794	12.5	5	126
Rubus idaeus	L.	3	1	—	982	982	982	10	5	126
Rubus idaeus	L.	4	1	—	4	4	4	20	4	111
Rubus idaeus	L.	4	2	—	3	29	16	4	5	176
Rubus idaeus	L.	4	1	—	44	44	44	25	5	101
Rubus idaeus	L.	4	2	—	27	117	72	0	5	157
Rubus idaeus	L.	4	5	—	79	426	198	5	5	220
Rubus idaeus	L.	4	6	—	38	1025	246	10	5	255
Rubus idaeus	L.	4	1	—	310	310	310	12.5	5	126
Rubus idaeus	L.	4	1	—	404	404	404	10	5	126
Rubus idaeus	L.	4	1	—	1016	1016	1016	5	5	80
Rubus idaeus	L.	4	2	—	292	2718	1505	10	5	150
Rubus saxatilis	L.	1	1	—	0	0	0	5	5	82
Rubus saxatilis	L.	1	1	—	0	0	0	10	5	150
Rubus saxatilis	L.	1	5	—	0	0	0	5	6	170
Rubus saxatilis	L.	1	2	—	0	0	0	12	5	171
Sanguisorba minor	Scop.	1	2	—	0	0	0	4	5	176
Sanguisorba minor	Scop.	1	1	—	0	0	0	5	5	69
Sanguisorba minor	Scop.	1	1	—	0	0	0	1	5	96
Sanguisorba minor	Scop.	1	1	—	0	0	0	13	5	183
Sanguisorba minor	Scop.	1	1	—	0	0	0	10	5	255
Sanguisorba minor	Scop.	1	1	—	10	10	10	6.5	7	182
Sanguisorba minor	Scop.	2	1	—	46	46	46	6.5	7	182
Sanguisorba minor	Scop.	2	1	—	1146	1146	1146	20	5	97
Sanguisorba minor	Scop.	3	1	>30	24	24	24	10	5	181
Sanguisorba minor	Scop.	4	1	—	5	5	5	5	5	122
Sanguisorba minor	Scop.	4	1	—	40	40	40	7	5	58

Species	Authority									
Sanguisorba minor	Scop.	4	2	—	32	56	44	6	5	181
Sanguisorba minor	Scop.	4	1	—	75	75	75	10	5	255
Sanguisorba minor	Scop.	4	1	—	85	85	85	10	5	110
Sanguisorba officinalis	L.	1	2	—	0	0	0	6.5	5	68
Sanguisorba officinalis	L.	1	2	—	0	0	0	13	5	68
Sanguisorba officinalis	L.	1	1	—	0	0	0	6.5	5	128
Sanguisorba officinalis	L.	1	7	—	0	0	0	10	5	149
Sanguisorba officinalis	L.	1	19	—	0	0	0	10	5	255
Sanguisorba officinalis	L.	1	2	—	0	96	48	10	5	252
Sibbaldia procumbens	L.	1	2	—	0	0	0	6	5	39
Sibbaldia procumbens	L.	4	2	—	20	33	27	6	5	39
Sorbus aria	(L.) Crantz	1	1	—	0	0	0	10	5	255
Sorbus aucuparia	L.	1	2	—	0	0	0	5	5	82
Sorbus aucuparia	L.	1	2	—	0	0	0	16	5	216
Sorbus aucuparia	L.	1	1	—	0	0	0	20	4	225
Sorbus aucuparia	L.	1	1	—	0	0	0	5	5	248
Sorbus aucuparia	L.	1	7	—	0	0	0	10	5	255
Sorbus aucuparia	L.	1	9	—	0	0	0	15	5	257
Sorbus aucuparia	L.	3	1	>5	0	0	0	2	3	83

Rubiaceae

Species	Authority									
Asperula cynanchia	—	1	1	—	0	0	0	5	5	122
Asperula cynanchia	L.	1	1	—	0	0	0	6	5	181
Asperula cynanchia	L.	2	2	—	8	43	26	6.5	7	182
Asperula cynanchia	L.	4	1	—	1020	1020	1020	8.5	4	254
Cruciata laevipes	Opiz	1	1	—	0	0	0	3	5	96
Cruciata laevipes	Opiz	1	2	—	0	0	0	10	5	110
Cruciata laevipes	Opiz	1	1	—	0	0	0	3	7	235
Galium aparine	L.	1	4	—	0	0	0	25	4	36
Galium aparine	L.	1	2	—	0	0	0	7	5	58
Galium aparine	L.	1	1	—	0	0	0	50	5	66
Galium aparine	L.	1	2	—	0	0	0	20	5	67
Galium aparine	L.	1	1	0	0	0	0	0	1	75
Galium aparine	L.	1	1	0	0	0	0	7.5	1	75
Galium aparine	L.	1	6	—	0	0	0	3	5	96
Galium aparine	L.	1	1	—	0	0	0	10	5	126
Galium aparine	L.	1	1	—	0	0	0	3	5	132
Galium aparine	L.	1	1	—	0	0	0	13	5	183
Galium aparine	L.	1	2	—	0	0	0	25	4	185
Galium aparine	L.	1	3	—	0	0	0	3	5	208
Galium aparine	L.	1	2	—	0	0	0	6	5	208
Galium aparine	L.	1	5	—	0	0	0	20	4	225
Galium aparine	L.	1	1	—	0	0	0	10	5	252
Galium aparine	L.	1	6	—	0	0	0	10	5	255
Galium aparine	L.	1	2	—	0	0	0	15	5	257
Galium aparine	L.	1	2	—	0	10	5	3	7	235
Galium aparine	L.	1	9	—	53	237	91	8	7	275

Species	Authority	Seed bank type	Number of records	Longevity (y)	Minimum density (seeds m⁻²)	Maximum density (seeds m⁻²)	Mean density (seeds m⁻²)	Depth (cm)	Method	Source code
Galium aparine	L	1	1	—	667	667	667	50	5	260
Galium aparine	L	2	1	2	0	0	0	30	3	14
Galium aparine	L	2	1	3	0	0	0	25	3	209
Galium aparine	L	2	1	>3	0	0	0	20	3	268
Galium aparine	L	2	3	>1	0	0	0	—	1	275
Galium aparine	L	2	2	1	0	0	0	—	1	275
Galium aparine	L	2	1	—	85	85	85	6	5	208
Galium aparine	L	2	2	—	92	96	94	8	7	275
Galium aparine	L	2	5	—	38	163	95	10	5	255
Galium aparine	L	2	1	—	172	172	172	40	5	219
Galium aparine	L	4	1	—	22	22	22	25	4	36
Galium aparine	L	4	6	—	20	53	33	30	4	13
Galium aparine	L	4	1	—	76	76	76	15	5	142
Galium aparine	L	4	7	—	6	241	109	15	4	30
Galium aparine	L	4	1	—	134	134	134	6	5	49
Galium aparine	L	4	7	—	75	375	197	10	5	255
Galium aparine	L	4	1	—	220	220	220	10	5	15
Galium aparine	L	4	2	—	100	400	220	3	5	96
Galium aparine	L	4	2	—	83	417	250	10	5	184
Galium aparine	L	4	1	—	260	260	250	10	5	140
Galium aparine	L	1	1	—	789	789	260	8	7	275
Galium boreale	L	1	7	—	0	0	789	10	5	150
Galium boreale	L	1	17	—	0	0	0	10	5	152
Galium boreale	L	1	5	—	0	0	0	10	5	255
Galium boreale	L	4	1	—	4	4	0	10	5	33
Galium boreale	L	4	1	—	38	38	4	10	5	255
Galium boreale	L	4	2	—	50	133	38	10	5	184
Galium mollugo	L	1	1	—	0	0	92	13	5	68
Galium mollugo	L	1	1	—	0	0	0	5	5	122
Galium mollugo	L	1	2	—	0	0	0	6.5	5	128
Galium mollugo	L	1	1	—	0	0	0	10	5	150
Galium mollugo	L	1	3	—	0	0	0	5	6	170
Galium mollugo	L	1	1	—	0	0	0	12	5	171
Galium mollugo	L	1	5	—	0	0	0	12	5	174
Galium mollugo	L	1	3	—	0	0	0	20	4	225
Galium mollugo	L	1	18	—	0	0	0	10	5	255
Galium mollugo	L	1	1	—	50	50	50	6.5	5	68
Galium mollugo	L	1	1	—	223	223	223	6.5	7	182
Galium mollugo	L	2	1	3	0	0	0	7.5	2	195
Galium mollugo	L	2	1	—	8	8	8	25	4	36
Galium mollugo	L	2	1	—	128	128	128	6	5	181
Galium mollugo	L	4	1	—	5	5	5	5	5	122

Species	Author									
Galium mollugo	L.	4	4	—	60	200	125	6.5	5	128
Galium mollugo	L.	4	8	—	38	2025	311	10	5	255
Galium odoratum	(L.) Scop.	1	4	—	0	0	0	6	5	49
Galium odoratum	(L.) Scop.	1	2	—	0	0	0	20	5	67
Galium odoratum	(L.) Scop.	1	1	—	0	0	0	3	5	96
Galium odoratum	(L.) Scop.	1	2	—	0	0	0	10	5	126
Galium odoratum	(L.) Scop.	1	1	—	0	0	0	12.5	5	126
Galium odoratum	(L.) Scop.	1	2	—	0	0	0	30	5	173
Galium odoratum	(L.) Scop.	1	1	—	0	0	0	4	5	176
Galium odoratum	(L.) Scop.	1	1	—	0	0	0	5	5	220
Galium odoratum	(L.) Scop.	1	1	—	0	0	0	20	4	225
Galium odoratum	(L.) Scop.	1	1	—	0	0	0	15	5	257
Galium palustre	L.	1	5	—	0	0	0	10	5	27
Galium palustre	L.	1	2	—	0	0	0	3	5	96
Galium palustre	L.	1	1	—	0	0	0	20	4	111
Galium palustre	L.	1	1	—	0	0	0	20	5	111
Galium palustre	L.	1	2	—	0	0	0	6.5	5	128
Galium palustre	L.	1	2	—	0	0	0	6	5	151
Galium palustre	L.	1	4	—	0	0	0	3	5	208
Galium palustre	L.	1	5	—	0	0	0	20	4	225
Galium palustre	L.	1	2	—	0	0	0	10	5	255
Galium palustre	L.	1	2	—	52	52	52	2	7	99
Galium palustre	L.	1	2	—	85	566	326	6	5	208
Galium palustre	L.	2	2	—	79	90	84	3	7	235
Galium palustre	L.	2	1	—	170	170	170	6	5	208
Galium palustre	L.	3	4	—	113	425	212	6	5	208
Galium palustre	L.	3	1	—	1026	1026	1026	30	5	107
Galium palustre	L.	3	2	—	572	6344	3458	10	5	27
Galium palustre	L.	4	1	—	6	6	6	5	5	213
Galium palustre	L.	4	2	—	75	100	88	10	5	255
Galium palustre	L.	4	2	—	36	145	91	16	5	216
Galium palustre	L.	4	1	—	117	117	117	3	5	96
Galium palustre	L.	4	3	—	25	300	136	5	5	98
Galium palustre	L.	4	2	—	151	529	340	17.8	5	155
Galium palustre	L.	4	5	—	198	764	504	3	5	208
Galium parisiense	L.	1	2	—	0	0	0	3	5	132
Galium parisiense	L.	4	2	—	38	38	38	3	5	132
Galium pumilum	Murray	1	1	—	0	0	0	5	5	69
Galium pumilum	Murray	1	1	—	0	0	0	20	5	97
Galium pumilum	Murray	1	4	—	0	0	0	10	5	255
Galium pumilum	Murray	4	4	—	50	188	107	10	5	255
Galium saxatile	L.	1	1	—	0	0	0	7.5	5	84
Galium saxatile	L.	1	2	—	0	0	0	5	5	92
Galium saxatile	L.	1	3	—	0	0	0	3	5	96
Galium saxatile	L.	1	1	—	60	60	60	10	5	255
Galium saxatile	L.	1	1	—	60	344	60	15	5	257
Galium saxatile	L.	1	1	—	344	344	344	30	5	45

Species	Authority	Seed bank type	Number of records	Longevity (y)	Minimum density (seeds m⁻²)	Maximum density (seeds m⁻²)	Mean density (seeds m⁻²)	Depth (cm)	Method	Source code
Galium saxatile	L.	2	1	—	22	22	22	3	7	235
Galium saxatile	L.	2	1	—	38	38	38	10	5	255
Galium saxatile	L.	2	1	—	53	53	53	9.5	5	84
Galium saxatile	L.	2	2	—	54	667	361	22.5	5	45
Galium saxatile	L.	2	1	—	409	409	409	15	5	45
Galium saxatile		2	3	—	100	3660	2033	15	5	257
Galium saxatile		3	1	>120	44	44	44	32	5	169
Galium saxatile	L.	4	1	—	11	11	11	3	7	235
Galium saxatile	L.	4	1	—	23	23	23	2	5	153
Galium saxatile	L.	4	1	—	29	29	29	5	5	233
Galium saxatile	L.	4	1	—	48	48	48	5	5	69
Galium saxatile	L.	4	3	—	56	173	124	5	5	92
Galium saxatile	L.	4	6	—	13	398	191	7	5	232
Galium saxatile	L.	4	12	—	177	1591	728	17.8	5	156
Galium saxatile	L.	4	6	—	113	1575	775	10	5	255
Galium saxatile	L.	4	1	—	1117	1117	1117	3	5	96
Galium saxatile	L.	4	2	—	953	1440	1197	5	5	123
Galium saxatile	L.	4	5	—	252	7855	2976	17.8	5	155
Galium sterneri	Ehrend.	1	1	—	6	6	0	3	5	96
Galium sterneri	Ehrend.	1	1	—	6	6	6	3	7	235
Galium sterneri	Ehrend.	2	1	—	5	5	5	3	7	235
Galium sterneri	Ehrend.	4	1	—	83	83	83	3	5	96
Galium tricornutum	Dandy	1	3	—	0	0	0	10	4	22
Galium tricornutum	Dandy	4	3	—	30	86	57	15	4	30
Galium tricornutum	Dandy	4	1	—	82	82	82	10	4	22
Galium trifidum	L.	2	1	—	28	28	28	4	5	217
Galium trifidum	L.	2	1	—	50	50	50	10	5	255
Galium trifidum	L.	4	1	—	5	5	5	5	5	166
Galium trifidum	L.	4	1	—	36	36	36	4	5	217
Galium triflorum	Michx.	1	2	—	0	0	0	12	5	171
Galium triflorum	Michx.	4	1	—	6	6	6	10	5	33
Galium uliginosum	L.	1	2	—	0	0	0	10	5	150
Galium uliginosum	L.	1	4	—	0	0	0	3	5	208
Galium uliginosum	L.	1	3	—	0	0	0	20	4	225
Galium uliginosum	L.	1	5	—	0	0	0	10	5	255
Galium uliginosum	L.	1	1	—	65	65	65	2	7	99
Galium uliginosum	L.	2	2	—	38	38	38	10	5	255
Galium uliginosum	L.	4	5	—	25	704	231	12	5	174
Galium uliginosum	L.	4	2	—	142	425	284	3	5	208*
Galium uliginosum	L.	4	7	—	50	825	306	10	5	255
Galium verum	L.	4	2	—	2079	2339	2209	50	5	216
Galium verum	L.	1	1	—	0	0	0	6	5	48

Galium verum	L.	1	1	—	0	0	0	5	6	67
Galium verum	L.	1	3	—	0	0	0	4	10	88
Galium verum	L.	1	2	—	0	0	0	5	10	110
Galium verum	L.	1	2	—	0	0	0	5	5	122
Galium verum	L.	1	8	—	0	0	0	5	10	150
Galium verum	L.	1	16	—	0	0	0	5	10	152
Galium verum	L.	1	1	—	0	0	0	5	2	153
Galium verum	L.	1	2	—	0	0	0	5	6	181
Galium verum	L.	1	1	—	0	0	0	7	6.5	182
Galium verum	L.	1	3	—	0	0	0	7	3	235
Galium verum	L.	1	1	—	0	0	0	5	5	250
Galium verum	L.	1	10	—	0	0	0	5	5	251
Galium verum	L.	1	7	—	0	0	0	5	10	255
Galium verum	L.	2	1	—	57	57	57	5	30	29
Galium verum	L.	2	1	—	75	75	75	5	10	162
Galium verum	L.	2	1	—	96	96	96	5	13	183
Galium verum	L.	4	1	—	112	112	112	5	12	162
Galium verum	L.	4	1	—	40	40	40	5	6	181
Galium verum	L.	4	1	—	50	50	50	5	10	255
Galium verum	L.	4	1	—	408	408	408	4	8.5	254
Galium verum	L.	4	2	—	400	500	450	5	15	249
Galium verum	L.	4	1	—	2350	2350	2350	5	3	96
Sherardia arvensis	L.	1	2	—	0	0	0	4	10	22
Sherardia arvensis	L.	1	1	—	0	0	0	4	25	36
Sherardia arvensis	L.	1	1	—	0	0	0	5	2	95
Sherardia arvensis	L.	2	1	>3	0	0	0	2	7.5	195
Sherardia arvensis	L.	4	1	—	33	33	33	4	30	13
Sherardia arvensis	L.	4	1	—	35	35	35	5	25	2
Sherardia arvensis	L.	4	1	—	77	77	77	5	17.8	155
Sherardia arvensis	L.	4	1	—	250	250	250	5	2	95
Ruppiaceae	—			—						
Ruppia cirrhosa	(Petagna) Grande	4	1	—	831	831	831	5	5	116
Ruppia maritima	L.	1	1	—	0	0	0	5	5	116
Ruppia maritima	L.	3	1	—	50000	50000	50000	5	20	210
Salicaceae	—			—						
Populus tremula	L.	1	1	—	0	0	0	5	2	82
Populus tremula	L.	1	2	—	0	0	0	5	5	82
Populus tremula	L.	1	1	—	0	0	0	5	10	152
Populus tremula	L.	1	1	—	0	0	0	5	10	255
Salix alba	L.	1	2	—	0	0	0	4	20	225
Salix alba	L.	3	7	—	313	655	1146	5	20	97
Salix arbuscula	L.	1	1	—	0	0	0	5	10	149
Salix aurita	L.	1	2	—	0	0	0	5	10	255
Salix aurita	L.	3	1	—	228	228	228	5	50	107
Salix caprea	L.	1	1	—	0	0	0	5	2	82

Species	Authority	Seed bank type	Number of records	Longevity (y)	Minimum density (seeds m⁻²)	Maximum density (seeds m⁻²)	Mean density (seeds m⁻²)	Depth (cm)	Method	Source code
Salix caprea	L.	1	1	—	0	0	0	5	5	82
Salix caprea	L.	1	1	—	0	0	0	10	5	149
Salix caprea	L.	1	6	—	0	0	0	10	5	255
Salix cinerea	L.	1	1	—	0	0	0	3	7	235
Salix fragilis	L.	1	1	—	0	0	0	3	5	96
Salix glauca	L.	1	3	—	0	0	0	6	5	39
Salix glauca	L.	1	1	—	0	0	0	10	5	63
Salix herbacea	L.	1	1	—	0	0	0	5	5	55
Salix herbacea	L.	1	2	—	0	0	0	5	4	91
Salix polaris	Wahlenb.	1	1	—	0	0	0	5	5	71
Salix repens	L.	1	3	—	0	0	0	10	5	149
Salix repens	L.	1	1	—	0	0	0	10	5	152
Salix repens	L.	1	1	—	0	0	0	2	5	153
Salix repens	L.	1	4	—	0	0	0	12	5	174
Salix repens	L.	4	1	—	15	15	15	12	5	174
Salix reticulata	L.	1	3	—	0	0	0	6	5	39
Salix reticulata	L.	1	1	—	0	0	0	5	5	55
Salix reticulata	L.	1	1	—	0	0	0	10	5	63
Salix reticulata	L.	1	1	—	0	0	0	5	5	71
Salix reticulata	L.	1	2	—	0	0	0	20	5	191
Salix viminalis	L.	1	1	—	0	0	0	10	5	255
Saxifragaceae										
Chrysosplenium oppositifolium	—	1	1	—	0	0	0	3	7	235
Chrysosplenium oppositifolium	L.	4	3	—	117	483	350	3	5	96
Chrysosplenium tetrandrum	(N. Lund) Th. Fries	1	1	—	205	205	205	16	5	130
Saxifraga aizoides	L.	1	1	—	0	0	0	5	5	55
Saxifraga cernua	L.	1	1	—	0	0	0	5	5	55
Saxifraga granulata	L.	1	1	—	0	0	0	6	5	67
Saxifraga granulata	L.	1	1	—	0	0	0	3	7	235
Saxifraga granulata	L.	1	1	—	0	0	0	10	5	255
Saxifraga granulata	L.	2	1	—	234	234	234	10	5	150
Saxifraga granulata	L.	2	7	—	63	1100	352	10	5	255
Saxifraga granulata	L.	2	2	—	858	3744	2301	10	5	152
Saxifraga granulata	L.	3	2	—	117	205	161	10	5	150
Saxifraga granulata	L.	4	1	—	63	63	63	10	5	255
Saxifraga hieracifolia	Waldst. & Kit. ex Willd.	1	1	—	167	167	167	10	5	63
Saxifraga oppositifolia	L.	1	2	—	0	0	0	5	5	55
Saxifraga oppositifolia	L.	4	3	—	5	84	33	2	5	73
Saxifraga paniculata	Miller	1	2	—	0	0	0	5	5	55
Saxifraga tridactylites	L.	1	1	—	0	0	0	1	5	96
Saxifraga tridactylites	L.	2	1	—	40	40	40	3	7	235

Species	Author									
Scrophulariaceae	—									
Antirrhinum majus	L.	3	1	—	60	60	60	15	5	257
Bartsia alpina	L.	1	2	—	0	0	0	5	5	55
Chaenorhinum minus	(L.) Lange	1	1	—	0	0	0	25	4	36
Chaenorhinum minus	(L.) Lange	2	1	—	6	6	6	5	5	122
Chaenorhinum minus	(L.) Lange	2	1	—	9	9	9	10	5	110
Chaenorhinum minus	(L.) Lange	2	4	—	293	522	368	15	5	272
Chaenorhinum minus	(L.) Lange	3	1	>30	58	58	58	6	5	211
Chaenorhinum minus	(L.) Lange	3	1	—	146	146	146	30	4	12
Chaenorhinum minus	(L.) Lange	4	18	—	20	406	78	30	4	13
Chaenorhinum minus	(L.) Lange	4	2	—	6	1163	585	15	4	30
Chaenorhinum minus	(L.) Lange	4	4	—	572	1377	836	15	5	272
Digitalis purpurea	L.	1	1	—	0	0	0	10	5	255
Digitalis purpurea	L.	1	1	—	0	0	0	15	5	257
Digitalis purpurea	L.	1	1	—	26	26	26	2	7	99
Digitalis purpurea	L.	2	1	>3	0	0	0	7.5	2	195
Digitalis purpurea	L.	2	1	—	20	20	20	8	5	108
Digitalis purpurea	L.	2	3	—	7	61	29	3	7	235
Digitalis purpurea	L.	2	2	—	50	50	50	3	5	96
Digitalis purpurea	L.	2	2	—	88	113	101	10	5	255
Digitalis purpurea	L.	2	1	—	114	114	114	15	5	32
Digitalis purpurea	L.	2	1	—	1800	1800	1800	15	5	257
Digitalis purpurea	L.	2	1	—	4256	4256	4256	10	5	93
Digitalis purpurea	L.	3	1	—	43	43	43	30	5	45
Digitalis purpurea	L.	3	1	—	169	169	169	12	4	51
Digitalis purpurea	L.	3	1	—	1257	1257	1257	16	4	51
Digitalis purpurea	L.	3	7	—	80	6140	1903	15	5	257
Digitalis purpurea	L.	4	2	—	101	2920	1511	17.8	5	155
Digitalis purpurea	L.	4	7	—	50	7138	2018	10	5	255
Digitalis purpurea	L.	4	1	—	2627	2627	2627	17.8	5	156
Euphrasia micrantha	Reichb.	2	1	>3	0	0	0	0	1	270
Euphrasia nemorosa	(Pers.) Wallr.	4	1	>2	28	28	28	10	5	110
Euphrasia nemorosa	(Pers.) Wallr.	1	1	—	0	0	0	25	4	36
Euphrasia officinalis	—	1	1	—	0	0	0	6	5	67
Euphrasia officinalis	—	1	2	—	0	0	0	5	5	122
Euphrasia officinalis	—	1	1	—	0	0	0	2	5	153
Euphrasia officinalis	—	2	1	—	97	97	97	30	5	45
Euphrasia pseudokerneri	Pugsley	4	1	—	27	27	27	1.5	5	121
Euphrasia rostkoviana	Hayne	1	1	—	0	0	0	6	5	181
Euphrasia rostkoviana	Hayne	1	1	—	8	8	8	6.5	7	182
Euphrasia rostkoviana	Hayne	2	1	—	80	80	80	6.5	7	182
Euphrasia stricta	D. Wolff ex J. Lehm.	1	1	—	0	0	0	6	5	181
Euphrasia stricta	D. Wolff ex J. Lehm.	2	1	>2	0	0	0	0	1	270
Euphrasia stricta	D. Wolff ex J. Lehm.	2	1	—	21	21	21	6	5	211
Kickxia elatine	(L.) Dumort.	2	1	—	10	10	10	25	4	36

Species	Authority	Seed bank type	Number of records	Longevity (y)	Minimum density (seeds m⁻²)	Maximum density (seeds m⁻²)	Mean density (seeds m⁻²)	Depth (cm)	Method	Source code
Kickxia elatine	(L.) Dumort.	2	1	—	102	102	102	15	5	142
Kickxia elatine	(L.) Dumort.	3	1	>5	0	0	0	7.5	2	195
Kickxia elatine	(L.) Dumort.	3	1	>30	23	23	23	6	5	211
Kickxia elatine	(L.) Dumort.	3	1	>6	49	49	49	15	5	62
Kickxia elatine	(L.) Dumort.	3	1	—	4080	4080	4080	12	5	81
Kickxia elatine	(L.) Dumort.	4	3	—	33	73	53	30	4	13
Kickxia spuria	(L.) Dumort.	1	1	—	0	0	0	10	4	22
Kickxia spuria	(L.) Dumort.	3	1	>5	0	0	0	30	3	14
Kickxia spuria	(L.) Dumort.	3	1	>5	0	0	0	7.5	2	195
Kickxia spuria	(L.) Dumort.	3	1	—	465	465	465	30	4	12
Kickxia spuria	(L.) Dumort.	4	1	—	204	204	204	25	4	136
Kickxia spuria	(L.) Dumort.	4	6	—	33	850	229	10	5	15
Kickxia spuria	(L.) Dumort.	4	34	—	20	1179	232	30	4	13
Linaria alpina	(L.) Miller	1	1	—	0	0	0	5	5	55
Linaria arenaria	DC.	1	1	—	0	0	0	3	5	132
Linaria vulgaris	Miller	1	1	—	0	0	0	3	5	96
Linaria vulgaris	Miller	1	10	—	0	0	0	10	5	255
Linaria vulgaris	Miller	1	1	—	3	3	3	3	7	235
Linaria vulgaris	Miller	1	1	—	53	53	53	8	7	275
Linaria vulgaris	Miller	2	5	>1	0	0	0	—	1	275
Linaria vulgaris	Miller	2	2	—	38	50	44	10	5	255
Linaria vulgaris	Miller	3	1	>5	0	0	0	7.5	2	195
Linaria vulgaris	Miller	4	3	—	175	175	175	10	5	255
Melampyrum arvense	L.	2	1	2	0	0	0	25	3	209
Melampyrum cristatum	L.	2	1	—	0	0	0	10	5	152
Melampyrum nemorosum	L.	1	1	—	0	0	0	5	6	170
Melampyrum pratense	L.	1	1	—	0	0	0	5	5	92
Melampyrum pratense	L.	1	2	—	0	0	0	5	6	170
Melampyrum pratense	L.	1	2	—	0	0	0	12	5	171
Melampyrum pratense	L.	1	9	—	0	0	0	15	5	257
Odontites vernus	(Bellardi) Dumort.	3	1	>5	0	0	0	7.5	2	195
Odontites vernus	(Bellardi) Dumort.	3	1	>11	0	0	0	25	3	209
Odontites vernus	(Bellardi) Dumort.	4	1	—	14	14	14	20	4	111
Odontites vernus	(Bellardi) Dumort.	4	4	—	11	321	137	15	4	30
Pedicularis palustris	L.	2	4	—	60	140	90	10	5	149
Pedicularis sylvatica	L.	1	2	—	0	0	0	5	5	222
Pedicularis sylvatica	L.	1	1	—	0	0	0	10	5	255
Rhinanthus alectorolophus	(Scop.) Pollich	2	1	>3	0	0	0	6	3	179
Rhinanthus alectorolophus	(Scop.) Pollich	2	1	1	0	0	0	25	3	209
Rhinanthus angustifolius	C. Gmelin	1	6	—	0	0	0	6	5	11
Rhinanthus angustifolius	C. Gmelin	1	1	—	0	0	0	6.5	5	68
Rhinanthus angustifolius	C. Gmelin	1	1	—	0	0	0	13	5	68

Species	Authority									
Rhinanthus angustifolius	C. Gmelin	1	14	—	0	0	0	6	5	99
Rhinanthus angustifolius	C. Gmelin	1	6	—	156	156	156	10	5	27
Rhinanthus angustifolius	C. Gmelin	1	4	—	728	1118	923	6	5	28
Rhinanthus angustifolius	C. Gmelin	4	2	—	156	156	156	6	5	99
Rhinanthus aristatus	Celak.	1	2	—	0	0	0	10	5	255
Rhinanthus aristatus	Celak.	1	1	—	56	56	56	6.5	7	182
Rhinanthus minor	L.	1	1	—	0	0	0	7	5	58
Rhinanthus minor	L.	1	2	—	0	0	0	10	5	149
Rhinanthus minor	L.	1	1	—	0	0	0	10	5	150
Rhinanthus minor	L.	1	3	—	0	0	0	10	5	255
Rhinanthus minor	L.	2	1	>3	0	0	0	6	3	179
Rhinanthus minor	L.	2	1	4	0	0	0	7.5	2	195
Rhinanthus minor	L.	2	1	—	14	14	14	10	5	252
Rhinanthus minor	L.	2	2	—	32	109	71	6.5	7	182
Rhinanthus minor	L.	2	1	—	200	200	200	10	5	149
Rhinanthus minor	L.	3	1	—	32	32	32	30	5	45
Rhinanthus minor	L.	4	3	—	227	2821	1226	17.8	5	155
Scrophularia auriculata	L.	2	1	3	8	8	8	7.5	2	195
Scrophularia auriculata	L.	3	2	—	100	160	130	3	7	235
Scrophularia auriculata	L.	1	1	—	0	0	0	15	5	257
Scrophularia nodosa	L.	2	2	—	2	2	2	10	5	255
Scrophularia nodosa	L.	2	2	—	5	29	17	15	5	32
Scrophularia nodosa	L.	2	1	—	50	50	50	3	7	235
Scrophularia nodosa	L.	3	1	—	56	56	56	10	5	255
Scrophularia nodosa	L.	3	1	—	60	60	60	32	5	169
Scrophularia nodosa	L.	3	1	—	60	60	60	30	5	173
Scrophularia nodosa	L.	3	2	—	22	101	62	15	5	257
Scrophularia nodosa	L.	3	1	—	112	112	112	12.5	5	126
Scrophularia nodosa	L.	3	1	—	178	178	178	20	5	67
Scrophularia nodosa	L.	4	1	—	50	50	50	16	5	168
Scrophularia nodosa	L.	4	2	—	83	83	83	10	5	110
Scrophularia nodosa	L.	4	1	—	58	130	94	3	5	96
Scrophularia nodosa	L.	4	2	—	101	101	101	5	5	220
Scrophularia nodosa	L.	4	1	—	163	163	163	10	5	126
Scrophularia umbrosa	Dumort.	2	1	—	388	388	388	10	5	255
Scrophularia umbrosa	Dumort.	4	1	—	0	0	0	10	5	255
Verbascum blattaria	L.	3	1	>100	0	0	0	—	3	124
Verbascum nigrum	L.	1	2	—	200	200	200	10	5	255
Verbascum nigrum	L.	2	1	—	875	875	875	30	5	162
Verbascum nigrum	L.	4	1	—	1020	1020	1020	10	5	162
Verbascum phoeniceum	L.	1	1	—	0	0	0	8.5	4	254
Verbascum thapsus	L.	2	1	—	0	0	0	11.5	5	134
Verbascum thapsus	L.	2	1	—	7	7	7	10	4	127
Verbascum thapsus	L.	2	1	—	32	32	32	3	5	59
Verbascum thapsus	L.	3	1	—	144	144	144	10	5	181
Verbascum thapsus	L.	3	1	>100	0	0	0	—	3	124
Verbascum thapsus	L.	3	1	>30	0	0	0	15	1	240

Species	Authority	Seed bank type	Number of records	Longevity (y)	Minimum density (seeds m⁻²)	Maximum density (seeds m⁻²)	Mean density (seeds m⁻²)	Depth (cm)	Method	Source code
Verbascum thapsus	L.	3	1	>39	0	0	0	50	1	240
Verbascum thapsus	L.	3	1	>39	0	0	0	100	1	240
Verbascum thapsus	L.	3	2	—	37	49	43	11.5	5	134
Verbascum thapsus	L.	3	1	—	160	160	160	15	5	257
Verbascum thapsus	L.	3	2	>40	2096	2096	2096	18	5	162
Verbascum thapsus	L.	4	1	—	1112	1112	1112	10	5	33
Veronica agrestis	L.	2	1	—	76	76	76	15	5	142
Veronica agrestis	L.	2	1	—	138	138	138	10	5	255
Veronica agrestis	L.	3	1	>10	355	355	355	30	5	29
Veronica agrestis	L.	3	1	—	431	431	431	30	5	29
Veronica agrestis	L.	4	1	—	7	7	7	20	4	111
Veronica agrestis	L.	4	1	—	86	86	86	30	4	13
Veronica agrestis	L.	4	2	—	714	1133	924	25	5	260
Veronica alpina	L.	1	1	—	0	0	0	5	5	55
Veronica alpina	L.	1	2	—	0	0	0	5	4	91
Veronica alpina	L.	4	2	—	49	393	221	10	5	9
Veronica anagallis-aquatica	L.	2	2	—	60	60	60	6.5	5	128
Veronica anagallis-aquatica	L.	4	2	—	83	150	117	3	5	96
Veronica arvensis	L.	1	2	—	0	0	0	25	4	36
Veronica arvensis	L.	1	1	—	0	0	0	6	5	67
Veronica arvensis	L.	1	1	—	0	0	0	6.5	5	68
Veronica arvensis	L.	1	1	0	0	0	0	5	5	69
Veronica arvensis	L.	1	1	—	0	0	0	0	1	75
Veronica arvensis	L.	1	2	—	0	0	0	1	5	96
Veronica arvensis	L.	1	1	—	0	0	0	3	5	132
Veronica arvensis	L.	1	1	—	32	32	32	100	5	219
Veronica arvensis	L.	1	2	—	53	53	53	8	7	275
Veronica arvensis	L.	1	1	—	60	60	60	8	5	25
Veronica arvensis	L.	1	1	—	72	72	72	13	5	68
Veronica arvensis	L.	2	1	>2	0	0	0	0	2	75
Veronica arvensis	L.	2	1	>2	0	0	0	5	1	75
Veronica arvensis	L.	2	1	>2	0	0	0	5	2	75
Veronica arvensis	L.	2	1	>3	0	0	0	20	3	268
Veronica arvensis	L.	2	1	—	6	6	6	25	4	36
Veronica arvensis	L.	2	2	—	32	40	36	12	5	174
Veronica arvensis	L.	2	1	—	96	96	96	5	5	69
Veronica arvensis	L.	2	3	—	50	200	104	10	5	255
Veronica arvensis	L.	2	3	—	150	1333	617	3	5	96
Veronica arvensis	L.	2	2	—	1014	4134	2574	10	5	152
Veronica arvensis	L.	2	2	—	1552	5789	3671	8	5	25
Veronica arvensis	L.	4	3	—	12	102	45	12	5	174

Species	Authority									
Veronica arvensis	L.	4	2	—	27	162	95	13.5	5	163
Veronica arvensis	L.	4	3	—	50	238	134	3	5	132
Veronica arvensis	L.	4	1	—	246	246	246	20	5	111
Veronica arvensis	L.	4	2	—	170	504	337	15	4	31
Veronica arvensis	L.	4	4	—	317	633	404	3	5	96
Veronica arvensis	L.	4	1	—	445	445	445	15	5	207
Veronica arvensis	L.	4	1	—	672	672	672	5	5	69
Veronica arvensis	L.	4	1	—	773	773	773	25	5	2
Veronica arvensis	L.	4	2	—	267	1733	1000	25	5	260
Veronica arvensis	L.	4	3	—	400	1700	1033	15	5	249
Veronica arvensis	L.	4	7	—	20	3510	1039	30	4	13
Veronica arvensis	L.	1	7	—	19	2123	1280	15	4	30
Veronica austriaca	L.	1	2	—	0	0	0	6	5	181
Veronica beccabunga	L.	2	3	—	0	78	26	2	7	99
Veronica beccabunga	L.	2	2	—	0	60	30	2	7	99
Veronica beccabunga	L.	2	1	—	38	38	38	10	5	255
Veronica beccabunga	L.	2	2	—	320	320	320	6.5	5	128
Veronica beccabunga	L.	2	2	—	260	572	416	6	5	99
Veronica beccabunga	L.	3	1	—	188	188	188	8	5	108
Veronica beccabunga	L.	4	1	—	40	40	40	30	4	13
Veronica beccabunga	L.	4	1	—	2067	2067	2067	3	5	96
Veronica catenata	Pennell	2	6	—	100	4000	1023	6.5	5	128
Veronica chamaedrys	L.	1	1	—	0	0	0	25	4	36
Veronica chamaedrys	L.	1	2	—	0	0	0	5	4	53
Veronica chamaedrys	L.	1	2	—	0	0	0	5	5	69
Veronica chamaedrys	L.	1	1	—	0	0	0	9	5	84
Veronica chamaedrys	L.	1	1	—	0	0	0	5	4	88
Veronica chamaedrys	L.	1	1	—	0	0	0	5	5	92
Veronica chamaedrys	L.	1	1	—	0	0	0	20	5	97
Veronica chamaedrys	L.	1	3	—	0	0	0	10	5	110
Veronica chamaedrys	L.	1	2	—	0	0	0	6.5	5	128
Veronica chamaedrys	L.	1	2	—	0	0	0	10	5	152
Veronica chamaedrys	L.	1	2	—	0	0	0	2	5	153
Veronica chamaedrys	L.	1	1	—	0	0	0	12	5	171
Veronica chamaedrys	L.	1	1	—	0	0	0	6	5	181
Veronica chamaedrys	L.	1	3	—	0	0	0	6.5	7	182
Veronica chamaedrys	L.	1	3	—	0	0	0	3	5	208
Veronica chamaedrys	L.	1	3	—	0	0	0	10	5	255
Veronica chamaedrys	L.	1	1	—	3	6	5	3	7	235
Veronica chamaedrys	L.	1	1	—	42	42	42	2	7	99
Veronica chamaedrys	L.	1	1	—	53	53	53	8	7	275
Veronica chamaedrys	L.	1	1	—	60	60	60	13	5	68
Veronica chamaedrys	L.	1	2	—	100	100	100	6.5	5	68
Veronica chamaedrys	L.	2	1	—	1	2	2	5	6	170
Veronica chamaedrys	L.	2	1	—	2	2	2	15	5	32
Veronica chamaedrys	L.	2	1	—	48	48	48	13	5	68
Veronica chamaedrys	L.	2	1	—	48	48	48	5	5	69

Species	Authority	Seed bank type	Number of records	Longevity (y)	Minimum density (seeds m⁻²)	Maximum density (seeds m⁻²)	Mean density (seeds m⁻²)	Depth (cm)	Method	Source code
Veronica chamaedrys	L	2	1	—	56	56	56	5	5	92
Veronica chamaedrys	L	2	5	—	38	250	100	10	5	255
Veronica chamaedrys	L	2	1	—	220	220	220	15	5	257
Veronica chamaedrys	L	3	1	—	75	75	75	6.5	5	68
Veronica chamaedrys	L	3	1	—	80	80	80	8	5	25
Veronica chamaedrys	L	3	1	—	80	80	80	20	5	173
Veronica chamaedrys	L	3	1	—	85	85	85	3	5	208
Veronica chamaedrys	L	3	1	—	260	260	260	30	5	173
Veronica chamaedrys	L	3	3	—	80	740	300	15	5	257
Veronica chamaedrys	L	3	1	—	753	753	753	30	5	29
Veronica chamaedrys	L	4	3	—	14	65	34	6	5	49
Veronica chamaedrys	L	4	1	—	60	60	60	6.5	5	128
Veronica chamaedrys	L	4	3	—	50	88	71	10	5	255
Veronica chamaedrys	L	4	2	—	177	181	179	5	5	122
Veronica chamaedrys	L	4	1	—	234	234	234	10	5	152
Veronica chamaedrys	L	4	2	—	71	540	306	5	5	92
Veronica hederifolia	L	1	1	—	0	0	0	30	5	66
Veronica hederifolia	L	1	1	—	0	0	0	50	5	66
Veronica hederifolia	L	1	1	—	0	0	0	15	5	142
Veronica hederifolia	L	1	1	—	0	0	0	25	4	185
Veronica hederifolia	L	1	1	—	53	53	53	8	7	275
Veronica hederifolia	L	2	1	4	0	0	0	2.5	2	203
Veronica hederifolia	L	2	1	4	0	0	0	7.5	2	203
Veronica hederifolia	L	2	1	>3	0	0	0	15	2	203
Veronica hederifolia	L	2	1	—	0	0	0	20	3	268
Veronica hederifolia	L	2	1	—	14	14	14	6	5	49
Veronica hederifolia	L	2	3	—	6	24	16	25	4	36
Veronica hederifolia	L	2	5	—	38	100	65	10	5	255
Veronica hederifolia	L	2	1	—	75	75	75	30	5	45
Veronica hederifolia	L	2	1	—	89	89	89	30	4	12
Veronica hederifolia	L	2	2	—	417	483	450	10	5	184
Veronica hederifolia	L	2	1	—	765	765	765	50	5	260
Veronica hederifolia	L	2	1	—	2288	2288	2288	8	5	25
Veronica hederifolia	L	3	1	>5	0	0	0	7.5	2	196
Veronica hederifolia	L	3	1	>5	0	0	0	2.5	1	203
Veronica hederifolia	L	3	1	>5	0	0	0	7.5	1	203
Veronica hederifolia	L	3	1	>5	0	0	0	15	1	203
Veronica hederifolia	L	3	1	>20	24	24	24	20	5	40
Veronica hederifolia	L	3	1	—	65	65	65	30	5	29
Veronica hederifolia	L	3	1	>10	205	205	205	30	5	29
Veronica hederifolia	L	4	1	—	16	16	16	25	4	36
Veronica hederifolia	L	4	1	—	26	26	26	10	4	22

Species	Authority									
Veronica hederifolia	L.	4	1	—	59	59	59	15	4	31
Veronica hederifolia	L.	4	7	—	33	165	90	10	5	15
Veronica hederifolia	L.	4	12	—	20	526	107	30	4	13
Veronica hederifolia	L.	4	2	—	229	294	262	25	5	260
Veronica hederifolia	L.	4	7	—	22	799	454	15	4	30
Veronica hederifolia	L.	4	1	—	741	741	741	15	5	207
Veronica montana	L.	1	1	—	0	0	0	3	5	96
Veronica montana	L.	1	1	—	3	3	3	3	7	235
Veronica montana	L.	2	1	—	19	19	19	15	5	32
Veronica montana	L.	2	1	—	80	80	80	15	5	257
Veronica montana	L.	3	3	—	80	100	87	15	5	257
Veronica officinalis	L.	1	2	—	0	0	0	5	5	19
Veronica officinalis	L.	1	1	—	0	0	0	6	5	49
Veronica officinalis	L.	1	1	—	0	0	0	3	5	59
Veronica officinalis	L.	1	1	—	0	0	0	5	5	69
Veronica officinalis	L.	1	1	—	0	0	0	16	5	216
Veronica officinalis	L.	1	1	—	0	0	0	15	5	257
Veronica officinalis	L.	1	2	—	0	42	21	12	5	171
Veronica officinalis	L.	2	5	—	0	7	3	5	6	170
Veronica officinalis	L.	2	1	—	5	5	5	0	5	157
Veronica officinalis	L.	2	2	—	7	9	8	5	5	19
Veronica officinalis	L.	2	1	—	25	25	25	15	5	32
Veronica officinalis	L.	2	2	—	29	43	36	5	5	220
Veronica officinalis	L.	2	2	—	38	388	213	10	5	255
Veronica officinalis	L.	2	1	—	1120	1120	1120	15	5	257
Veronica officinalis	L.	3	2	>100	67	78	73	32	5	169
Veronica officinalis	L.	3	1	—	1819	1819	1819	9	5	84
Veronica officinalis	L.	4	1	—	10	10	10	6	5	49
Veronica officinalis	L.	4	2	—	10	37	24	5	5	19
Veronica officinalis	L.	4	1	—	26	26	26	5	5	123
Veronica officinalis	L.	4	5	—	38	188	90	10	5	255
Veronica officinalis	L.	4	4	—	101	225	138	17.8	5	155
Veronica officinalis	L.	2	2	—	123	291	207	5	5	189
Veronica peregrina	L.	4	1	—	101	101	101	12	5	187
Veronica peregrina	L.	4	6	—	27	568	209	13.5	5	163
Veronica persica	Poiret	1	3	—	0	0	0	25	4	36
Veronica persica	Poiret	1	1	—	0	0	0	50	5	66
Veronica persica	Poiret	1	1	—	0	0	0	3	5	96
Veronica persica	Poiret	1	1	—	0	0	0	15	5	257
Veronica persica	Poiret	1	3	1	53	105	88	8	7	275
Veronica persica	Poiret	2	1	>3	0	0	0	20	1	219
Veronica persica	Poiret	2	1	—	0	0	0	20	3	268
Veronica persica	Poiret	2	1	—	28	28	28	15	5	272
Veronica persica	Poiret	2	2	—	50	300	175	3	5	96
Veronica persica	Poiret	2	10	—	75	825	368	10	5	255
Veronica persica	Poiret	2	3	—	199	5352	2706	8	5	25

Species	Authority	Seed bank type	Number of records	Longevity (y)	Minimum density (seeds m⁻²)	Maximum density (seeds m⁻²)	Mean density (seeds m⁻²)	Depth (cm)	Method	Source code
Veronica persica	Poiret	3	1	>5	0	0	0	7.5	2	196
Veronica persica	Poiret	3	1	>5	0	0	0	2.5	1	203
Veronica persica	Poiret	3	1	>5	0	0	0	2.5	2	203
Veronica persica	Poiret	3	1	>5	0	0	0	7.5	1	203
Veronica persica	Poiret	3	1	>5	0	0	0	7.5	2	203
Veronica persica	Poiret	3	1	>5	0	0	0	15	1	203
Veronica persica	Poiret	3	1	>5	0	0	0	15	2	203
Veronica persica	Poiret	3	1	>10	614	614	614	30	5	29
Veronica persica	Poiret	3	1	—	1552	1552	1552	8	5	25
Veronica persica	Poiret	3	1	—	2260	2260	2260	30	5	29
Veronica persica	Poiret	3	1	—	5660	5660	5660	12	5	81
Veronica persica	Poiret	4	1	—	45	45	45	10	4	22
Veronica persica	Poiret	4	1	—	50	50	50	25	4	36
Veronica persica	Poiret	4	1	—	50	50	50	3	5	96
Veronica persica	Poiret	4	2	—	20	99	60	15	5	272
Veronica persica	Poiret	4	2	—	102	153	128	15	5	142
Veronica persica	Poiret	4	17	—	20	2171	340	30	4	13
Veronica persica	Poiret	4	1	—	480	480	480	25	4	3
Veronica persica	Poiret	4	2	—	17	1064	541	15	4	31
Veronica persica	Poiret	4	2	—	350	878	614	25	4	185
Veronica persica	Poiret	4	1	—	825	825	825	20	4	111
Veronica persica	Poiret	4	5	—	463	3088	1920	10	5	255
Veronica persica	Poiret	4	1	—	3211	3211	3211	15	5	207
Veronica polita	Fries	1	1	—	0	0	0	25	4	36
Veronica polita	Fries	2	1	—	53	53	53	20	5	111
Veronica polita	Fries	3	1	>20	56	56	56	16	5	168
Veronica polita	Fries	3	2	>40	89	100	95	24	5	168
Veronica polita	Fries	3	1	>15	222	222	222	24	5	168
Veronica polita	Fries	4	5	—	9	360	86	15	4	30
Veronica praecox	All.	1	2	—	0	0	0	10	5	255
Veronica scutellata	L.	1	1	—	0	0	0	16	5	216
Veronica scutellata	L.	1	1	—	26	26	26	2	7	99
Veronica scutellata	L.	1	3	—	255	255	255	6	5	208
Veronica scutellata	L.	2	1	—	80	80	80	6.5	5	128
Veronica scutellata	L.	3	1	—	85	85	85	6	5	208
Veronica scutellata	L.	4	1	—	36	36	36	16	5	216
Veronica scutellata	L.	4	2	—	42	92	67	5	5	98
Veronica scutellata	L.	4	2	—	58	4640	2349	12	4	230
Veronica scutellata	L.	4	2	—	2079	4158	3119	50	5	216
Veronica serpyllifolia	L.	1	6	—	0	0	0	25	4	36
Veronica serpyllifolia	L.	1	3	—	0	0	0	5	4	53
Veronica serpyllifolia	L.	1	1	—	0	0	0	3	5	59

Species										
Veronica serpyllifolia	L.	1	1	—	0	0	0	6.5	5	68
Veronica serpyllifolia	L.	1	1	—	0	0	0	13	5	68
Veronica serpyllifolia	L.	1	4	—	0	0	0	5	5	69
Veronica serpyllifolia	L.	1	1	—	0	0	0	10	5	255
Veronica serpyllifolia	L.	1	1	—	172	172	172	30	5	45
Veronica serpyllifolia	L.	2	1	—	3	3	3	15	5	32
Veronica serpyllifolia	L.	2	1	—	44	44	44	5	5	244
Veronica serpyllifolia	L.	2	1	—	53	53	53	16	5	23
Veronica serpyllifolia	L.	2	1	—	71	71	71	5	5	222
Veronica serpyllifolia	L.	2	12	—	38	550	140	10	5	255
Veronica serpyllifolia	L.	2	2	—	249	308	279	8	5	25
Veronica serpyllifolia	L.	2	1	—	377	377	377	30	5	29
Veronica serpyllifolia	L.	2	1	—	721	721	721	30	5	45
Veronica serpyllifolia	L.	2	1	—	833	833	833	20	5	97
Veronica serpyllifolia	L.	2	1	—	1216	1216	1216	6	5	151
Veronica serpyllifolia	L.	3	1	>5	0	0	0	7.5	2	195
Veronica serpyllifolia	L.	3	2	>40	44	56	50	24	5	168
Veronica serpyllifolia	L.	3	1	—	75	75	75	30	5	29
Veronica serpyllifolia	L.	3	1	—	80	80	80	8	5	108
Veronica serpyllifolia	L.	3	3	>100	44	144	81	32	5	169
Veronica serpyllifolia	L.	3	1	>20	89	89	89	24	5	168
Veronica serpyllifolia	L.	3	2	—	117	321	219	10	5	150
Veronica serpyllifolia	L.	3	3	—	300	300	300	20	5	4
Veronica serpyllifolia	L.	3	8	—	183	1830	655	30	5	45
Veronica serpyllifolia	L.	3	1	—	700	700	700	20	5	69
Veronica serpyllifolia	L.	3	3	—	149	1482	726	8	5	25
Veronica serpyllifolia	L.	4	1	—	10	10	10	2	5	153
Veronica serpyllifolia	L.	4	2	—	43	129	86	3	5	59
Veronica serpyllifolia	L.	4	1	—	130	130	130	20	4	111
Veronica serpyllifolia	L.	4	4	—	48	288	168	5	5	69
Veronica serpyllifolia	L.	4	4	—	323	323	323	6	5	151
Veronica serpyllifolia	L.	4	2	—	205	526	366	10	5	150
Veronica serpyllifolia	L.	4	1	—	533	533	533	3	5	96
Veronica serpyllifolia	L.	4	3	—	126	2670	1284	17.8	5	155
Veronica spicata	L.	1	1	—	0	0	0	10	5	152
Veronica spicata	L.	1	1	—	0	0	0	8.5	4	254
Veronica spicata	L.	4	2	—	234	390	312	10	5	152
Veronica verna	L.	1	1	—	0	0	0	3	5	132
Solanaceae										
Atropa belladonna	L.	1	1	—	0	0	0	10	5	181
Atropa belladonna	L.	2	2	—	17	48	33	6	5	49
Atropa belladonna	L.	3	1	—	67	67	67	16	5	168
Datura stramonium	L.	1	2	>2	0	0	0	5	5	189
Datura stramonium	L.	2	1	>3	0	0	0	20	1	219
Datura stramonium	L.	2	1	>3	0	0	0	2.5	1	224
Datura stramonium	L.	2	1	—	0	0	0	10.2	1	224

Species	Authority	Seed bank type	Number of records	Longevity (y)	Minimum density (seeds m⁻²)	Maximum density (seeds m⁻²)	Mean density (seeds m⁻²)	Depth (cm)	Method	Source code
Datura stramonium	L.	3	1	>9	0	0	0	22.5	1	214
Datura stramonium	L.	3	1	8	0	0	0	22.5	2	214
Datura stramonium	L.	3	1	>30	0	0	0	15	1	240
Datura stramonium	L.	3	1	>39	0	0	0	50	1	240
Datura stramonium	L.	3	1	>39	0	0	0	100	1	240
Hyoscyamus niger	L.	2	1	—	337	337	337	12	5	162
Hyoscyamus niger	L.	3	1	>5	0	0	0	7.5	2	195
Hyoscyamus niger	L.	3	1	>90	79	79	79	55	5	161
Hyoscyamus niger	L.	3	1	>660	119	119	119	27	5	161
Hyoscyamus niger	L.	3	1	>40	180	180	180	18	5	162
Hyoscyamus niger	L.	3	1	>75	351	351	351	22	5	162
Hyoscyamus niger	L.	3	1	>75	2188	2188	2188	17	5	162
Hyoscyamus niger	L.	3	1	>35	2473	2473	2473	65	5	161
Lycopersicon esculentum	Miller	2	1	1	0	0	0	15	1	240
Lycopersicon esculentum	Miller	2	1	1	0	0	0	50	1	240
Lycopersicon esculentum	Miller	2	1	1	0	0	0	100	1	240
Nicotiana tabacum	L.	3	1	>30	0	0	0	15	1	240
Nicotiana tabacum	L.	3	1	>39	0	0	0	50	1	240
Nicotiana tabacum	L.	3	1	>39	0	0	0	100	1	240
Nicotiana tabacum	L.	3	1	—	53	53	53	16	5	23
Solanum dulcamara	L.	1	1	—	0	0	0	3	5	96
Solanum dulcamara	L.	1	1	—	0	0	0	15	5	257
Solanum dulcamara	L.	1	1	—	10	10	10	3	7	235
Solanum dulcamara	L.	2	1	3	0	0	0	7.5	2	195
Solanum dulcamara	L.	3	1	>6	7	7	7	15	5	62
Solanum dulcamara	L.	3	1	—	22	22	22	12.5	5	126
Solanum dulcamara	L.	4	1	—	3	3	3	10	5	33
Solanum nigrum	L.	1	2	—	0	0	0	10	4	22
Solanum nigrum	L.	1	7	—	53	105	60	8	7	275
Solanum nigrum	L.	2	1	—	14	14	14	15	5	32
Solanum nigrum	L.	2	1	—	30	30	30	5	5	244
Solanum nigrum	L.	2	1	—	43	43	43	15	5	175
Solanum nigrum	L.	2	1	—	112	112	112	27	5	162
Solanum nigrum	L.	2	1	—	226	226	226	65	5	162
Solanum nigrum	L.	2	1	—	250	250	250	17	5	162
Solanum nigrum	L.	2	4	—	375	738	548	15	5	272
Solanum nigrum	L.	2	2	—	875	875	875	10	5	162
Solanum nigrum	L.	2	1	—	1750	1750	1750	30	5	162
Solanum nigrum	L.	3	1	>5	0	0	0	7.5	2	195
Solanum nigrum	L.	3	1	>11	0	0	0	25	3	209
Solanum nigrum	L.	3	1	>30	0	0	0	15	1	240
Solanum nigrum	L.	3	1	>39	0	0	0	50	1	240

Species	Author									
Solanum nigrum	L.	3	1	>39	0	0	0	100	1	240
Solanum nigrum	L.	3	1	—	7	7	7	15	5	62
Solanum nigrum	L.	3	1	—	12	12	12	70	5	219
Solanum nigrum	L.	3	1	>20	20	20	20	20	5	40
Solanum nigrum	L.	3	1	>20	75	75	75	15	5	162
Solanum nigrum	L.	3	1	—	129	129	129	15	5	175
Solanum nigrum	L.	3	1	>25	2188	2188	2188	17	5	162
Solanum nigrum	L.	4	1	—	4	4	4	20	4	111
Solanum nigrum	L.	4	11	—	20	1298	335	30	4	13
Solanum nigrum	L.	4	1	—	612	612	612	23	5	202
Solanum nigrum	L.	4	1	—	800	800	800	15	5	205
Solanum nigrum	L.	4	4	—	341	2049	1202	15	5	272
Solanum nigrum	L.	4	1	—	2594	2594	2594	15	5	207
Solanum sarachoides	Sendtner	3	1	>5	0	0	0	7.5	2	195
Sparganiaceae	—									
Sparganium erectum	L.	1	1	—	0	0	0	3	5	96
Sparganium erectum	L.	1	2	—	0	0	0	6.5	5	128
Sparganium erectum	L.	1	3	—	0	0	0	20	4	225
Tiliaceae	—									
Tilia cordata	Miller	1	1	—	0	0	0	4	5	176
Typhaceae	—									
Typha latifolia	L.	1	1	—	3	3	3	3	7	235
Typha latifolia	L.	2	1	—	2950	2950	2950	10	5	131
Typha latifolia	L.	2	2	—	5910	19300	12605	32	5	131
Typha latifolia	L.	3	1	—	2	2	2	10	4	127
Typha latifolia	L.	3	1	—	117	117	117	10	5	150
Typha latifolia	L.	3	1	—	1443	1443	1443	12.5	5	126
Typha latifolia	L.	3	1	—	9510	9510	9510	10	5	131
Typha latifolia	L.	4	1	—	100	100	100	6.5	5	128
Typha latifolia	L.	4	2	—	1200	80400	40800	12	4	230
Typha × glauca	Godron	2	6	—	1950	16863	6408	35	5	247
Ulmaceae	—									
Ulmus glabra	Hudson	1	1	—	0	0	0	20	4	111
Ulmus glabra	Hudson	1	1	—	0	0	0	20	5	111
Ulmus procera	Salisb.	1	1	—	0	0	0	3	7	235
Umbelliferae (Apiaceae)	—									
Aegopodium podagraria	L.	1	2	—	0	0	0	4	5	176
Aegopodium podagraria	L.	1	1	—	0	0	0	5	5	220
Aegopodium podagraria	L.	1	4	—	0	0	0	20	4	225
Aegopodium podagraria	L.	1	3	—	0	0	0	10	5	255
Aegopodium podagraria	L.	2	5	1	0	0	0	—	1	275
Aethusa cynapium	L.	1	1	—	0	0	0	25	4	36

Species	Authority	Seed bank type	Number of records	Longevity (y)	Minimum density (seeds m⁻²)	Maximum density (seeds m⁻²)	Mean density (seeds m⁻²)	Depth (cm)	Method	Source code
Aethusa cynapium	L	2	1	>3	0	0	0	10	3	135
Aethusa cynapium	L	2	1	>3	0	0	0	25	3	135
Aethusa cynapium	L	2	1	—	112	112	112	27	5	162
Aethusa cynapium	L	2	1	—	150	150	150	30	5	162
Aethusa cynapium	L	2	9	—	38	888	164	10	5	255
Aethusa cynapium	L	2	3	—	102	458	272	15	5	142
Aethusa cynapium	L	2	1	—	439	439	439	52	5	162
Aethusa cynapium	L	2	4	—	75	1104	539	10	5	162
Aethusa cynapium	L	2	1	—	625	625	625	37	5	162
Aethusa cynapium	L	2	2	—	750	1625	1188	17	5	162
Aethusa cynapium	L	3	1	>5	0	0	0	30	3	14
Aethusa cynapium	L	3	3	>5	0	0	0	7.5	2	193
Aethusa cynapium	L	3	1	>20	54	54	54	20	5	40
Aethusa cynapium	L	3	1	—	290	290	290	30	4	12
Aethusa cynapium	L	4	12	—	20	493	118	30	4	13
Aethusa cynapium	L	4	7	—	24	411	126	15	4	30
Aethusa cynapium	L	4	2	—	75	250	163	10	5	255
Aethusa cynapium	L	4	2	—	360	1010	685	10	5	15
Ammi majus	L	1	1	—	0	0	0	10	4	22
Ammi majus	L	2	1	—	1400	1400	1400	7	4	50
Ammi majus	L	4	2	—	2300	6100	4200	7	4	50
Angelica sylvestris	L	1	1	—	0	0	0	6	5	48
Angelica sylvestris	L	1	1	—	0	0	0	20	5	67
Angelica sylvestris	L	1	1	—	0	0	0	1	5	96
Angelica sylvestris	L	1	4	—	0	0	0	3	5	96
Angelica sylvestris	L	1	1	—	0	0	0	6.5	5	128
Angelica sylvestris	L	1	8	—	0	0	0	10	5	149
Angelica sylvestris	L	1	1	—	0	0	0	5	6	170
Angelica sylvestris	L	1	3	—	0	0	0	12	5	174
Angelica sylvestris	L	1	7	—	0	0	0	3	5	208
Angelica sylvestris	L	1	3	—	0	0	0	6	5	208
Angelica sylvestris	L	1	7	—	0	0	0	20	4	225
Angelica sylvestris	L	1	1	—	0	0	0	5	5	248
Angelica sylvestris	L	1	9	—	0	0	0	10	5	255
Angelica sylvestris	L	1	2	—	3	3	3	3	7	235
Angelica sylvestris	L	1	5	—	53	158	100	8	7	275
Angelica sylvestris	L	2	1	3	0	0	0	7.5	2	193
Angelica sylvestris	L	2	5	>1	0	0	0	—	1	275
Angelica sylvestris	L	2	1	—	3	3	3	5	6	170
Angelica sylvestris	L	3	1	>5	0	0	0	7.5	2	193
Angelica sylvestris	L	4	1	—	9	9	9	12	5	174
Angelica sylvestris	L	4	1	—	83	83	83	3	5	96

Angelica sylvestris	L.	4	1	—	85	85	85	3	5	208
Anthriscus caucalis	M. Bieb.	1	3	—	0	0	0	3	5	132
Anthriscus caucalis	M. Bieb.	3	1	>5	0	0	0	7.5	2	195
Anthriscus caucalis	M. Bieb.	4	1	—	217	217	217	3	5	96
Anthriscus sylvestris	(L.) Hoffm.	1	2	—	0	0	0	6	5	11
Anthriscus sylvestris	(L.) Hoffm.	1	1	—	0	0	0	6	5	99
Anthriscus sylvestris	(L.) Hoffm.	1	5	—	0	0	0	10	5	150
Anthriscus sylvestris	(L.) Hoffm.	1	9	—	0	0	0	20	4	225
Anthriscus sylvestris	(L.) Hoffm.	1	1	—	0	0	0	10	5	252
Anthriscus sylvestris	(L.) Hoffm.	1	12	—	0	0	0	10	5	255
Anthriscus sylvestris	(L.) Hoffm.	1	3	—	53	263	175	8	7	275
Anthriscus sylvestris	(L.) Hoffm.	2	2	2	0	0	0	7.5	2	193
Anthriscus sylvestris	(L.) Hoffm.	2	1	3	0	0	0	7.5	2	275
Anthriscus sylvestris	(L.) Hoffm.	2	1	>1	0	0	0	—	1	275
Anthriscus sylvestris	(L.) Hoffm.	2	4	1	700	700	700	6	5	162
Anthriscus sylvestris	(L.) Hoffm.	4	1	—	88	88	88	10	5	150
Apium graveolens	L.	3	1	>39	0	0	0	100	1	240
Apium graveolens	L.	3	1	16	0	0	0	15	1	240
Apium graveolens	L.	3	1	30	0	0	0	50	1	240
Apium nodiflorum	(L.) Lagasca	4	1	—	50	50	50	3	5	96
Berula erecta	(Hudson) Cov.	1	2	—	0	0	0	6.5	5	128
Berula erecta	(Hudson) Cov.	1	1	—	0	0	0	20	4	225
Berula erecta	(Hudson) Cov.	1	1	—	0	0	0	3	7	235
Berula erecta	(Hudson) Cov.	1	1	—	26	26	26	2	7	99
Berula erecta	(Hudson) Cov.	2	1	—	156	156	156	6	5	99
Berula erecta	(Hudson) Cov.	2	4	—	60	2260	900	6.5	5	128
Berula erecta	(Hudson) Cov.	4	2	—	60	760	410	6.5	5	128
Bupleurum falcatum	L.	1	1	—	0	0	0	6	5	67
Bupleurum subovatum	Link ex Sprengel	1	1	—	0	0	0	10	4	22
Carum carvi	L.	1	1	—	0	0	0	5	5	69
Carum carvi	L.	1	3	—	0	0	0	10	5	150
Caucalis platycarpos	L.	1	1	—	0	0	0	6	5	67
Caucalis platycarpos	L.	2	1	2	0	0	0	25	3	209
Chaerophyllum aureum	L.	1	2	—	0	0	0	10	5	255
Chaerophyllum aureum	L.	2	2	—	38	50	44	10	5	255
Chaerophyllum hirsutum	L.	4	1	—	6	6	6	5	5	213
Chaerophyllum temulum	L.	1	2	—	0	0	0	7	5	58
Chaerophyllum temulum	L.	1	1	—	53	53	53	8	7	275
Chaerophyllum temulum	L.	2	1	3	0	0	0	7.5	2	195
Chaerophyllum temulum	L.	2	4	>1	0	0	0	—	1	275
Chaerophyllum temulum	L.	2	1	1	0	0	0	—	1	275
Cicuta virosa	L.	1	1	—	0	0	0	5	5	248
Conium maculatum	L.	1	1	—	0	0	0	15	5	142
Conium maculatum	L.	3	3	>5	0	0	0	7.5	2	193
Conopodium majus	(Gouan) Loret	1	1	—	0	0	0	7	5	232
Conopodium majus	(Gouan) Loret	1	2	—	0	0	0	15	5	257

Species	Authority	Seed bank type	Number of records	Longevity (y)	Minimum density (seeds m^{-2})	Maximum density (seeds m^{-2})	Mean density (seeds m^{-2})	Depth (cm)	Method	Source code
Conopodium majus	(Gouan) Loret	2	1	1	0	0	0	7.5	2	193
Conopodium majus	(Gouan) Loret	2	1	—	269	269	269	30	5	29
Crithmum maritimum	L.	4	3	—	25	133	72	5	5	98
Daucus carota	L.	1	1	—	0	0	0	5	5	19
Daucus carota	L.	1	3	—	0	0	0	20	5	97
Daucus carota	L.	1	1	—	0	0	0	6	5	181
Daucus carota	L.	1	3	—	0	0	0	20	4	225
Daucus carota	L.	2	1	>3	0	0	0	10	5	255
Daucus carota	L.	2	1	>3	0	0	0	20	1	61
Daucus carota	L.	2	5	—	0	0	0	6	3	179
Daucus carota	L.	2	1	>1	0	0	0	—	1	275
Daucus carota	L.	2	1	—	37	37	37	11.5	5	134
Daucus carota	L.	2	1	—	112	112	112	30	5	219
Daucus carota	L.	2	2	—	101	137	119	6	5	211
Daucus carota	L.	2	1	—	162	162	162	16	5	23
Daucus carota	L.	3	1	20	313	313	313	20	5	97
Daucus carota	L.	3	1	>5	0	0	0	25	1	125
Daucus carota	L.	3	1	>20	0	0	0	7.5	2	195
Daucus carota	L.	3	1	>12	33	33	33	24	5	168
Daucus carota	L.	3	2	>35	36	36	36	50	5	219
Daucus carota	L.	3	1	—	89	89	89	24	5	168
Daucus carota	L.	4	1	—	63	63	63	10	5	255
Daucus carota	L.	4	1	—	80	80	80	10	5	33
Daucus carota	L.	4	1	—	80	80	80	10	5	86
Daucus carota	L.	4	1	—	112	112	112	5	5	69
Daucus carota	L.	4	2	—	24	440	232	6	5	181
Daucus carota	L.	4	1	—	252	252	252	6	5	67
Eryngium campestre	L.	1	1	—	0	0	0	5	5	69
Eryngium campestre	L.	1	1	—	0	0	0	8.5	4	254
Falcaria vulgaris	Bernh.	1	2	—	0	0	0	6	5	67
Foeniculum vulgare	Miller	1	2	—	0	0	0	10	4	22
Heracleum mantegazzianum	Sommier & Levier	1	6	—	0	0	0	3	5	208
Heracleum sphondylium	L.	1	1	—	0	0	0	6	5	11
Heracleum sphondylium	L.	1	2	—	0	0	0	6	5	49
Heracleum sphondylium	L.	1	13	—	0	0	0	3	5	96
Heracleum sphondylium	L.	1	5	—	0	0	0	20	5	97
Heracleum sphondylium	L.	1	1	—	0	0	0	10	5	110
Heracleum sphondylium	L.	1	1	—	0	0	0	12.5	5	126
Heracleum sphondylium	L.	1	2	—	0	0	0	6.5	5	128
Heracleum sphondylium	L.	1	1	—	0	0	0	15	5	142
Heracleum sphondylium	L.	1	1	—	0	0	0	10	5	149
Heracleum sphondylium	L.	1	7	—	0	0	0	20	4	225

Species	Authority									
Heracleum sphondylium	L.	1	20	—	0	0	0	10	5	255
Heracleum sphondylium	L.	1	4	—	1	3	0	3	7	235
Heracleum sphondylium	L.	2	1	1	0	0	0	7.5	2	193
Heracleum sphondylium	L.	2	1	2	0	0	0	7.5	2	193
Heracleum sphondylium	L.	3	1	>5	0	0	0	7.5	2	193
Heracleum sphondylium	L.	3	1	5	0	0	0	7.5	5	193
Heracleum sphondylium	L.	4	1	—	9	9	9	10	5	110
Hydrocotyle vulgaris	L.	1	1	—	0	0	0	40	5	11
Hydrocotyle vulgaris	L.	1	1	—	0	0	0	3	5	96
Hydrocotyle vulgaris	L.	1	1	—	0	0	0	5	5	248
Hydrocotyle vulgaris	L.	1	1	—	8	8	8	3	7	235
Hydrocotyle vulgaris	L.	1	2	—	26	26	26	2	7	99
Hydrocotyle vulgaris	L.	1	1	—	32	32	32	30	5	45
Hydrocotyle vulgaris	L.	1	1	—	208	208	208	6	5	11
Hydrocotyle vulgaris	L.	3	1	—	416	416	416	10	5	27
Hydrocotyle vulgaris	L.	3	2	—	2081	3933	228	30	5	107
Hydrocotyle vulgaris	L.	4	1	—	203	203	203	17.8	5	155
Ligusticum mutellina	(L.) Crantz	1	2	—	0	0	0	5	4	91
Myrrhis odorata	(L.) Scop.	2	1	2	0	0	0	7.5	2	195
Myrrhis odorata	(L.) Scop.	3	1	>5	0	0	0	30	3	14
Oenanthe fistulosa	L.	1	3	—	0	0	0	3	5	208
Oenanthe fistulosa	L.	1	1	—	0	0	0	20	4	225
Oenanthe fistulosa	L.	1	3	—	311	311	311	6	5	208
Pastinaca sativa	L.	2	1	2	0	0	0	7.5	2	193
Pastinaca sativa	L.	2	5	>1	0	0	0	—	1	275
Pastinaca sativa	L.	3	1	10	0	0	0	15	1	240
Pastinaca sativa	L.	3	1	10	0	0	0	50	1	240
Pastinaca sativa	L.	3	1	16	0	0	0	100	1	240
Peucedanum cervaria	(L.) Lapeyr.	1	1	—	0	0	0	10	5	255
Peucedanum palustre	(L.) Moench	1	2	—	0	0	0	16	5	216
Pimpinella major	(L.) Hudson	1	1	—	0	0	0	6	5	48
Pimpinella major	(L.) Hudson	1	3	—	0	0	0	10	4	88
Pimpinella major	(L.) Hudson	1	7	—	0	0	0	20	5	97
Pimpinella major	(L.) Hudson	1	1	—	0	0	0	6.5	5	128
Pimpinella major	(L.) Hudson	1	1	—	0	0	0	10	5	149
Pimpinella major	(L.) Hudson	1	9	—	0	0	0	5	5	251
Pimpinella major	(L.) Hudson	1	1	—	0	0	0	10	5	255
Pimpinella major	(L.) Hudson	1	1	—	3	3	3	3	7	235
Pimpinella major	(L.) Hudson	2	5	>1	0	0	0	—	1	275
Pimpinella saxifraga	L.	1	1	—	0	0	0	6	5	67
Pimpinella saxifraga	L.	1	1	—	0	0	0	5	5	69
Pimpinella saxifraga	L.	1	2	—	0	0	0	3	5	96
Pimpinella saxifraga	L.	1	3	—	0	0	0	20	5	97
Pimpinella saxifraga	L.	1	3	—	0	0	0	10	5	150
Pimpinella saxifraga	L.	1	12	—	0	0	0	10	5	152
Pimpinella saxifraga	L.	1	2	—	0	0	0	6	5	181
Pimpinella saxifraga	L.	1	21	—	0	0	0	10	5	255

Species	Authority	Seed bank type	Number of records	Longevity (y)	Minimum density (seeds m^{-2})	Maximum density (seeds m^{-2})	Mean density (seeds m^{-2})	Depth (cm)	Method	Source code
Pimpinella saxifraga	L.	1	2	—	3	3	3	3	7	235
Pimpinella saxifraga	L.	1	1	—	18	18	18	6.5	7	182
Pimpinella saxifraga	L.	1	2	—	0	1617	809	6	5	48
Pimpinella saxifraga	L.	2	1	3	0	0	0	7.5	2	195
Pimpinella saxifraga	L.	4	1	—	32	32	32	6	5	181
Pimpinella saxifraga	L.	4	1	—	306	306	306	8.5	4	254
Sanicula europaea	L.	1	1	—	0	0	0	7	5	58
Sanicula europaea	L.	1	1	—	0	0	0	20	5	67
Sanicula europaea	L.	1	2	—	0	0	0	4	5	176
Sanicula europaea	L.	1	1	—	0	0	0	10	5	181
Sanicula europaea	L.	2	1	3	0	0	0	7.5	2	195
Scandix pecten-veneris	L.	1	3	—	0	0	0	10	4	22
Scandix pecten-veneris	L.	4	6	—	32	728	343	15	4	30
Scandix pecten-veneris	L.	4	1	—	490	490	490	10	4	22
Selinum carvifolia	(L.) L.	1	1	—	0	0	0	5	5	69
Selinum carvifolia	(L.) L.	1	5	—	0	0	0	10	5	149
Selinum carvifolia	(L.) L.	1	3	—	0	0	0	10	5	255
Selinum carvifolia	(L.) L.	4	2	—	38	38	38	10	5	255
Seseli annuum	L.	1	1	—	0	0	0	8.5	4	254
Silaum silaus	(L.) Schinz & Thell.	1	1	—	0	0	0	10	5	252
Sison amomum	L.	2	1	4	0	0	0	7.5	2	193
Sison amomum		3	1	>5	0	0	0	7.5	2	193
Sison amomum		3	1	5	0	0	0	7.5	2	193
Sium latifolium	L.	1	1	—	0	0	0	5	5	248
Sium latifolium	L.	4	1	—	1308	1308	1308	50	5	216
Smyrnium olusatrum	L.	2	1	1	0	0	0	7.5	2	193
Smyrnium olusatrum	L.	4	1	—	60	60	60	30	4	13
Torilis arvensis	(Hudson) Link	4	7	—	13	243	108	15	4	30
Torilis japonica	(Houtt.) DC.	1	1	—	0	0	0	3	5	96
Torilis japonica	(Houtt.) DC.	1	1	—	0	0	0	13	5	183
Torilis japonica	(Houtt.) DC.	1	3	—	105	342	245	8	7	275
Torilis japonica	(Houtt.) DC.	2	5	>1	0	0	0	—	1	275
Torilis japonica	(Houtt.) DC.	3	3	>5	0	0	0	7.5	2	193
Torilis japonica	(Houtt.) DC.	4	1	—	50	50	50	3	5	96
Torilis japonica	(Houtt.) DC.	4	2	—	3556	3986	3771	5	5	16
Torilis nodosa	(L.) Gaertner	1	1	—	0	0	0	10	4	22
Torilis nodosa	(L.) Gaertner	1	1	—	0	0	0	10	5	140
Urticaceae	—			—	—	—	—	—	—	—
Urtica dioica	L.	1	2	—	0	0	0	6	5	11
Urtica dioica	L.	1	2	—	0	0	0	5	5	69
Urtica dioica	L.	1	2	—	0	0	0	3	5	96

Urtica dioica	L.	1	1	—	0	0	0	15	5	142
Urtica dioica	L.	1	5	—	0	0	0	20	4	225
Urtica dioica	L.	1	7	—	0	0	0	10	5	255
Urtica dioica	L.	1	5	—	0	172	52	2	7	99
Urtica dioica	L.	1	1	—	104	104	104	6	5	28
Urtica dioica	L.	1	1	—	170	170	170	6	5	208
Urtica dioica	L.	1	1	—	313	313	313	20	5	97
Urtica dioica	L.	2	1	>2	0	0	0	20	1	219
Urtica dioica	L.	2	1	—	4	4	4	10	4	127
Urtica dioica	L.	2	2	—	12	17	15	3	7	235
Urtica dioica	L.	2	2	—	75	175	125	40	5	162
Urtica dioica	L.	2	1	—	130	130	130	6	5	28
Urtica dioica	L.	2	15	—	50	500	138	3	5	96
Urtica dioica	L.	2	1	—	156	156	156	10	5	27
Urtica dioica	L.	2	2	—	156	156	156	6	5	99
Urtica dioica	L.	2	1	—	180	180	180	10	5	149
Urtica dioica	L.	2	1	—	198	198	198	6	5	208
Urtica dioica	L.	2	2	—	188	230	209	10	5	252
Urtica dioica	L.	2	1	—	337	337	337	57	5	162
Urtica dioica	L.	2	1	—	533	533	533	10	5	184
Urtica dioica	L.	2	4	—	112	1311	534	12	5	162
Urtica dioica	L.	2	1	—	700	700	700	6	5	162
Urtica dioica	L.	2	2	—	112	1311	712	27	5	162
Urtica dioica	L.	2	1	—	752	752	752	20	5	162
Urtica dioica	L.	2	5	—	275	875	755	10	5	162
Urtica dioica	L.	2	1	—	780	780	780	10	5	152
Urtica dioica	L.	2	2	—	300	1750	1025	15	5	162
Urtica dioica	L.	2	3	—	359	2096	1517	18	5	162
Urtica dioica	L.	2	1	—	1577	1577	1577	11	5	162
Urtica dioica	L.	2	9	—	112	2188	1735	17	5	162
Urtica dioica	L.	2	1	—	2180	2180	2180	15	5	257
Urtica dioica	L.	2	1	—	3070	3070	3070	22	5	162
Urtica dioica	L.	2	1	—	4400	4400	4400	20	4	225
Urtica dioica	L.	2	13	—	140	49076	10354	8	7	275
Urtica dioica	L.	3	3	>5	0	0	0	7.5	2	198
Urtica dioica	L.	3	2	>8	78	104	91	6	5	28
Urtica dioica	L.	3	1	—	156	156	156	6	5	99
Urtica dioica	L.	3	1	—	237	237	237	30	5	45
Urtica dioica	L.	3	1	—	311	311	311	6	5	208
Urtica dioica	L.	3	1	—	417	417	417	20	5	97
Urtica dioica	L.	3	1	—	478	478	478	8	5	108
Urtica dioica	L.	3	1	—	600	600	600	20	5	69
Urtica dioica	L.	3	3	—	120	1560	607	15	5	257
Urtica dioica	L.	3	1	—	731	731	731	10	5	150
Urtica dioica	L.	3	2	—	156	2548	1352	65	5	27
Urtica dioica	L.	3	1	>35	2473	2473	2473		5	161
Urtica dioica	L.	4	2	—	1	3	2	4	5	176

Species	Authority	Seed bank type	Number of records	Longevity (y)	Minimum density (seeds m⁻²)	Maximum density (seeds m⁻²)	Mean density (seeds m⁻²)	Depth (cm)	Method	Source code
Urtica dioica	L.	4	1	—	4	4	4	5	5	166
Urtica dioica	L.	4	2	—	10	49	30	15	5	62
Urtica dioica	L.	4	1	—	40	40	40	5	5	213
Urtica dioica	L.	4	1	—	48	48	48	5	5	69
Urtica dioica	L.	4	3	—	9	72	48	10	5	110
Urtica dioica	L.	4	1	—	67	67	67	10	5	184
Urtica dioica	L.	4	3	—	58	83	69	5	5	98
Urtica dioica	L.	4	3	—	96	329	231	6	5	49
Urtica dioica	L.	4	2	—	173	312	243	7	5	58
Urtica dioica	L.	4	5	—	63	525	255	10	5	255
Urtica dioica	L.	4	2	—	101	722	412	5	5	220
Urtica dioica	L.	4	3	—	83	1450	989	3	5	96
Urtica dioica	L.	4	6	—	1600	9600	4533	20	4	225
Urtica dioica	L.	4	3	—	2264	6509	4924	3	5	208
Urtica dioica	L.	4	5	—	526	67979	23819	10	5	150
Urtica urens	L.	1	1	—	0	0	0	3	5	96
Urtica urens	L.	2	1	—	2	2	2	5	5	244
Urtica urens	L.	2	1	—	188	188	188	17	5	162
Urtica urens	L.	2	1	—	200	200	200	10	5	162
Urtica urens	L.	2	3	—	88	563	271	10	5	255
Urtica urens	L.	3	1	>5	0	0	0	7.5	2	196
Urtica urens	L.	3	1	>5	0	0	0	2.5	1	203
Urtica urens	L.	3	1	>5	0	0	0	2.5	2	203
Urtica urens	L.	3	1	>5	0	0	0	7.5	1	203
Urtica urens	L.	3	1	>5	0	0	0	7.5	2	203
Urtica urens	L.	3	1	>5	0	0	0	15	1	203
Urtica urens	L.	3	1	>5	0	0	0	15	2	203
Urtica urens	L.	4	1	—	84	84	84	20	4	111
Urtica urens	L.	4	1	—	400	400	400	15	5	205
Urtica urens	L.	4	1	—	525	525	525	23	5	202
Urtica urens	L.	4	1	—	22749	22749	22749	15	5	207
Valerianaceae	—			—						
Valeriana dioica	L.	1	2	—	0	0	0	6.5	5	128
Valeriana dioica	L.	1	7	—	0	0	0	10	5	149
Valeriana dioica	L.	1	6	—	0	0	0	3	5	208
Valeriana dioica	L.	1	3	—	0	0	0	10	5	255
Valeriana officinalis	L.	1	1	—	0	0	0	6	5	48
Valeriana officinalis	L.	1	1	—	0	0	0	20	5	67
Valeriana officinalis	L.	1	1	—	0	0	0	5	4	88
Valeriana officinalis	L.	1	1	—	0	0	0	6	5	181
Valeriana officinalis	L.	1	2	—	0	0	0	20	4	225

Species	Author									
Valeriana officinalis	L.	1	2	—	0	0	0	5	5	248
Valeriana officinalis	L.	1	1	—	7	7	7	3	5	235
Valeriana officinalis	L.	1	1	—	52	52	52	2	7	99
Valerianella coronata	(L.) DC.	1	2	—	0	0	0	25	7	36
Valerianella coronata	(L.) DC.	1	1	—	0	0	0	6	4	67
Valerianella coronata	(L.) DC.	1	1	—	0	0	0	3	5	132
Valerianella coronata	(L.) DC.	2	1	—	50	50	50	10	5	255
Valerianella coronata	(L.) DC.	2	1	—	128	128	128	5	5	69
Valerianella coronata	(L.) DC.	4	6	—	46	340	156	30	4	13
Valerianella dentata	(L.) Pollich	3	3	—	0	0	0	25	4	36
Valerianella dentata	(L.) Pollich	3	1	>30	28	28	28	6	5	211
Valerianella locusta	(L.) Laterr.	4	2	—	50	67	59	3	5	96
Valerianella rimosa	Bast.	1	1	—	0	0	0	25	4	36
Verbenaceae	—									
Verbena officinalis	—	1	1	—	0	0	0	6	5	181
Verbena officinalis	L.	2	1	—	35	35	35	15	5	272
Verbena officinalis	L.	4	2	—	20	26	23	30	4	13
Verbena officinalis	L.	4	3	—	20	64	37	15	5	272
Violaceae	—									
Viola arvensis	Murray	1	8	—	0	0	0	25	4	36
Viola arvensis	Murray	1	1	—	0	0	0	7	4	50
Viola arvensis	Murray	1	1	—	0	0	0	3	5	96
Viola arvensis	Murray	2	1	>2	0	0	0	0	1	75
Viola arvensis	Murray	2	1	>2	0	0	0	0	2	75
Viola arvensis	Murray	2	1	>2	0	0	0	5	1	75
Viola arvensis	Murray	2	1	>2	0	0	0	5	2	75
Viola arvensis	Murray	2	1	—	34	34	34	6	5	49
Viola arvensis	Murray	2	1	—	126	126	126	17.8	5	155
Viola arvensis	Murray	2	1	—	224	224	224	38	5	186
Viola arvensis	Murray	2	14	—	75	1275	391	10	5	255
Viola arvensis	Murray	3	1	>5	0	0	0	2.5	1	203
Viola arvensis	Murray	3	1	>5	0	0	0	2.5	2	203
Viola arvensis	Murray	3	1	>5	0	0	0	7.5	1	203
Viola arvensis	Murray	3	1	>5	0	0	0	7.5	2	203
Viola arvensis	Murray	3	1	>5	0	0	0	15	1	203
Viola arvensis	Murray	3	1	>5	0	0	0	15	2	203
Viola arvensis	Murray	3	1	>20	26	26	26	20	5	40
Viola arvensis	Murray	3	1	—	175	175	175	17.8	5	155
Viola arvensis	Murray	3	1	>8	237	237	237	30	5	45
Viola arvensis	Murray	3	1	>6	312	312	312	30	5	45
Viola arvensis	Murray	4	2	—	16	144	80	15	4	31
Viola arvensis	Murray	4	1	—	150	150	150	3	5	96
Viola arvensis	Murray	4	1	—	272	272	272	15	5	207
Viola arvensis	Murray	4	7	—	88	2150	881	10	5	255
Viola arvensis	Murray	4	1	—	1225	1225	1225	20	5	118

Species	Authority	Seed bank type	Number of records	Longevity (y)	Minimum density (seeds m⁻²)	Maximum density (seeds m⁻²)	Mean density (seeds m⁻²)	Depth (cm)	Method	Source code
Viola arvensis	Murray	4	25	—	20	10476	1390	30	4	13
Viola arvensis	Murray	4	5	—	833	2667	1633	20	5	74
Viola arvensis	Murray	4	3	—	900	4100	3000	7	4	50
Viola arvensis	Murray	4	2	—	320	5920	3120	25	4	3
Viola arvensis	Murray	4	1	—	6103	6103	6103	25	5	2
Viola canina	L.	1	1	—	0	0	0	10	5	150
Viola canina	L.	1	1	—	0	0	0	12	5	171
Viola canina	L.	1	1	—	0	0	0	6.5	7	182
Viola canina	L.	1	2	—	0	0	0	10	5	255
Viola canina	L.	2	5	—	1	23	7	5	6	170
Viola canina	L.	2	1	—	32	32	32	30	5	45
Viola canina	L.	2	2	—	263	1113	688	10	5	255
Viola canina	L.	3	1	>40	80	80	80	20	5	173
Viola canina	L.	3	2	—	205	292	249	10	5	150
Viola canina	L.	4	6	—	50	1525	648	10	5	255
Viola canina	L.	4	2	—	526	1169	848	10	5	150
Viola collina	Besser	3	1	—	13125	13125	13125	20	5	97
Viola hirta	L.	1	1	—	0	0	0	20	5	97
Viola hirta	L.	1	6	—	0	0	0	10	5	152
Viola hirta	L.	1	2	—	0	0	0	3	7	235
Viola hirta	L.	1	2	—	0	0	0	10	5	255
Viola hirta	L.	2	1	—	24	24	24	6	5	181
Viola hirta	L.	2	1	—	312	312	312	10	5	152
Viola hirta	L.	4	1	—	92	92	92	5	5	122
Viola hirta	L.	4	1	—	352	352	352	6	5	181
Viola hirta	L.	4	2	—	858	1170	1014	10	5	152
Viola lutea	Hudson	4	1	—	227	227	227	17.8	5	155
Viola mirabilis	L.	3	1	—	20	20	20	10	5	172
Viola odorata	L.	4	1	—	17	17	17	7	5	58
Viola palustris	L.	1	1	—	0	0	0	6	5	11
Viola palustris	L.	1	1	—	0	0	0	40	5	11
Viola palustris	L.	1	1	—	0	0	0	6	5	28
Viola palustris	L.	1	2	—	0	0	0	12	5	174
Viola palustris	L.	1	2	—	0	0	0	10	5	255
Viola palustris	L.	1	1	—	3	3	3	3	7	235
Viola palustris	L.	2	1	—	9	9	9	12	5	174
Viola palustris	L.	4	3	—	52	137	84	12	5	174
Viola palustris	L.	4	1	—	203	203	203	17.8	5	155
Viola palustris	L.	4	1	—	310	310	310	16	5	216
Viola reichenbachiana	Jordan ex Boreau	1	1	—	0	0	0	20	5	67
Viola reichenbachiana	Jordan ex Boreau	3	1	—	528	528	528	20	5	67
Viola riviniana	Reichb.	1	1	—	0	0	0	6	5	48

Species	Authority									
Viola riviniana	Reichb.	1	2	—	0	0	0	6	5	49
Viola riviniana	Reichb.	1	2	—	0	0	0	5	5	92
Viola riviniana	Reichb.	1	2	—	0	0	0	3	5	96
Viola riviniana	Reichb.	1	1	—	0	0	0	2	5	153
Viola riviniana	Reichb.	1	3	—	0	0	0	3	7	235
Viola riviniana	Reichb.	1	2	—	0	0	0	15	5	257
Viola riviniana	Reichb.	1	1	—	49	49	49	12	4	51
Viola riviniana	Reichb.	4	1	—	7	7	7	15	5	32
Viola riviniana	Reichb.	4	3	—	58	79	67	5	5	220
Viola riviniana	Reichb.	4	1	—	191	191	191	9	5	84
Viola riviniana	Reichb.	2	1	—	88	88	88	10	5	255
Viola tricolor	L.	2	1	—	160	160	160	5	5	69
Viola tricolor	L.	2	3	—	486	5111	2495	50	5	260
Viola tricolor	L.	3	1	>11	0	0	0	25	3	209
Viola tricolor	L.	3	1	—	581	581	581	30	5	29
Viola tricolor	L.	3	1	—	1200	1200	1200	20	5	69
Viola tricolor	L.	4	1	—	42	42	42	10	5	15
Viola tricolor	L.	4	4	—	200	471	306	25	5	260
Zannichelliaceae	—	—	—	—	—	—	—	—	—	—
Zannichellia palustris	L.	1	1	—	0	0	0	5	5	177
Zannichellia palustris	L.	2	5	—	20	90	45	4	5	217
Zannichellia palustris	L.	2	2	—	89	266	178	5	5	177
Zannichellia palustris	L.	2	1	—	364	364	364	5	5	116
Zannichellia palustris	L.	4	1	—	133	133	133	5	5	177
Zannichellia palustris	L.	4	1	—	484	484	484	10	5	87
Zannichellia palustris	L.	4	5	—	182	3143	1756	5	5	116
Zosteraceae	—	—	—	—	—	—	—	—	—	—
Zostera marina	L.	1	1	—	0	0	0	—	—	100
Zostera marina	L.	1	1	—	0	0	0	5	5	116
Zostera noltii	Hornem.	1	1	—	0	0	0	20	5	67
Zostera noltii	Hornem.	1	1	—	0	0	0	—	—	100
Zostera noltii	Hornem.	1	1	—	0	0	0	6.5	5	128

Data sources and references cited in the text

6

Source Code

1. Abrams, M. D. (1988). Effect of burning regime on buried seed banks and canopy coverage in a Kansas tallgrass prairie. *Southwestern Naturalist*, **33**, 65–70.

2. Albrecht, H. (1993). *Modeluntersuchung und Literaturauswertung zum Diasporenvorrat gefährdeter Wildkräuter in Ackerböden.* MS Internationales Symposium Flora und Fauna der Acker und Weinberge, Kommer 1992.

3. Albrecht, H. and Bachtaler, G. (1988). Die Segatalflora zwei bayerischer Ackerstandorte 1986/87 im Vergleich zu Untersuchungergebnissen von 1955/56 bzw. 1965. *Zeitschrift für Pflanzenkrankheit und Pflanzenschutz*, **11**, 163–174.

4. Altena, H.J. (1986). Onderzoek naar de aanwezige zaadvoorraad in de bodem van het proefobject 'De Veenkampen'. Internal Report Centre for Agrobiological Research, Wageningen (mimeo).

5. Arai, M. and Kataoka, T. (1956). Ecological studies on *Alopecurus aequalis Sobol. Proceedings of the Crop Science Society of Japan*, **24**, 319–323.

6. Archer, K. A. and Rochester, I. J. (1982). Numbers and germination characteristics of white clover seed recovered from soils on the northern Tablelands of New South Wales after drought. *Journal of the Australian Institute of Agricultural Science*, **48**, 99–101.

7. Archibold, O. W. (1978). Buried viable propagules as a factor in post-fire regeneration in northern Saskatchewan. *Canadian Journal of Botany*, **57**, 54–58.

8. Archibold, O. W. (1981). Buried viable propagules in native prairie and adjacent agricultural sites in central Saskatchewan. *Canadian Journal of Botany*, **59**, 701–706.

9. Archibold, O. W. (1984). A comparison of seed reserves in arctic, subarctic and alpine soils. *Canadian Field-Naturalist*, **98**, 337–344.

10. Archibold, O. W. and Hume, L. (1983). A preliminary survey of seed input into fallow fields in Saskatchewan. *Canadian Journal of Botany*, **61**, 1216–1221.

    Bakker, D. (1960). *Senecio congestus* (R.Br.) DC in the lake Isselmeerpolders. *Acta Botanica Neerlandica*, **9**, 235–259.

    Bakker, J. P. (1989). *Nature management by grazing and cutting.* Dordrecht: Kluwer.

    Bakker, J. P., Bakker, E. S., Rosén, E., Verweij, G. L. and Bekker, R. M. (1996). Soil seed bank composition along a gradient from dry alvar grassland to *Juniperus* scrub. *Journal of Vegetation Science*, **7**, 165–176.

11. Bakker, J.P., Bos, A.F., Hoogveld, J. and Muller, H.J. (1991). The role of the seed bank in restoration management of semi-natural grasslands. In *Terrestrial and aquatic ecosystems; perturbation and recovery*, ed. O. Ravera pp. 449–455. New York: Ellis Horwood. (+ unpublished data.)

12. Barralis, G. and Chadoeuf, R. (1980). Etude de la dynamique d'une communeauté adventice. I. Evolution de la flore adventice au secours du

cycle végétatif d'une culture. *Weed Research*, **20**, 231–237.

13. Barralis, G. and Chadoeuf, R. (1987). Potentiel semencier des terres arables. *Weed Research*, **27**, 417–424. (+ unpublished data.)

14. Barralis, G., Chadoeuf, R. and Lonchamp, J.P. (1988). Longevité des semences de mauvais herbes annuelles dans un sol cultivé. *Weed Research*, **28**, 407–418.

15. Barralis, G. and Salin, D. (1973). Relations entre flore potentielle et flore réelle dans quelques types de sol de Côte-d'Or. *IV. Colloque international sur l'écologie, la biologie et la systématique des mauvaises herbes*, 94–101.

16. Baskin, J. M. and Baskin, C. C. (1975). Ecophysiology of seed dormancy and germination in *Torilis japonica* in relation to its life cycle strategy. *Bulletin of the Torrey Botanical Club*, **102**, 67–72.

17. Baskin, J. M. and Baskin, C. C. (1989). Seasonal changes in the germination responses of buried seeds of *Barbarea vulgaris. Canadian Journal of Botany*, **67**, 2131–2134.

18. Baskin, J. M. and Baskin, C. C. (1990). Seed germination ecology of poison hemlock, *Conium maculatum. Canadian Journal of Botany*, **68**, 2018–2024.

19. Beatty, S. W. (1991). Colonization dynamics in a mosaic landscape: the buried seed pool. *Journal of Biogeography*, **18**, 553–563.

Benoit, D. L., Kenkel, N. C. and Cavers, P. B. (1989). Factors influencing the precision of soil seed bank estimates. *Canadian Journal of Botany*, **67**, 2833–2840.

Benz, W., Koch, W. & Moosman, A. (1984). Ein Extraktionsverfahren zur Bestimmung des Unkrautsamenpotentials im Boden. *Zeitschrift für Pflanzenkrankheit und Pflanzenschutz*, **10**, 106–114.

20. Bernhardt, K-G. (1992). Besiedlungsstrategieen an sandigen Extremstandorten im Tidebereich. *Flora*, **187**, 272–281.

21. Bernhardt, K-G. and Handke, P. (1992). Successional dynamics on newly created saline marsh soils. *Ekologia (CSFR)*, **11**, 139–152.

22. Bernhardt, K-G. and Hurka, H. (1989). Dynamik des

Samenspeichers in einigen Mediterranen Kulturböden. *Weed Research*, **29**, 247–254.

Bernhardt, K-G. and Poschlod, P. (1993). Diasporenbanken im Boden als Vegetationsbestandteil. Teil 1: Europa. *Excerpta Botanica* Section B, **29**, 241–260.

Bertiller, M. B. and Coronato, F. (1994). Seed bank patterns of *Festuca pallescens* in semiarid Patagonia (Argentina): a possible limit to bunch reestablishment. *Biodiversity and Conservation*, **3**, 57–67.

23. Bigwood, D. W. and Inouye, D. W. (1988). Spatial pattern analysis of seed banks: an improved method and optimized sampling. *Ecology*, **69**, 487–507.

24. Bostock, S. J. (1976). Seed germination strategies of five perennial weeds. *Oecologia*, **31**, 113–126.

25. Boutin, C. and Harper, J. L. (1991). A comparative study of the population dynamics of five species of *Veronica* in natural habitats. *Journal of Ecology*, **79**, 199–221.

26. Bowes, G. G. and Thomas, A. G. (1976). Longevity of leafy spurge seeds in the soil following various control programs. *Journal of Range Management*, **31**, 137–140.

27. Brands, R. and Hoekstra, E. (1980). *De invloed van beheerexperimenten op de kieming en vestiging van plantesoorten in grasland*. Internal Report Department of Plant Ecology, University Groningen (mimeo). (see Bakker 1989).

28. Brandsma, O. (1984). *De zaadbank in relatie met de samenstelling en structuur van de vegetatie op het Westerholt*. Internal Report Department of Plant Ecology, University Groningen / Research Institute for Nature Management, Leersum (mimeo). (see Bakker 1989).

29. Brenchley, W. E. (1918). Buried weed seeds. *Journal of Agricultural Science*, **9**, 1–31.

30. Brenchley, W. E. and Warington, K. (1930). The weed seed population of arable soil. I. Numerical estimation of viable seeds and observations on their natural dormancy. *Journal of Ecology*, **18**, 235–272.

31. Brenchley, W. E. and Warington, K. (1933). The weed

seed population of arable soil. II. Influence of crop, soil and methods of cultivation upon the relative abundance of viable seeds. *Journal of Ecology*, **21**, 103–127.

32. Brown, A. H. F. and Oosterhuis, L. (1981). The role of buried seeds in coppicewoods. *Biological Conservation*, **21**, 19–38.

33. Brown, D. (1992). Estimating the composition of a forest seed bank: a comparison of the seed extraction and seedling emergence methods. *Canadian Journal of Botany*, **70**, 1603–1612.

34. Brown, E. O. and Porter, R. H. (1942). The viability and germination of seeds of *Convolvulus arvensis* L. and other perennial weeds. *Iowa Agricultural Experiment Station Research Bulletin*, **294**, 473–504.

35. Bruch, E. C. (1961). 1932–California Department of Agriculture buried seed project–1960. *California Department of Agriculture Bulletin*, **50**, 29–30.

35. Goss, W. L. (1939). Germination of buried weed seeds. *Californian Department of Agriculture Monthly Bulletin*, **28**, 132–135.

36. Büchli, M. (1936). Ökologie der Ackerunkräuter der Nordostschweiz. *Beiträge geobotanische Landesaufnahmen der Schweiz*, **19**, 1–354

37. Budd, A. C., Chepil, W. S. and Doughty, J. L. (1954). Germination of weed seeds. III. The influence of crops and fallow on the weed seed population of the soil. *Canadian Journal of Agricultural Science*, **34**, 18–27.

38. Burnside, O. C., Fenster, C. R., Evetts, L. L. and Mumm, R. R. (1981). Germination of exhumed weed seed in Nebraska. *Weed Science*, **29**, 577–586.

39. Chambers, J. C. (1993). Seed and vegetation dynamics in an alpine herbfield: effects of disturbance type. *Canadian Journal of Botany*, **71**, 471–485.

40. Chancellor, R. J. (1986). Decline of arable weed seeds during 20 years in soil under grass and the periodicity of seedling emergence after cultivation. *Journal of Applied Ecology*, **23**, 631–637.

41. Chapman, D. F. and Anderson, C. B. (1987). Natural re-seeding and *Trifolium repens* demography in grazed hill pastures. I. Flowerhead appearance and fate,

and seed dynamics. *Journal of Applied Ecology*, **24**, 1025–1035.

42. Charlton, J. F. L. (1977). Establishment of pasture legumes in North Island hill country. I. Buried seed populations. *New Zealand Journal of Experimental Agriculture*, **5**, 211–214.

43. Cheam, A. H. (1987). Longevity of *Bromus diandrus* Roth. seed in soil at three sites in Western Australia. *Plant Protection Quarterly*, **2**, 137–139.

44. Chepil, W. S. (1946). Germination of weed seeds. I. Longevity, periodicity of germination, and vitality of seeds in cultivated soil. *Scientific Agriculture*, **26**, 307–346.

45. Chippindale, H. G. and Milton, W. E. J. (1934) . On the viable seeds present in the soil beneath pastures. *Journal of Ecology*, **22**, 508–531.

Clark, D. L. and Wilson, M. V. (1994). Heat-treatment effects on seed bank species of an old-growth douglas-fir forest. *Northwest Science*, **68**, 1–5.

46. Conn, J. S. and Farris, M. L. (1987). Seed viability and dormancy of 17 weed species after 21 months in Alaska. *Weed Science*, **35**, 524–529.

47. Conn, J. S., Cochrane, C. L. and Delapp, J. A. (1984). Soil seed bank changes after forest clearing and agricultural use in Alaska. *Weed Science*, **32**, 343–347.

48. Cresswell, E. G. (1982). The developmental origin and ecological consequences of seed germination responses to light. PhD Thesis, University of Sheffield.

49. Csontos, P. (1991). (unpublished)

50. D'Angela, E., Facelli, J. M. and Jacobo, E. (1988). The role of the permanent soil seed bank in early stages of a post-agricultural succession in the Inland pampa, Argentina. *Vegetatio*, **74**, 39–45.

51. Darby, C. D. (1987). The dynamics of buried seed banks beneath woodlands, with particular reference to *Hypericum pulchrum*. PhD Thesis, Plymouth Polytechnic.

52. Dawson, J. H. and Bruns, V. F. (1975). Longevity of barnyardgrass, green foxtail, and yellow foxtail seeds in soil. *Weed Science*, **5**, 437–440.

53. Delpech, R. (1969). Essai d'estimation du stock de

semences viables dans l'horizon superficiel du sol d'une vieille prairie pâturée de l'Anxois (Côte-d'Or). *III. Colloque international sur l'écologie, la biologie et la systématique des mauvaises herbes*, 80–86.

54. Dickie, J. B. Gajjar, K. H., Birch, P. and Harris, J. A. (1988). The survival of viable seeds in stored topsoil from opencast coal workings and its implications for site restoration. *Biological Conservation*, **43**, 257–265.

55. Diemer, M. and Prock, S. (1993). Estimates of alpine seed bank size in two central European and one Scandinavian subarctic plant communities. *Arctic and Alpine Research*, **25**, 194–200.

56. Diemont, W. H. (1990). Seedling emergence after sod cutting in grass heath. *Journal of Vegetation Science*, **1**, 129–132.

57. Dijkstra, M. (1985). *Enkele mogelijke oorzaken van verschillen in vegetatie tussen de onbeweide en de door jongvee beweide Oosterkwelder, Schiermonnikoog*. Internal Report Department of Plant Ecology, University Groningen (mimeo). (see Bakker 1989).

58. Donelan, M. and Thompson, K. (1980). Distribution of buried viable seeds along a successional series. *Biological Conservation*, **17**, 297–311.

59. Dore, W. G. and Raymond, L. C. (1942). Pasture studies XXIV. Viable seeds in pasture soil and manure. *Scientific Agriculture*, **23**, 69–79.

60. Dorph-Peterson, K. (1910). Kurze Mitteilungen über Keimungsuntersuchungen mit Samen verscheidener wildwachsenden Pflanzen. *Jahresbericht der Vereiningung der Vertreter der Angewandten Botanik*, **8**, 239–247.

61. Dorph-Peterson, K. (1924). Examinations of the occurrence and vitality of various weed seed species under different conditions, made at the Danish State Seed Testing Station during the years 1896–1923. *Report of the 4th International Seed Testing Congress*, 124–138.

62. Douglas, G. (1965). The weed flora of chemically-renewed lowland swards. *Journal of the British Grassland Society*, **20**, 91–100.

63. Ebersole, J. J. (1988). Role of the seed bank in providing

colonizers on a tundra disturbance in Alaska. *Canadian Journal of Botany*, **67**, 466–471.

64. Egley, G. H. and Chandler, J. M. (1978). Germination and viability of weed seeds after 2.5 years in a 50 year buried seed study. *Weed Science*, **26**, 230–239.

64. Egley, G. H. and Chandler, J. M. (1983). Longevity of weed seeds after 5.5 years in the Stoneville 50-year buried seed study. *Weed Science*, **31**, 264–270.

Elliott, J. M. (1977). *Some methods for the statistical analysis of samples of benthic invertebrates*, 2nd edn. Freshwater Biological Association Scientific Publication No. 25.

Fay, P. K. & Olsen, W. A. (1978). Technique for separating weed seeds from soil. *Weed Science*, **26**, 530–533.

Fenner, M. (1985). *Seed ecology*. London: Chapman & Hall.

65. Fernandez-Quintanilla, C., Navarrete, L., Andujar, J. L. G., Fernandez, A. and Sanchez, M. J. (1986). Seedling recruitment and age-specific survivorship and reproduction in populations of *Avena sterilis* L. ssp. *ludoviciana* (Durieu) Nyman. *Journal of Applied Ecology*, **23**, 945–955.

Finlayson, C. M., Cowie, I. D. & Bailey, B. J. (1990). Sediment seed banks on the Magela Creek floodplain, northern Australia. *Aquatic Botany*, **38**, 163–176.

66. Fischer, A. (1983). Wildkrautvegetation der Weinberge des Rheingaus (Hessen); Gesellschaften, Abhängigkeit von modernen Bewirtschaftungsmethoden, Aufgaben des Naturschutzes. *Phytocoenologia*, **11**, 335–383.

67. Fischer, A. (1987). *Untersuchungen zur Populationsdynamik am Beginn von Sekundärsukzessionen*. Berlin: Cramer.

Fitter, A. H. and Peat, H. J. (1994). The ecological flora database. *Journal of Ecology*, **82**, 415–425.

Fitter, R., Fitter, A. and Blamey, M. (1985). *The wild flowers of Britain and Northern Europe*, 4th edn. London: Collins.

Fitter, R., Fitter, A. and Farrer, A. (1984). *Collins guide to the grasses, sedges, rushes and ferns of Britain and Northern Europe.* London: Collins.

68. Fix, K. and Poschlod, P. (1993). Extensivierung von Grünlandstandorten am Beispiel Wackershofen (Landkreis Schwäbischen Hall; Gipskeuper). Bedeutung von Nährstoffstatus und Diasporenbank. *Verhandlungen Gesellschaft für Ökologie,* **22,** 39–45.

69. Foerster, E. (1956). Ein Beitrag zur Kenntnis der Selbstverjüngung von Dauerweiden. *Zeitschrift für Acker- und Pflanzenbau,* **100,** 273–301.

70. Forbes, N. (1963). The survival of wild oat seeds under a long ley. *Experimental Husbandry,* **9,** 10–13.

71. Fox, J. F. (1983). Germinable seed banks of interior Alaskan tundra. *Arctic and Alpine Research,* **15,** 405–411.

72. Frank, R. M. and Safford, L. O. (1970). Lack of viable seeds in the forest floor after clearcutting. *Journal of Forestry,* **75,** 776–778.

73. Freedman, B., Hill, N., Svoboda, J. and Henry, G. (1982). Seed banks and seedling occurrence in a high arctic oasis at Alexandra Fjord, Ellesmere Island, Canada. *Canadian Journal of Botany,* **60,** 2112–2118.

74. Froud-Williams, R. J., Chancellor, R. J. and Drennan, D. S. H. (1983). Influence of cultivation regime upon buried weed seeds in arable cropping systems. *Journal of Applied Ecology,* **20,** 199–208.

75. Froud-Williams, R. J., Chancellor, R. J. and Drennan, D. S. H. (1984). The effects of seed burial and soil disturbance on emergence and survival of arable weeds in relation to minimal cultivation. *Journal of Applied Ecology,* **21,** 629–641.

76. Fyles, J. W. (1989). Seed bank populations in upland coniferous forests in central Alberta. *Canadian Journal of Botany,* **67,** 274–278.

Galinato, M. I. & Van der Valk, A. G. (1986). Seed germination traits of annuals and emergents recruited during drawdowns in the Delta Marsh, Manitoba, Canada. *Aquatic Botany,* **26,** 89–102.

77. Gardner, G. (1977). The reproductive capacity of

Fraxinus excelsior on the Derbyshire Limestone. *Journal of Ecology,* **65,** 107–118.

78. Gartner, B. L., Chapin, S. F. and Shaver, G. R. (1983). Demographic patterns of seedling establishment and growth of native graminoids in an Alaskan tundra disturbance. *Journal of Applied Ecology,* **20,** 965–980.

79. Gleischner, J. A. and Appleby, A. P. (1989). Effect of depth and duration of seed burial on ripgut brome (*Bromus rigidus*). *Weed Science,* **37,** 68–72.

Goyeau, H. and Fablet, G. (1982). Étude du stock de semences de mauvaises herbes dans le sol: le problème de l'échantillonage. *Agronomie* (Paris), **2,** 545–552.

80. Graber, R. E. and Thompson, D. F. (1978). *Seeds in the organic layers and soil of four beech–birch–maple stands.* Forest Service Research Paper NE-401. Broomall, PA: US Department of Agriculture. 8 pp.

81. Graham, D. J. and Hutchings, M. J. (1988a). Estimation of the seed bank of a chalk grassland ley established on former arable land. *Journal of Applied Ecology,* **25,** 241–252.

81. Graham, D. J. and Hutchings, M. J. (1988b). A field investigation of germination from the seed bank of a chalk grassland ley on former arable land. *Journal of Applied Ecology,* **25,** 253–263.

Granados, F. L. and Torres, L. G. (1993). Seed bank and other demographic parameters of broomrape (*Orobanche crenata* Forsk.) populations in faba bean (*Vicia faba* L.). *Weed Research,* **33,** 319–327.

82. Granström, A. (1982). Seed banks in five boreal forest stands originating between 1810 and 1963. *Canadian Journal of Botany,* **60,** 1815–1821.

83. Granström, A. (1987). Seed viability of fourteen species during four years of storage in a forest soil. *Journal of Ecology,* **75,** 321–331.

84. Granström, A. (1988). Seed banks at six open and afforested heathland sites in southern Sweden. *Journal of Applied Ecology,* **25,** 297–306.

85. Granström, A. and Fries, C. (1985). Depletion of viable seeds of *Betula pubescens* and *Betula verrucosa* sown

onto some north Swedish forest soils. *Canadian Journal of Forest Research*, **15**, 1176–1180.

Grime, J. P., Mason, G., Curtis, A. V., Rodman, J., Band, S. R., Mowforth, M. A., Neal, A. M. and Shaw, S. C. (1981). A comparative study of germination characteristics in a local flora. *Journal of Ecology*, **69**, 1017–1059.

Gross, K. L. (1990). A comparison of methods for estimating seed numbers in the soil. *Journal of Ecology*, **78**, 1079–1093.

86. Guyot, L. (1960). Sur la présence dans les terres cultivées et incultes de semences dormantes des espèces adventices. *Bulletin du Service de la Carte Phytogéographique Série B*, **5**, 197–254.

87. Haag, R. W. (1983). Emergence of seedlings of aquatic macrophytes from lake sediments. *Canadian Journal of Botany*, **61**, 148–156.

88. Ham, M. (1980). *Enkele methoden voor het bepalen van de zaadflora van grondmonsters van proeven in de Tielerwaard en op de Ossekampen.* Internal Report, Centre for Agrobiological Research, Wageningen (mimeo).

Harley, J. L. and Harley, E. L. (1986). A checklist of mycorrhizas in the British flora. *New Phytologist*, suppl. **105**, 1–102.

Harper, J. L. (1977). *Population biology of plants.* London: Academic Press.

89. Harradine, A. R. (1986). Seed longevity and seedling establishment of *Bromus diandrus* R. *Weed Research*, **26**, 173–180.

90. Hartman, J. M. (1988). Recolonization of small disturbance patches in a New England salt marsh. *American Journal of Botany*, **75**, 1625–1631.

91. Hatt, M. (1991). Samenvorrat von zwei alpinen Böden. *Berichte Geobotanisches Institut ETH Stiftung Rübel Zürich*, **57**, 41–47.

Haukos, D. A. and Smith, L. M. (1993). Seed-bank composition and predictive ability of field vegetation in playa lakes. *Wetlands*, **13**, 32–40.

92. Hester, A. J., Gimingham, C. H. and Miles, J. (1991). Succession from heather moorland to birch woodland. III. Seed availability, germination and early growth. *Journal of Ecology*, **79**, 329–334.

93. Hill, M. O. and Stevens, P. A. (1981). The density of viable seed in soils of forest plantations in upland Britain. *Journal of Ecology*, **69**, 693–709.

94. Hill, N. M., Patriquin, D. G. and Van der Kloet, S. P. (1989). Weed seed bank and vegetation at the beginning and end of the first cycle of a 4-course crop rotation with minimal weed control. *Journal of Applied Ecology*, **26**, 233–246.

95. Hobbs, R. J. and Mooney, H. A. (1986). Community changes following shrub invasion of grassland. *Oecologia*, **70**, 508–513.

96. Hodgson, J. G. and Grime, J. P. (1979). (unpublished).

Hodgson, J. G. and Grime, J. P. (1990). The role of dispersal mechanisms, regenerative strategies and seed banks in the vegetation dynamics of the British landscape. In *Species dispersal in agricultural habitats*, ed. R. G. H. Bunce and D. C. Howard, pp. 65–81. London: Belhaven.

Hodgson, J. G., Grime, J. P., Hunt, R. and Thompson, K. (1995). *The electronic comparative plant ecology.* London: Chapman and Hall.

97. Hofman, R. and Griffioen, C. (1980). *Het Zaadkapitaal onder enkele graslanden in Zuid-Limburg (Gerendal).* Internal Report Department of Vegetation Science and Plant Ecology. University Utrecht (mimeo).

98. Hofstede, R., Bakker, J.P. and Van Diggelen, R. (1991). Het Friesche Veen: mogelijkheden voor vegetatieontwikkeling. *De Levende Natuur*, **92**, 94–100.

99. Hoogveld, J. and Muller, H. J. M. (1985). *Zaadkapitaal en kieming en vestiging van ingezaaid zaad in vijf graslanden bij het Anloërdiepje.* Internal Report Department of Plant Ecology, University Groningen (mimeo). (see Bakker 1989).

100. Hootsmans, M. J. M., Vermaat, J. E. and Van Vierssen, W. (1987). Seed-bank development, germination and early seedling survival of two seagrass species from the Netherlands: *Zostera marina* L. and *Zostera noltii* Hornem. *Aquatic Botany*, **28**, 275–286.

101. Hughes, J. W. and Fahey, T. J. (1991). Colonization dynamics of herbs and shrubs in a disturbed northern hardwood forest. *Journal of Ecology*, **79**, 605–616.

102. Hulbert, L. C. (1955). Ecological studies of *Bromus tectorum* and other annual bromegrasses. *Ecological Monographs*, **25**, 181–213.

Hutchings, M. J. (1986). Plant population biology. In *Methods in plant ecology*, 2nd edn. ed. P. D. Moore and S. B. Chapman, pp. 377–435. Oxford: Blackwell.

103. Hutchings, M. J. and Russell, P. J. (1989). The seed regeneration dynamics of an emergent salt marsh. *Journal of Ecology*, **77**, 615–637.

104. Hyde, O. C. and Suckling, F. E. T. (1953). Dormant seeds of clovers and other legumes in agricultural soils. *New Zealand Journal of Science and Technology*, **34**, 375–385.

105. Ingersoll, C. A. and Wilson, M. V. (1993). Buried propagule bank of a high subalpine site: microsite variation and comparisons with aboveground vegetation. *Canadian Journal of Botany*, **71**, 712–717.

106. Isaac, L. A. (1935). Life of Douglas fir seed in the forest floor. *Journal of Forestry*, **33**, 61–66.

107. Janiesch, P., Mellin, C., Von Lemm, R. and Wolf, D. (1991). Die Aktievierung van Samenbanken ehemaliger Feuchtwiesen und -Wälder als Grundlage für Renaturierungen. *Verhandlungen Gesellschaft für Ökologie*, **20**, 347–350.

108. Jayasingam, T. (1983). Distribution of viable seeds in soil of a permanent pasture and their relation to surface vegetation. (unpublished)

109. Jefferies, R. L., Davy, A. J. and Rudmik, T. (1981). Population biology of the salt marsh annual *Salicornia europaea* agg. *Journal of Ecology*, **69**, 17–31.

110. Jefferson, R. G. and Usher, M. B. (1987). The seed bank in soils of disused chalk quarries in the Yorkshire Wolds, England: implications for conservation management. *Biological Conservation*, **42**, 287–302.

111. Jensen, H. A. (1969). Content of buried seeds in arable soil in Denmark and its relation to the weed population. *Dansk Botanisk Arkiv*, **27**, 9–56.

112. Jerling, L. (1983). Composition and viability of the seed bank along a successional gradient on a Baltic seashore meadow. *Holarctic Ecology*, **6**, 150–156.

113. Joenje, W. (1979). Plant succession and nature conservation of newly embanked tidal flats in the Lauwerszeepolder. In *Ecological processes in coastal environments*, ed. R. L. Jefferies and A. J. Davy, pp. 617–634. Oxford: Blackwell.

114. Johnston, A., Smoliak, S. and Stringer, P. W. (1969). Viable seed populations in Alberta prairie topsoils. *Canadian Journal of Plant Science*, **49**, 75–82.

115. Kasahara, Y., Nishi, K. and Ueyama, Y. (1967). Studies on the germination of seeds and their growth in rush (*Juncus effusus* L. var. *decipiens* Buchen.) and weeds, buried for about 50 years. *Hikobia*, **5**, 91–103.

116. Kautsky, L. (1990). Seed and tuber banks of aquatic macrophytes in the Askö area, northern Baltic proper. *Holarctic Ecology*, **13**, 143–148.

117. Keddy, P. A. and Reznicek, A. A. (1982). The role of seed banks in the persistence of Ontario's coastal plain flora. *American Journal of Botany*, **69**, 13–22.

118. Kees, H. (1986). Einfluss zehnjähringer Unkrautbekämpfung mit 4 unterschiedlichen Intensitätsstufen unter Berücksichtigung der wirtschaftlichen Schadenschwelle auf Unkrautflora und Unkrautsamensvorrat in Boden. *Proceedings EWRS Symposium on Economic Weed Control, Stuttgart*, pp. 399–406.

119. Kellman, M. C. (1970). The viable seed content of some forest soil in coastal British Columbia. *Canadian Journal of Botany*, **48**, 1383–1385.

120. Kellman, M. C. (1974). Preliminary seed budgets for two plant communities in coastal British Columbia. *Journal of Biogeography*, **1**, 123–133.

121. Kelly, D. (1989). Demography of short-lived plants in chalk grassland. I. Life cycle variation in annuals and strict biennials. *Journal of Ecology*, **77**, 747–769.

Kent, D. H. (1992). *List of vascular plants of the British Isles*. London: Botanical Society of the British Isles.

Kiirikki, M. (1993). Seed bank and vegetation

succession in abandoned fields in Karkali Nature Reserve, southern Finland. *Annales Botanici Fennici*, **30**, 139–152.

122. King, T. J. (1969) (unpublished)

123. King, T. J. (1976). The viable seed contents of anthill and pasture soil. *New Phytologist*, 77, 143–147. (+ unpublished data.)

124. Kivilaan, A. and Bandurski, R. S. (1981). The one hundred-year period for Dr Beal's seed viability experiment. *American Journal of Botany*, **68**, 1290–1292.

125. Kjær, A. (1940). Germination of buried and dry stored seeds. I. 1934–1939. *Proceedings of the International Seed Testing Association*, **12**, 167–190.

125. Kjær, A. (1948). Germination of buried and dry stored seeds. II. 1934–1944. *Proceedings of the International Seed Testing Association*, **14**, 19–26.

125. Madsen, S. B. (1962). Germination of buried and dry stored seeds III. 1934–1960. *Proceedings of the International Seed Testing Association*, **27**, 920–928.

126. Kjellson, G. (1992). Seed banks in Danish deciduous forests: species composition, seed influx and distribution pattern in soil. *Ecography*, **15**, 86–100.

127. Kramer, N. B. and Johnson, F. D. (1986). Mature forest seed banks of three habitat types in central Idaho. *Canadian Journal of Botany*, **65**, 1961–1966.

128. Krause, U., Poschlod, P. and Kapfer, A. (1993). Pflege- und Ausbaumassnahmen an Gräben der Singener Aach-Niederung. Ihre Auswirkungen auf Vegetation und Diasporenbank. In *Biologie semiaquatischer Lebensräume – Aspekte der Populationsbiologie* ed. K. G. Bernhardt, H. Hurka, and P. Poschlod. pp. 19–39. Solingen: Verlag Natur und Wissenschaft.

Kropác, Z. (1966). Estimation of weed seeds in arable soil. *Pedobiologia*, **6**, 105–128.

129. Kropác, Z., Havranek, T. and Dobry, J. (1986). Effect of duration and depth of burial on seed survival of *Avena fatua* in arable soil. *Folia Geobotanica Phytotaxonomica*, **21**, 249–262.

130. Leck, M. A. (1980). Germination in Barrow, Alaska, tundra soil cores. *Arctic and Alpine Research*, **12**, 343–349.

Leck, M. A., Parker, V. T. and Simpson, R. L. (eds.) (1989). *Ecology of soil seed banks.* London: Academic press.

131. Leck, M. A. and Simpson, R. L. (1987). Seed bank of a freshwater tidal wetland: turnover and relationship to vegetation change. *American Journal of Botany*, **74**, 360–370.

Leck, M. A. and Simpson, R. L. (1992). Effect of oil on recruitment from the seed bank of two tidal freshwater wetlands. *Wetlands Ecology and Management*, **1**, 223–231.

132. Levassor, C., Ortega, M. and Peco, B. (1990). Seed bank dynamics of Mediterranean pastures subjected to mechanical disturbance. *Journal of Vegetation Science*, **1**, 339–344. (+ unpublished data.)

133. Lewis, J. (1973). Longevity of crop and weed seeds: survival after 20 years in soil. *Weed Research*, **13**, 179–191.

Linder, C. R. and Schmitt, J. (1994). Assessing the risks of transgene escape through time and crop–wild hybrid peresistence. *Molecular Ecology*, **3**, 23–30.

134. Livingston, R. B. & Allessio, M. L. (1968). Buried viable seed in successional field and forest stands, Harvard Forest, Massachusetts. *Bulletin of the Torrey Botanical Club*, **95**, 58–69.

135. Lonchamp, J.P., Chadoeuf, R. and Barralis, G. (1984). Evolution de la capacité de germination des semences de mauvaises herbes enfouies dans le sol. *Agronomie*, **4**, 671–682.

Lonsdale, W. M. (1993). Losses from the seed bank of *Mimosa pigra*: soil micro-organisms vs. temperature fluctuations. *Journal of Applied Ecology*, **30**, 654–660.

136. Lopez, C., Abramovsky, P. Verdier, J.L. and Mamarot, J. (1988). Estimation du stock semencier dans le cadre d'un essai étudiant l'influence des systèmes culturaux sur l'évolution de la flore adventice. *Weed Research*, **28**, 215–221.

137. Lush, W. M. (1988). Biology of *Poa annua* in a temperate zone of golf putting green (*Agrostis stolonifera/Poa annua*). II. The seed bank. *Journal of Applied Ecology*, **25**, 989–997.

138. Mack, R. N. (1976). Survivorship of *Cerastium atrovirens* at Abberffraw, Anglesey. *Journal of Ecology*, **64**, 309–312.

139. Major, J. and Pyott, W. T. (1966). Buried viable seeds in two California bunchgrass sites and their bearing on the definition of a flora. *Vegetatio*, **13**, 253–282.

   Malone, C. R. (1967). A rapid method for enumeration of viable seeds in soil. *Weeds*, **15**, 381–382.

140. Marañon, T. and Bartolome, J. W. (1989). Seed and seedling populations in two contrasted communities: open grassland and oak (*Quercus agrifolia*) understory in California. *Acta Oecologia/Oecologia Plantarum*, **10**, 147–158.

141. Marquis, D. A. (1975). Seed storage and germination under northern hardwood forests. *Canadian Journal of Forest Research*, **5**, 478–484.

142. Marshall, E. J. P. (1989). Distribution patterns of plants associated with arable field edges. *Journal of Applied Ecology*, **26**, 247–257.

   Marshall, E . J. P. and Arnold, G. M. (1994). Weed seed banks in arable fields under contrasting pesticide regimes. *Annals of Applied Biology*, **125**, 349–360.

143. Matlack, G. R. and Good, R. E. (1990). Spatial heterogeneity in the soil seed bank of a mature Coastal Plain forest. *Bulletin of the Torrey Botanical Club*, **117**, 143–152.

144. McGraw, J. B. (1980). Seed bank size and distribution of seeds in cottongrass tussock tundra, Eagle Creek, Alaska. *Canadian Journal of Botany*, **58**, 1607–1611.

145. McGraw, J. B. (1987). Seed bank properties of an Appalachian sphagnum bog and a model of the depth distribution of viable seeds. *Canadian Journal of Botany*, **65**, 2028–2035.

146. McGraw, J. B., Vavrek, M. C. and Bennington, C. C. (1991). Ecological genetic variation in seed banks I. Establishment of a time transect. *Journal of Ecology*, **79**, 617–625.

147. McIntyre, S. (1985). Seed reserves in temperate Australian rice fields following pasture rotation and continuous cropping. *Journal of Applied Ecology*, **22**, 875–884.

148. Meredith, T. C. (1985). Factors affecting recruitment from the seed bank of sedge (*Cladium mariscus*) dominated communities at Wicken Fen, Cambridgeshire, England. *Journal of Biogeography*, **12**, 463–472.

149. Mika, V. (1978). Der Vorrat am Keimfähigen Samen im südböhmischen Niedermoorböden. *Zeitschrift für Acker- und Pflanzenbau*, **146**, 222–234.

   Milberg, P. (1990). Hur länge kan ett frö leva? *Svensk Botanisk Tidskrift*, **84**, 323–352.

150. Milberg, P. (1992). Seed bank in a 35 year old experiment with different treatments of a semi-natural grassland. *Acta Oecologica*, **13**, 743–752.

151. Milberg, P. (1993). Seed bank and seedlings emerging after soil disturbance in a wet semi-natural grassland in Sweden. *Annales Botanici Fennici*, **30**, 9–13. (+ unpublished data.)

   Milberg, P. (1994). Germination ecology of the endangered grassland biennial *Gentianella campestris*. *Biological Conservation*, **70**, 287–290.

152. Milberg, P. and Hansson, M. L. (1994). Soil seed bank and species turnover in a limestone grassland. *Journal of Vegetation Science*, **5**, 35–42.

   Milberg, P. and Persson, T. S. (1994). Soil seed bank and species recruitment in road verge grassland vegetation. *Annales Botanici Fennici*, **31**, 155–162.

153. Miles, J. (1973). Natural recolonization of experimentally bared soil in Callunetum in NE Scotland. *Journal of Ecology*, **61**, 399–412.

154. Miller, G. R. and Cummins, R. P. (1987). Role of buried viable seeds in the recolonization of disturbed ground by heather (*Calluna vulgaris* [L.] Hull) in the Cairngorm mountains, Scotland, UK. *Arctic and Alpine Research*, **19**, 396–401.

155. Milton, W. E. J. (1936). The buried viable seeds of enclosed and unenclosed hill land. *Bulletin of the Welsh Plant Breeding Station*, **14**, 58–73.

156. Milton, W. E. J. (1939). The occurrence of buried viable

seeds in soils at different elevations and on a salt marsh, *Journal of Ecology*, **27**, 149–159.

Milton, W. E. J. (1943). The buried viable seed content of a midland calcareous clay soil. *Empire Journal of Experimental Agriculture*, **11**, 155–167.

Milton, W. E. J. (1948). The buried viable-seed content of upland soils in Montgomeryshire. *Empire Journal of Experimental Agriculture*, **16**, 163–177.

Moore, R. P. (1972). Tetrazolium staining for assessing seed quality. In *Seed ecology. Proceedings of the 19th Easter school in Agricultural Science, University of Nottingham*, ed. W. Heydeker, pp. 347–366. London: Butterworths.

157. Morash, R. and Freedman, B. (1983). Seed banks in several recently clear-cut and mature hardwood forests in Nova Scotia. *Proceedings of the Nova Scotian Institute of Science*, **33**, 85–94.

158. Morin, H. and Payette, S. (1987). Buried seed populations in the montane, subalpine and alpine belts of Mont Jacques-Cartier, Québec. *Canadian Journal of Botany*, **66**, 101–107.

159. Moss, S. R. (1985). The survival of *Alopecurus myosuroides* Huds. seeds in soil. *Weed Research*, **25**, 201–211.

Myerscough, P. J. and Whitehead, F. H. (1966). Comparative biology of *Tussilago farfara* L., *Chamaenerion angustifolium* (L.) Scop., *Epilobium montanum* L. and *Epilobium adenocaulon* Hausskn. *New Phytologist*, **65**, 192–210.

Nakagoshi, N. (1985). Buried viable seeds in temperate forests. In *The population structure of vegetation*, ed. J. White, pp. 551–570. Dordrecht: Junk.

Netherlands Bureau of Statistics (1994). *Botanical database*, Voorburg: Netherlands Bureau of Statistics.

160. Nicholson, A. and Keddy, P. A. (1983). The depth profile of a shoreline seedbank in Matchedash Lake, Ontario. *Canadian Journal of Botany*, **61**, 3293–3296.

161. Odum, S. (1965). Germination of ancient seeds. *Dansk Botanisk Arkiv*, **24**(2), 70 pp.

162. Odum, S. (1974). Seeds in ruderal soils, their longevity and contribution to the flora of disturbed ground in Denmark. *12th British Weed Control Conference*, 1131–1144.

162. Odum, S. (1978). *Dormant seeds in Danish ruderal soils.* Denmark: Royal Veterinary and Agricultural University, Horsholm Arboretum.

Olmsted, N. W. and Curtis, J. D. (1947). Seeds of the forest floor. *Ecology*, **28**, 49–52.

163. Oosting, H. T. and Humphreys, M. E. (1940). Buried viable seeds in a successional series of old field and forest soils. *Bulletin of the Torrey Botanical Club*, **67**, 253–273.

164. Panetta, F. D. (1985). Population studies on Pennyroyal mint (*Mentha pulegium*). II. Seed banks. *Weed Research*, **25**, 311–316.

165. Pavone, L. V. and Reader, R. J. (1982). The dynamics of seed bank size and seed state of *Medicago lupulina*. *Journal of Ecology*, **70**, 537–547.

166. Pederson, R. L. (1981). Seed bank characteristics of the Delta marsh, Manitoba: applications for wetland management. In *Selected Proceedings Midwest Conference on Wetland Values and Management*, ed B. Richardson, pp 61–69. Navarre, Manitoba: Freshwater Society.

Perring, F. H. and Walters, S. M. (1962). *Atlas of the British Flora*. London: Nelson.

167. Pessala, B. (1978). Longevity of *Avena fatua* seeds in the field. Proceedings of the *19th Swedish Weed Control Conference*, C14–C24.

168. Peter, A. (1893). Culturversuche mit 'ruhenden' Samen. *Nachrichte Gesellschaft Wissenschaft Göttingen Mitteilung*, **17**, 673–691.

169. Peter, A. (1894). Culturversuche mit 'ruhenden' Samen II. *Nachrichte Gesellschaft Wissenschaft Göttingen Mitteilung*, **4**, 373–393.

170. Petrov, V. V. (1977). Reserve of viable plant seeds in the uppermost soil layer beneath the canopies of coniferous and small-leaved forests. *Vestnik Moskovskogo Universiteta, Biologiya*, **32**, 33–40.

171. Petrov, V. V. (1981). Amount of dormant viable plant

seeds in the soil of coniferous forest types. *Vestnik Moskovskogo Universiteta, Biologya*, 36, 3–8.

172. Petrov, V. V. (1987). New data on seed bank in the soil under broad-leaved deciduous forests. *Vestnik Moskovskogo Universiteta, Biologya*, 42, 55–59.

173. Petrov, V. V. and Palkina, T. A. (1983). The content of dormant viable seeds in the soil of a broad-leaved forest and a spruce plantation. *Vestnik Moskovskogo Universiteta, Biologya*, 38, 31–35.

174. Pfadenhauer, J. and Maas, D. (1987). Samenpotential in Niedermoorböden des Alpenvorlandes bei Grünlandnutzung unterschiedlicher Intensität. *Flora*, 179, 85–97.

175. Pipal, F. J. (1916). Weed seeds in the soil. *Proceeding of the Indiana Academy of Science*, 368–377.

176. Piroznikow, E. (1983). Seed bank in the soil of stabilized ecosystem of a deciduous forest (Tilio–Carpinetum) in the Bialowieza National Park. *Ekologia Polska*, 31, 145–172.

177. Poiani, K. A. and Johnson, W. C. (1988). Evaluation of the emergence method in estimating seed bank composition of prairie wetlands. *Aquatic Botany*, 32, 91–97.

177. Poiani, K. A. and Johnson, W. C. (1989). Effect of hydroperiod on seed bank composition in semipermanent prairie wetlands. *Canadian Journal of Botany*, 67, 856–864.

178. Pons, T. L. (1989). Dormancy, germination and mortality of seeds buried in heathland and inland sand dunes. *Acta Botanica Neerlandica*, 38, 327–335.

179. Pons, T. L. (1991). Dormancy, germination and mortality of seeds in a chalk-grassland flora. *Journal of Ecology*, 79, 765–780.

180. Poschlod, P. (1990). *Vegetationsentwicklung in abgetorften Hochmooren des bayerischen Alpenvorlandes unter besonderer Berücksichtigung standortskundlicher und populationsbiologische Faktoren.* Berlin: Cramer.

Poschlod, P. (1993). Die Dauerhaftigkeit von generativen Diasporenbanken in Böden von Kalkmagerrasenpflanzen und deren Bedeutung für den botanischen Arten- und Biotopschutz. *Verhandlungen der Gesellschaft für Ökologie*, 22, 229–240.

181. Poschlod, P., Deffner, A., Beier, B. and Grunicke, U. (1991). Untersuchungen zur Diasporenbank von Samenpflanzen auf beweideten, gemähten, brachgefallenen und aufgeforsteten Kalkmagerrasenstandorten. *Verhandlungen Gesellschaft für Ökologie*, 20, 893–904.

182. Poschlod, P. and Jackel, A.K. (1993). Untersuchungen zur Dynamik von generativen Diasporenbanken von Samenpflanzen in Kalkmagerrasen. I. Jahreszeitliche Dynamik des Diasporenregens und der Diasporenbank auf zwei Kalkmagerrasenstandorten der Schwäbischen Alp. *Flora*, 188, 49–71. (+ unpublished data.)

183. Poschlod, P. and Jordan, S. (1992). Wiederbesiedlung eines aufgeforsteten Kalkmagerrasenstandortes nach Rodung. *Zeitschrift für Ökologie und Naturschutz*, 1, 119–139.

184. Pratt, D. W., Black, R. A. and Zamora, B. A. (1984). Buried viable seed in a ponderosa pine community. *Canadian Journal of Botany*, 62, 44–52.

Priestley, D. A. (1986). *Seed aging.* Ithaca: Cornell University Press.

185. Pulcher, M. and Hurle, K. (1984). Unkrautflora und Unkrautsamenvorrat in Weizenmonokultur bei unterschiedlicher Pflanzenschutzintenistät. *Zeitschrift für Pflanzenkrankheit und Pflanzenschutz*, 10, 51–61.

186. Putensen, H. (1882). Untersuchungen über die im Ackerboden enthaltenen Unkrautsämereien. *Hannoverischs Land – und Forstwirtschaftliches Verein Blatt*, 21, 514–524.

187. Rabinowitz, D. (1981). Buried viable seeds in a North American tall grass prairie: the resemblance of their abundance and composition to dispersing seeds. *Oikos*, 36, 191–195.

188. Rampton, H. H. and Ching, T. M. (1966). Longevity and dormancy in seeds of several cool-season grasses and legumes buried in soil. *Agronomy Journal*, 58, 220–222.

188. Rampton, H. H. and Ching, T. M. (1970). Persistence of crop seeds in soil. *Agronomy Journal*, 62, 272–277.

189. Raynal, D. J. and Bazzaz, F. A. (1973). Establishment of early successional plant populations on forest and prairie soil. *Ecology*, **54**, 1335-1341.

190. Reeves, T. G., Code, G. R. and Piggin, C. M. (1981). Seed production and longevity, seasonal emergence, and phenology of wild radish, (*Raphanus raphanistrum* L.). *Australian Journal of Experimental Agriculture and Animal Husbandry*, **21**, 524-530.

191. Roach, D. A. (1983). Buried seed and standing vegetation in two adjacent tundra habitats, northern Alaska. *Oecologia*, **60**, 359-364.

192. Roberts, H. A. (1964). Emergence and longevity in cultivated soil of seeds of some annual weeds. *Weed Research*, **4**, 296-307.

193. Roberts, H. A. (1979). Periodicity of seedling emergence and seed survival in some Umbelliferae. *Journal of Applied Ecology*, **16**, 195-201.

Roberts, H. A. (1981). Seed banks in soils. In *Advances in Applied Biology 6*, ed. T. H. Coaker, pp. 1-55. London: Academic Press.

194. Roberts, H. A. (1986). Persistence of seeds of some grass species in cultivated soil. *Grass and Forage Science*, **41**, 273-276.

195. Roberts, H. A. (1986). Seed persistence in soil and seasonal emergence in plant species from different habitats. *Journal of Applied Ecology*, **23**, 638-656.

196. Roberts, H. A. and Boddrell, J. E. (1983). Seed survival and periodicity of emergence in eight species of Cruciferae. *Annals of Applied Biology*, **103**, 301-309.

197. Roberts, H. A. and Boddrell, J. E. (1984). Seed survival and periodicity of seedling emergence in four weedy species of Papaver. *Weed Research*, **24**, 195-200.

198. Roberts, H. A. and Boddrell, J. E. (1984). Seed survival and seasonal emergence of seedlings of some ruderal plants. *Journal of Applied Ecology*, **21**, 617-628.

199. Roberts, H. A. and Boddrell, J. E. (1985). Seed survival and seasonal emergence in some species of *Geranium, Ranunculus* and *Rumex. Annals of Applied Biology*, **107**, 231-238.

200. Roberts, H. A. and Boddrell, J. E. (1985). Seed survival and seasonal pattern of emergence in some Leguminosae. *Annals of Applied Biology*, **106**, 125-132.

201. Roberts, H. A. and Chancellor, R. J. (1979). Periodicity of seedling emergence and achene survival in some species of *Carduus, Cirsium* and *Onopordum. Journal of Applied Ecology*, **16**, 641-647.

202. Roberts, H. A. and Dawkins, P. A. (1967). Effect of cultivation on the numbers of viable weed seeds in soil. *Weed Research*, **7**, 290-301.

203. Roberts, H. A. and Feast, P. M. (1972). Fate of seeds of some annual weeds in different depths of cultivated and undisturbed soil. *Weed Research*, **12**, 316-324.

204. Roberts, H. A. and Neilson, J. E. (1980). Seed survival and periodicity of seedling emergence in some species of *Atriplex, Chenopodium, Polygonum* and *Rumex. Annals of Applied Biology*, **94**, 111-120.

205. Roberts, H. A. and Neilson, J. E. (1981). Changes in the soil seed bank of four longterm crop/herbicide experiments. *Journal of Applied Ecology*, **18**, 661-668.

206. Roberts, H. A. and Neilson, J. E. (1981). Seed survival and periodicity of seedling emergence in twelve weedy species of Compositae. *Annals of Applied Biology*, **97**, 325-334.

Roberts, H. A. and Ricketts, M. E. (1979). Quantitative relationships between the weed flora after cultivation and the seed population in the soil. *Weed Research*, **19**, 269-275.

207. Roberts, H. A. and Stokes, F. G. (1966). Studies on the weeds of vegetable crops. VI. Seed populations of soil under commercal cropping. *Journal of Applied Ecology*, **3**, 181-190.

208. Rosenthal, G. (1992). *Erhaltung und Regeneration von Feuchtwiesen*. Berlin: Cramer. (+ unpublished data.)

Röttele, M. and Koch, W. (1981). Verteilung van Unkrautsamen im Boden und Konsequenzen für die Bestimmung der Samendichte. *Zeitschrift für Pflanzenkrankheit und Pflanzenschutz. Sonderheft*, **9**, 383-391.

Rusch, G. (1992). Spatial pattern of seedling recruitment at two different scales in a limestone grassland. *Oikos*, **63**, 139–146.

209. Salzmann, R. (1954). Untersuchungen über die Lebensdauer von Unkrautsamen im Boden. *Mitteilungen der Schweizerische Landwirtschaft*, **10**, 170–176.

210. Schat, H. (1983). Germination ecology of some dune slack pioneers. *Acta Botanica Neerlandica*, **32**, 203–212. (+ unpublished data.)

211. Schenkeveld, A. J. and Verkaar, H. J. (1984). The ecology of short-lived forbs in chalk grasslands: distribution of germinative seeds and its significance for seedling emergence. *Journal of Biogeography*, **11**, 251–260.

212. Schmid, B. (1986). Colonizing plants with persistent seeds and persistent seedlings (*Carex flava* group). *Botanica Helvetica*, **96**, 19–26.

213. Schwabe, A. (1991). Zur Wiederbesiedlung von Auenwald-Vegetationskomplexen nach Hochwasser-Ereignissen: Bedeutung der Diasporen-Verdriftung, der generativen und vegetativen Etablierung. *Phytocoenologia*, **20**, 65–94.

214. Schwerzel, P. J. and Thomas, P. E. L. (1979). Effects of cultivation frequency on the survival of seeds of six weeds commonly found in Zimbabwe Rhodesia. *Zimbabwe Rhodesia Agricultural Journal*, **76**, 195–199.

215. Simpson, R. L., Leck, M. A. and Parker, V. T. (1985). The comparative ecology of *Impatiens capensis* Meerb. (Balsaminaceae) in central New Jersey. *Bulletin of the Torrey Botanical Club*, **112**, 295–311.

216. Skoglund, J. (1990). *Seed banks, seed dispersal and regeneration processes in wetland areas.* Comprehensive summaries of Uppsala Dissertations from the Faculty of Science 253, 33 pp.

217. Smith, L. M. and Kadlec, J. A. (1983). Seed banks and their role during drawdown of a North American marsh. *Journal of Applied Ecology*, **20**, 673–684.

218. Smits, A. J. M., Van Avesaath, P. H. and Van der Velde, G. (1990). Germination requirements and seed banks of some nymphaeid macrophytes: *Nymphaea alba* L., *Nuphar lutea* (L.) Sm. and *Nymphoides peltata* (Gmel.) O. Kuntze. *Freshwater Biology*, **24**, 315–326.

219. Snell, K. (1912). Über das Vorkommen von Keimfähigen Unkrautsamen im Boden. *Landwirtschaftliches Jahrbuch*, **43**, 323–347.

220. Staaf, H., Jonsson, M. and Olsen, L. G. (1987). Buried germinative seeds in mature beech forests with different herbaceous vegetation and soil types. *Holarctic Ecology*, **10**, 268–277.

221. Sterk, A. A., Van Duykeren, A., Hogervorst, J. and Verbeek, E. D. M. (1982). Demographic studies of *Anthyllis vulneraria* L. in the Netherlands. II. Population density fluctuations and adaptations to arid conditions, seed populations, seedling mortality and influence of the biocenosis on demographic features. *Acta Botanica Neerlandica*, **31**, 11–40.

Stewart, A., Pearman, D. A. and Preston, C. D. (1994). *Scarce plants in Britain.* Peterborough: Joint Nature Conservation Committee.

222. Stieperaere, H. and Timmerman, C. (1983). Viable seeds in the soils of some parcels of reclaimed and unreclaimed heath in the Flemish district (Northern Belgium). *Bulletin de la Société Royale de Botanique de Belgique*, **116**, 62–73. (+ unpublished data.)

223. Stoa, T. E. (1933). Persistence of viability of sweet clover seed in a cultivated soil. *Journal of the American Society of Agronomists*, **25**, 177–181.

224. Stoller, E. W. and Wax, L. M. (1974). Dormancy changes and the fate of some annual weed seeds in the soil. *Weed Science*, **22**, 151–155.

225. Storch, H. Bernhardt, K.G. and Hurka, H. (1990). Die Vegetation der Uferböschungen an Vorfluren im Abhängigkeit zur Pflegemahd und unter Berücksichtigung des Samenspeichers. *Verhandlungen Gesellschaft für Ökologie*, **19**, 410–416. (+ unpublished data.)

226. Symonides, E. (1978). Numbers, distribution and specific composition of diaspores in the soils of the plant association Spergulo-Corynephoretum. *Ekologia Polska*, **26**, 11–122.

227. Synnes, O. M. (1984). Effect of temperature, light, stratification, KNO_3 and storage on germination of fresh *Alopecurus geniculatus* L. seeds, and emergence

and longevity of seed and nodes in different soil depths. *Meldinger fra Norges Landbruchshogskole*, **63**(5), 14 pp.

228. Takabayashi, M. (1984). Dynamics of upland weed seeds and the ecological control methods. *Bulletin of the National Agricultural Research Center*, **2**, 120–123.

229. Taylor, G. B. and Ewing, M. A. (1988). Effect of depth of burial on the longevity of hard seeds of subterranean clover and annual medics. *Australian Journal of Experimental Agriculture*, **28**, 77–82.

230. Ter Heerdt, G. N. J. and Drost, H. J. (1994). Potential for the development of marsh vegetation from the seed bank after a drawdown. *Biological Conservation*, **67**, 1–11.

Ter Heerdt, G. N. J., Verweij, G. L., Bekker, R.M. and Bakker, J. P. (1996). An improved method for seed bank analysis: seedling emergence after removing the soil by sieving. *Functional Ecology*, **10**, 144–151.

231. Thompson, A. and Makepeace, W. (1983). Short note: longevity of buried ragwort (*Senecio jacobaea* L.) seed. *New Zealand Journal of Experimental Agriculture*, **11**, 251–260.

232. Thompson, K. (1985). Buried seed banks as indicators of seed output along an altitudinal gradient. *Journal of Biological Education*, **19**, 137–140. (+ unpublished data.)

233. Thompson, K. (1986). Small-scale heterogeneity in the seed bank of an acidic grassland. *Journal of Ecology*, **74**, 733–738.

Thompson, K. (1992) The functional ecology of seed banks. In *Seeds: the ecology of regeneration in plant communities*, ed. M. Fenner, pp. 231–258. Wallingford, UK: CAB International.

Thompson, K. (1993) Seed persistence in soil. In *Methods in comparative plant ecology*, ed. G. A. F. Hendry and J. P. Grime, pp. 199–202. London: Chapman and Hall.

Thompson, K., Band, S. R. and Hodgson, J. G. (1993). Seed size and shape predict persistence in soil. *Functional Ecology*, **7**, 236–241.

234. Thompson, K., Green, A. and Jewels, A. M. (1994).

Seeds in soil and worm casts from a neutral grassland. *Functional Ecology*, **8**, 29–35.

235. Thompson, K. and Grime, J. P. (1979). Seasonal variation in the seed banks of herbaceous species in ten contrasting habitats. *Journal of Ecology*, **67**, 893–921. (+ unpublished data.)

236. Thurston, J. M. (1961). The effect of depth of burying and frequency of cultivation on survival and germination of wild oats (*Avena fatua* L. and *A. ludoviciana* Dur.). *Weed Research*, **1**, 19–31.

237. Thurston, J. M. (1966). Survival of seeds of wild oats (*Avena fatua* L. and *Avena ludoviciana* Dur.) and charlock (*Sinapis arvensis* L.) in soil under leys. *Weed Research*, **6**, 67–80.

238. Timmons, F. (1948). Duration of viability of bindweed seed under field conditions and experimental results in the control of bindweed seedlings. *Agronomy Journal*, **41**, 130–133.

239. Tingey, D. C. (1961). Longevity of seeds of wild oats, winter rye, and wheat in cultivated soil. *Weeds*, **9**, 607–611.

240. Toole, E. H. and Brown, E. (1946). Final results of the Duvel buried seed experiment. *Journal of Agricultural Research*, **72**, 201–210.

241. Tsuyuzaki, S. (1991). Survival characteristics of buried seeds 10 years after the eruption of the Usu volcano in northern Japan. *Canadian Journal of Botany*, **69**, 2251–2256.

Tsuyuzaki, S. (1994). Fate of plants from buried seeds on Volcano Usu, Japan, after the 1977–1978 eruptions. *American Journal of Botany*, **81**, 395–399.

242. Ungar, I. A. (1988). A significant seed bank for *Spergularia marina* (Caryophyllaceae). *Ohio Journal of Science*, **88**, 200–202.

243. Ungar, I.A. (1991). Seed germination responses and the seed bank dynamics of the halophyte *Spergularia marina*. In *Proceedings International Symposium Jodhpur India*, ed. D. N. Sen and S. Mohammed, pp. 81–86.

244. Van Altena, S.C. and Minderhoud, J.W. (1972).

Keimfähige Samen von Gräsern und Kräutern in der Narbeschicht der Niederländischen Weiden. *Zeitschrift für Acker- und Pflanzenbau*, **136**, 95–109.

245. Van Breemen, A. M. M. (1984). Comparative germination ecology of three short-lived mono carpic Boraginaceae. *Acta Botanica Neerlandica*, **33**, 283–305.

246. Van der Toorn, J. and Ten Hove, H. J. (1982). On the ecology of *Cotula coronopifolia* L. and *Ranunculus sceleratus* L. II. Experiments on germination, seed longevity and seedling survival. *Oecologia Plantarum*, **3**, 409–418.

247. Van der Valk, A. G. and Davis, C. B. (1979). A reconstruction of the recent vegetational history of a prairie marsh, Eagle Lake, Iowa, from its seed bank. *Aquatic Botany*, **6**, 29–51.

248. Van der Valk, A. G. and Verhoeven, J. T. A. (1988). Potential role of seed banks and understorey vegetation in restoring quaking fens from floating forests. *Vegetatio*, **76**, 3–13.

249. Van Dijk, H. F. G. and Sykora, K. V. (1982). Onderzoek naar de aanwezigheid van kiemkrachtig zaad in de bodem van twee Noordlimburgse natuurgebieden. *De Levende Natuur*, **84**, 147–152.

250. Van Gool, C. R. and Van der Hoog, C. A. (1986). De zaadvoorraad in grasland. Internal Report Department of Graslandkunde, University Wageningen (mimeo).

Van Tooren, B. F. and During, H. J. (1988). Viable plant diaspores in the guts of earthworms. *Acta Botanica Neerlandica*, **37**, 181–194.

251. Van Zeist, C. (1978). *De invloed van het maaitijdstip op de vegetatieve en generatieve vermeerdering van graslandplanten, en op de samenstelling van de vegetatie.* Internal Report, Centre for Agrobiological Research, Wageningen (mimeo).

252. Vegelin, K. (1993). *Longevity of seeds and its importance for restoration of species-rich grasslands.* Internal report, Department of Plant Ecology, University of Groningen (mimeo).

253. Vieno, M., Komulainen, M. and Neuvonen, S. (1993). Seed bank composition in a subarctic pine–birch forest in Finnish Lapland: natural variation and the effect of simulated acid rain. *Canadian Journal of Botany*, **71**, 379–384.

254. Virágh, K. and Gerencsér, L. (1988). Seed bank in the soil and its role during secondary successions induced by some herbicides in a perennial grassland community. *Acta Botanica Hungarica*, **34**, 77–121.

255. Von Borstel, U. (1974). Untersuchungen zur Vegetationsentwicklung auf ökologisch verschiedenen Grünland- und Ackerbrachen hessischer Mittelgebirge. PhD Thesis, University of Giessen.

Vyvey, Q. (1986). Kiemkrachtige zaden in de bodem: betekenis voor het natuurbehoud. *Biologisch Jaarboek Dodonaea*, **54**, 116–130.

Vyvey, Q. (1989a). Bibliographical review on buried viable seeds in the soil. *Excerpta Botanica* Section B, **26**, 311–320.

Vyvey, Q. (1989b). Bibliographical review on buried viable seeds in the soil. *Excerpta Botanica* Section B, **27**, 1–52.

256. Waldron, L. R. (1904). Weed studies: vitality and growth of buried weed seed. *North Dakota Agricultural Experimental Station Bulletin*, **62**, 439–446.

257. Warr, S. J. (1991). Woodland ground flora and seed banks in South-West England. PhD Thesis, Polytechnic South West, Plymouth, UK.

257. Warr, S. J., Kent, M. and Thompson, K. (1994). Seed bank composition and variability in five woodlands in south-west England. *Journal of Biogeography*, **21**, 151–168.

258. Watanabe, Y. and Hirokawa, F. (1971). Longevity of buried weed seeds. *Weed Research* (Japan), **11**, 40–43.

259. Watkinson, A. R. (1978). The demography of a sand dune annual: *Vulpia fasciculata*. II. The dynamics of seed populations. *Journal of Ecology*, **66**, 35–44.

260. Wehsarg, O. (1912). *Das Unkraut im Ackerboden. Arbeiten der Deutschen Landwirtschafts Gesellschaft. Heft 226.* Berlin: Deutsche Landwirtschafts Gesellschaft.

261. Welling, C. H. and Becker, R. L. (1990). Seed bank dynamics of *Lythrum salicaria* L.: implications for

control of this species in North America. *Aquatic Botany*, **38**, 303–309.

262. Welling, C. H., Pederson, R. L. and Van der Valk, A. G. (1988). Temporal patterns in recruitment from the seed bank during drawdowns in a prairie wetland. *Journal of Applied Ecology*, **25**, 999–1007.

263. Wendel, G. W. (1972). *Longevity of black cherry seed in the forest floor*. Upper Darby, PA: US Department of Agriculture Forest Service Research Note NE-149. 4 pp.

    Westhoff, V. and Van der Maarel, E. (1978). The Braun–Blanquet approach. In *Classification of plant communities*. 2nd edn, ed. R. H. Whittaker, pp. 287–399. The Hague: Junk.

264. Whipple, S. A. (1978). The relationship of buried, germinating seeds to vegetation in an old-growth Colorado subalpine forest. *Canadian Journal of Botany*, **56**, 1505–1509.

265. Willems, J. H. (1988). Soil seed bank and regeneration of a *Calluna vulgaris* community after forest clearing. *Acta Botanica Neerlandica*, **37**, 313–320.

    Willems, J. H. and Huijsmans, K. G. A. (1994). Vertical seed dispersal by earthworms: a quantitative approach. *Ecography*, **17**, 124–130.

266. Williams, E. D. (1978). Germination and longevity of seeds of *Agropyron repens* L. Beauv. and *Agrostis gigantea* Roth. in soil in relation to different cultivation regimes. *Weed Research*, **18**, 129–138.

267. Williams, E. D. (1984). Changes during three years in the size and composition of the seed bank beneath a long-term pasture as influenced by defoliation and fertilizer regime. *Journal of Applied Ecology*, **21**, 603–615.

    Williams, J. T. (1969). Biological flora of the British Isles. *Chenopodium rubrum*. *Journal of Ecology*, **57**, 831–841.

268. Wilson, B. J. and Lawson, H. M. (1992). Seedbank persistence and seedling emergence of seven weed species in autumn-sown crops following a single year's seeding. *Annals of Applied Biology*, **120**, 105–116.

269. Wisheu, I. C. and Keddy, P. A. (1991). Seed banks of a rare wetland plant community: distribution patterns and effects of human-induced disturbance. *Journal of Vegetation Science*, **2**, 181–188.

270. Yeo, P. F. (1961). Germination, seedlings and the formation of haustoria in *Euphrasia*. *Watsonia*, **5**, 11–22.

271. Yli-Vakkuri, P. (1963). Experimental studies on the emergence and initial development of tree seedlings in spruce and pine stands. *Acta Forrestalia Fennica*, **75**, 1–110.

272. Zanin, G., Berti, A. and Zuin, M.C. (1989). Estimation du stock semencier d'un sol labouré ou en semis direct. *Weed Resaerch*, **29**, 407–417.

    Zar, J. H. (1984). *Biostatistical analysis*. Prentice-Hall.

273. Zhang, L. (1983). Vegetation ecology and population biology of *Fritillaria meleagris* L. at the Kungsangen Nature Reserve, Eastern Sweden. *Acta Phytogeographica Suecica*, **73**, 1–92.

274. Zorner, P. S., Zimdahl, R. L. and Schweizer, E. E. (1984). Effect of depth and duration of seed burial on Kochia (*Kochia scoparia*). *Weed Science*, **32**, 602–607.

275. Zwaenepoel, A. (1992). Beheer en typologie van wegbermen in Vlaanderen. PhD thesis, University of Gent. (+ unpublished data.)

This book is accompanied by an electronic version of the database on a standard 3.5 inch disk. The electronic version is identical to the printed version, with the single exception of an extra 'Family' field, which was omitted from the printed version in order to save space.

The disk contains a single file: seedbank.csv. This comma-delimited format should be accessible by all modern spreadsheet packages, including Microsoft Excel, Borland Quattro Pro, Lotus 1-2-3 and dBase. Users of modern Macintosh computers should also experience no difficulty. However, the publishers can make no guarantees about the functioning of the file. In the event of a problem please contact the authors.